Continuous Biopharmaceutical Processes

Chromatography, Bioconjugation, and Protein Stability

This innovative reference provides a coherent and critical view on the potential benefits of a transition from batch to continuous processes in the biopharmaceutical industry, with the main focus on chromatography. It also covers the key topics of protein stability and protein conjugation, addressing the chemical reaction and purification aspects together with their integration. This book offers a fine balance between theoretical modeling and illustrative case studies, between fundamental concepts and applied examples from the academic and industrial literature. Scientists interested in the design of biopharmaceutical processes will find useful practical methodologies, in particular for single-column and multicolumn chromatographic processes.

David Pfister and Lucrèce Nicoud were both doctorate students of Professor Morbidelli at ETH Zurich and graduated in 2015. Their PhD theses were focused on protein conjugation and protein aggregation, respectively, and their research also dealt with the fundamental understanding of protein chromatography. Before joining ETH Zurich, they both obtained a master's degree in chemical engineering from École Nationale Supérieure des Industries Chimiques (Nancy, France). Lucrèce Nicoud is the recipient of a fellowship from the Swiss National Science Foundation financing her current research on the crystallization of pharmaceutical ingredients at the Massachusetts Institute of Technology (Cambridge, MA, USA). David Pfister is head of software development at Ypso-Facto (Nancy, France), where he is in particular developing the software ChromWorks, which has been used to perform the large majority of the simulations of chromatographic processes shown in this book.

Massimo Morbidelli is professor at ETH Zurich (Switzerland) since 1996 and at Politecnico di Milano (Italy) since 1991. His research focuses on the production, purification, conjugation, and aggregation of therapeutic proteins. Other research activities carried out in his group deal with polymer and colloid science. Professor Morbidelli received the Excellence in Process Development Research Award from the American Institute of Chemical Engineers in 2017 and the Separation Science and Technology Award from the American Chemical Society in 2018 for his work on the continuous production and purification of monoclonal antibodies.

Cambridge Series in Chemical Engineering

Books in the Series

Baldea and Daoutidis, *Dynamics and Nonlinear Control of Integrated Process Systems*

Chamberlin, *Radioactive Aerosols*

Chau, *Process Control: A First Course with MATLAB*

Cussler, *Diffusion: Mass Transfer in Fluid Systems, Third Edition*

Cussler and Moggridge, *Chemical Product Design, Second Edition*

De Pablo and Schieber, *Molecular Engineering Thermodynamics*

Deen, *Introduction to Chemical Engineering Fluid Mechanics*

Denn, *Chemical Engineering: An Introduction*

Denn, *Polymer Melt Processing: Foundations in Fluid Mechanics and Heat Transfer*

Dorfman and Daoutidis, *Numerical Methods with Chemical Engineering Applications*

Duncan and Reimer, *Chemical Engineering Design and Analysis: An Introduction, Second Edition*

Fan, *Chemical Looping Partial Oxidation Gasification, Reforming, and Chemical Syntheses*

Fan and Zhu, *Principles of Gas-Solid Flows*

Fox, *Computational Models for Turbulent Reacting Flows*

Franses, *Thermodynamics with Chemical Engineering Applications*

Leal, *Advanced Transport Phenomena: Fluid Mechanics and Convective Transport Processes*

Lim and Shin, *Fed-Batch Cultures: Principles and Applications of Semi-Batch Bioreactors*

Litster, *Design and Processing of Particulate Products*

Marchisio and Fox, *Computational Models for Polydisperse Particulate and Multiphase Systems*

Mewis and Wagner, *Colloidal Suspension Rheology*

Morbidelli, Gavriilidis, and Varma, *Catalyst Design: Optimal Distribution of Catalyst in Pellets, Reactors, and Membranes*

Nicoud, *Chromatographic Processes: Modeling, Simulation and Design*

Noble and Terry, *Principles of Chemical Separations with Environmental Applications*

Orbey and Sandler, *Modeling Vapor-Liquid Equilibria: Cubic Equations of State and Their Mixing Rules*

Pfister, Nicoud, and Morbidelli, *Continuous Biopharmaceutical Processes: Chromatography, Bioconjugation, and Protein Stability*

Petyluk, *Distillation Theory and Its Applications to Optimal Design of Separation Units*

Ramkrishna and Song, *Cybernetic Modeling for Bioreaction Engineering*

Rao and Nott, *An Introduction to Granular Flow*

Russell, Robinson, and Wagner, *Mass and Heat Transfer: Analysis of Mass Contactors and Heat Exchangers*

Schobert, *Chemistry of Fossil Fuels and Biofuels*

Shell, *Thermodynamics and Statistical Mechanics*

Sirkar, *Separation of Molecules, Macromolecules and Particles: Principles, Phenomena and Processes*

Slattery, *Advanced Transport Phenomena*

Varma, Morbidelli, and Wu, *Parametric Sensitivity in Chemical Systems*

Continuous Biopharmaceutical Processes

Chromatography, Bioconjugation, and Protein Stability

DAVID PFISTER
Ypso-Facto

LUCRÈCE NICOUD
Massachusetts Institute of Technology

MASSIMO MORBIDELLI
ETH Zurich

CAMBRIDGE
UNIVERSITY PRESS

CAMBRIDGE
UNIVERSITY PRESS

Shaftesbury Road, Cambridge CB2 8EA, United Kingdom

One Liberty Plaza, 20th Floor, New York, NY 10006, USA

477 Williamstown Road, Port Melbourne, VIC 3207, Australia

314–321, 3rd Floor, Plot 3, Splendor Forum, Jasola District Centre, New Delhi – 110025, India

103 Penang Road, #05–06/07, Visioncrest Commercial, Singapore 238467

Cambridge University Press is part of Cambridge University Press & Assessment, a department of the University of Cambridge.

We share the University's mission to contribute to society through the pursuit of education, learning and research at the highest international levels of excellence.

www.cambridge.org
Information on this title: www.cambridge.org/9781108420228

DOI: 10.1017/9781108332897

First published 2018

A catalogue record for this publication is available from the British Library

Library of Congress Cataloging-in-Publication data
Names: Pfister, David, 1988– author. | Nicoud, Lucrèce, 1988– author. | Morbidelli, Massimo, 1954– author.
Title: Continuous biopharmaceutical processes : chromatography, bioconjugation and protein stability / David Pfister, Lucrèce Nicoud, Massimo Morbidelli.
Other titles: Cambridge series in chemical engineering. Description: Cambridge ; New York, NY : Cambridge University Press, 2018. | Series: Cambridge series in chemical engineering | Includes bibliographical references and index.
Identifiers: LCCN 2018016124 | ISBN 9781108420228 (hardback : alk. paper)
Subjects: | MESH: Proteins–chemistry | Proteins–therapeutic use | Chromatography–methods | Biopharmaceutics–methods
Classification: LCC RM301.55 | NLM QU 55 | DDC 615.7–dc23
LC record available at https://lccn.loc.gov/2018016124

ISBN 978-1-108-42022-8 Hardback

Contents

Preface

Galvanized by the success of continuous technologies in various industrial sectors, the biopharmaceutical industry is currently reconsidering its production strategies. The traditional batch processes are being reassessed and challenged by new ones operated in continuous mode. Such transition requires joint efforts from scientists and engineers to identify clearly the potential advantages of continuous processes. People with diverse backgrounds like biologists, chemists, or regulatory authorities may have their own way to look at the picture. As chemical engineers, we focus on designing efficient and robust unit operations. The production of biopharmaceuticals involves diverse steps from protein expression to protein purification and formulation, which are all associated with a specific domain of expertise. Here, we put emphasis on three main aspects: chromatography, bioconjugation, and protein stability.

In Chapter 1, we provide a general introduction to biopharmaceutical processes, including a short description of the single-unit operations typically encountered.

Chapters 2–5 are dedicated to chromatography, which is extensively used for the purification of therapeutic proteins. More precisely, Chapters 2 and 3 provide the fundamentals of protein chromatography and of continuous countercurrent chromatographic processes, whereas applications to the capture and polishing steps are presented in Chapters 4 and 5, respectively.

Chapter 6 deals with bioconjugation, which is of growing interest to improve certain drug characteristics such as potency, stability, or *in vivo* circulation half-life. We discuss both the kinetics of the conjugation reaction and the purification of the conjugated proteins, where chromatography plays a major role.

Finally, Chapter 7 addresses the question of protein stability. We concentrate on protein aggregation, which occurs all along the manufacturing process, for instance during viral inactivation, filtration, and purification by chromatography.

We aspired to make this book understandable and enjoyable both to academic researchers as well as scientists and engineers active in the biopharmaceutical industry. We provide the reader with practical methods, either empirical or model based, that are then illustrated with numerous case studies. We address the three topics covered in this book in the frame of a common methodology, which relies on the combination of experimental data and mechanistic models to describe the physical phenomena at stake. We build up on those models to design, simulate, and compare various processes, be they batch or continuous. In particular, all the simulations of chromatographic processes were performed with the software ChromWorks (Ypso-Facto, Nancy, France).

This book is the result of years of research and fruitful cooperation with many persons and institutions. The first thanks go to ETH Zurich with all its employees, students, and professors, who provide the perfect environment for any scientific adventure one may imagine and who made the writing of this book possible. In the frame of the Continuous Education Program at ETH Zurich, a short course on continuous chromatography for biotherapeutics is offered every year since 2015, where some of the material of this monograph is used. David Pfister and Lucrèce Nicoud heartily thank Roger-Marc Nicoud for many stimulating discussions and for critically reviewing the content of this book. Massimo Morbidelli is in debt to Hervé Broly, Michael Schulte, and Jonathan Souquet at Merck for the interesting discussions and the continuous support of our research program, as well as to the colleagues at ChromaCon for the close collaboration in the development of new concepts in biochromatography over the last several years. Finally, probably our biggest thanks go to all the research fellows of our research group, and particularly to the doctoral students who, with their dedication and talent, contributed to deepen our knowledge of the subjects covered in this book.

Notation

Symbols

Symbol	Description	Common Units
a_{VD}	first coefficient of the van Deemter equation	$cm^2\ min^{-1}$
b_{VD}	second coefficient of the van Deemter equation	cm
B	second virial coefficient	$L\ g^{-1}$
$\mathbf{B}$	matrix of binary coefficients in the conjugation model	
$\bar{c}_A$	local lumped concentration of A in the particles	$g\ L^{-1}$ or $mol\ L^{-1}$
c_{VD}	third coefficient of the van Deemter equation	min
c_{VD}^{perf}	c_{VD} for a perfusive medium	min
C_A	concentration of A in the liquid phase	$g\ L^{-1}$ or $mol\ L^{-1}$
C_A^*	C_A at thermodynamic equilibrium	$g\ L^{-1}$ or $mol\ L^{-1}$
$\overline{C}_A$	lumped concentration of A averaged over the particle volume	$g\ L^{-1}$ or $mol\ L^{-1}$
$\overline{C}_A^*$	$\overline{C}_A$ at thermodynamic equilibrium	$g\ L^{-1}$ or $mol\ L^{-1}$
$\overline{\overline{C}}_A$	concentration of A bound to the solid surface averaged over the particle volume	$g\ L^{-1}$ or $mol\ L^{-1}$
$\overline{\overline{C}}_A^*$	$\overline{\overline{C}}_A$ at thermodynamic equilibrium	$g\ L^{-1}$ or $mol\ L^{-1}$
C_A^F	C_A in the feed or at time 0	$g\ L^{-1}$ or $mol\ L^{-1}$
$\overline{C}_A^F$	$\overline{C}_A$ in equilibrium with the concentration C_A^F	$g\ L^{-1}$ or $mol\ L^{-1}$
C_A^k	C_A in the mixing cell k	$g\ L^{-1}$ or $mol\ L^{-1}$
$\overline{C}_A^k$	$\overline{C}_A$ in mixing cell k	$g\ L^{-1}$ or $mol\ L^{-1}$
C_A^P	C_A in the permeate	$g\ L^{-1}$ or $mol\ L^{-1}$
C_A^R	C_A in the retentate	$g\ L^{-1}$ or $mol\ L^{-1}$
c_i	concentration of the conjugated protein with the index i	$mol\ L^{-1}$
$\boldsymbol{c}_i^{\star}$	concentration vector of conjugated proteins involved in the formation of the species i	$mol\ L^{-1}$
C_j	concentration of conjugated proteins with a DoC equal to j or of aggregates containing j units	$mol\ L^{-1}$
C_M	modifier concentration	$g\ L^{-1}$ or $mol\ L^{-1}$
C_M^F	modifier concentration in the feed	$g\ L^{-1}$ or $mol\ L^{-1}$

C_M^0	C_M^F at time 0 for a gradient elution	$g\ L^{-1}$ or $mol\ L^{-1}$
CU	capacity utilization	
CV_A	retention volume of A normalized by the column volume	
CV^F	feed volume normalized by the column volume	
d_p	particle diameter	μm
d_f	fractal dimension	
D_A	molecular diffusion coefficient	$cm^2\ min^{-1}$
$\grave{D}_A$	diffusion coefficient of A in the intraparticle liquid	$cm^2\ min^{-1}$
$\overline{D}_A$	lumped diffusion coefficient of A in the particles	$cm^2\ min^{-1}$
$\overline{\overline{D}}_A$	diffusion coefficient of A on the solid surface	$cm^2\ min^{-1}$
D_{ax}	axial dispersion coefficient	$cm^2\ min^{-1}$
D_{ax}^{pseudo}	axial dispersion coefficient of a pseudosolute	$cm^2\ min^{-1}$
DV	number of diafiltration volumes	
g	gradient slope	$g\ L^{-1}\ min^{-1}$ or $mol\ L^{-1}\ min^{-1}$
D_{col}	column diameter	cm
h_k	coefficient in the Kozeny–Kármán equation	
J	number of mixing cells	
$k_{ads,A}$	adsorption rate constant of A	$mL\ mol^{-1}\ min^{-1}$
k_{app}	apparent reaction rate constant	$L^{n_{app}-1}\ mol^{n_{app}-1} min^{-1}$
k_B	Boltzmann constant	$m^2\ kg\ s^{-2}\ K^{-1}$
k_d	deactivation rate constant of the add-on molecule	min^{-1}
k_D	diffusion interaction coefficient	$L\ g^{-1}$
k_f	prefactor in fractal scaling	
$k_{i,j}$	aggregation rate constant between aggregates containing i and j units	min^{-1}
k_j^i	conjugation rate constant on the site i for species with a DoC equal to j	$L\ mol^{-1}\ min^{-1}$
k^i	intrinsic conjugation rate constant of the site i with the add-on molecule	$L\ mol^{-1}\ min^{-1}$
k^{iso}	conjugation rate constant under iso-reactivity assumption	$L\ mol^{-1}\ min^{-1}$
$\boldsymbol{k_j}$	column vector of the k_j^i	$L\ mol^{-1}\ min^{-1}$
k_S	sedimentation interaction coefficient	$L\ g^{-1}$
k_U	unfolding rate constant	min^{-1}
$\grave{K}_A$	partition coefficient of A	
$\tilde{K}_A$	adsorption equilibrium constant of A	L/g
$\tilde{K}_{0,A}$	parameter describing the variations of $\tilde{K}_A$ with C_M	
$\overline{K}_A$	lumped Henry coefficient of A	
$\overline{K}_A^F$	value of $\overline{K}_A$ under loading conditions	
$\overline{K}_A^R$	value of $\overline{K}_A$ under recovery conditions	
$\overline{K}^{pseudo}$	lumped Henry coefficient of a pseudosolute	

$\langle \overline{K} \rangle$	mean lumped Henry coefficient of a distribution of solutes	
$\overline{\overline{K}}_A$	Henry coefficient of A	
$\overline{\overline{K}}_{0,A}$	parameter describing the variation of $\overline{\overline{K}}_A$ with C_M	
$K_{i,j}$	aggregation equilibrium constant between aggregates containing i and j units	
L_{col}	column length	cm
L'_{col}	length of compressed bed	cm
L_{pore}	length of a pore	μm
m	dimensionless flow ratio ($Q/\overline{Q}$)	
$\langle m \rangle$	average value of the m parameters in the different zones	
m^D	dimensionless desorbent flow rate ($Q^D/\overline{Q}$)	
m^E	dimensionless extract flow rate ($Q^E/\overline{Q}$)	
m^F	dimensionless feed flow rate ($Q^F/\overline{Q}$)	
m^i	value of m in zone i ($Q^i/\overline{Q}$)	
m^R	dimensionless raffinate flow rate ($Q^R/\overline{Q}$)	
MW_A	molecular weight of A	g mol^{-1}
$\langle MW \rangle$	average molecular weight of the aggregates (including the monomer)	g mol^{-1}
$\langle MW^{agg} \rangle$	average molecular weight of the aggregates (excluding the monomer)	g mol^{-1}
n	number of conjugation sites	
n^i	number of columns in zone i	
n^F	number of columns in the load zone	
n^R	number of columns in the recovery/regeneration zone	
n^T	total number of columns	
n_{app}	apparent reaction order	
n_A	number of moles of A	mol
n_A^F	initial value of n_A	mol
N	number of theoretical plates	
N^{min}	minimum required number of plates	
N_A	Avogadro number	mol^{-1}
$\overline{\overline{N}}$	saturation capacity	g L^{-1} or mol L^{-1}
$\overline{N}$	lumped saturation capacity	g L^{-1} or mol L^{-1}
P^F	pressure in the feed of a filtration system	bar
P^P	pressure in the permeate	bar
P^R	pressure in the retentate	bar
Pe_c	column Péclet number	
Pe_p	particle Péclet number	
Pe_i	intraparticle Péclet number	
q_A	mass of solute A	g

$\dot{q}_A$	production requirement	kg day^{-1}
q_A^i	mass of solute A in column i	g
Q	fluid flow rate	mL min^{-1}
$\langle Q \rangle$	average flow rate in the different zones	mL min^{-1}
Q^D	desorbent flow rate	mL min^{-1}
Q^E	extract flow rate	mL min^{-1}
Q^F	feed flow rate	mL min^{-1}
Q_A^F	volumetric feed flow rate of A	mL min^{-1}
$Q^{F,DC}$	feed flow rate during the disconnected phase	mL min^{-1}
$Q^{F,IC}$	feed flow rate during the interconnected phase	mL min^{-1}
Q^i	value of Q in zone/operation i	mL min^{-1}
Q^R	raffinate flow rate	mL min^{-1}
$\overline{Q}$	lumped solid flow rate	mL min^{-1}
r_{pore}	average pore radius	nm
R_A	hydrodynamic radius of A	nm
$\langle R \rangle$	average radius of the aggregates	nm
Re	Reynolds number	
r_i	rate of formation of the conjugated protein with the index i	mol L^{-1} s^{-1}
$\mathcal{R}$	reaction rate	mol L^{-1} s^{-1}
$\mathcal{R}_j$	net rate of formation of a conjugated protein of DoC equal to j or of an aggregate containing j units	mol L^{-1} s^{-1}
s_A	sensitivity coefficient of A in HIC/RPC	
S_A	separation factor of A	
S_A^i	value of S_A in zone i	
S_{col}	column cross section	cm^2
$\mathcal{S}_A$	sieving coefficient of A in filtration	
Sc	Schmidt number	
Sh	Sherwood number	
t	time	min
t_0	zero-retention time	min
t_0^T	total zero-retention time in a multicolumn process	min
t_{BT}	breakthrough time	min
t^{BSTR}	time to reach a given conversion in a BSTR	min
$t_{c,A}^{beg}$	beginning of the collection window of A	min
$t_{c,A}^{end}$	end of the collection window of A	min
$t_{c,A/B}$	cut-time between the solutes A and B	min
$t_{e,A}$	characteristic time for external mass transfer of A	min
$t_{i,A}$	characteristic time for internal mass transfer of A	min
$t_{i,A}^F$	value of $t_{i,A}$ under loading conditions	min
$t_{i,A}^R$	value of $t_{i,A}$ under recovery conditions	min

t_i^{pseudo}	characteristic time for internal mass transfer of a pseudosolute	min
$t_{m,A}$	overall characteristic time for mass transfer of A	min
$t_{r,A}$	retention time or mean position of the front of A	min
T	temperature	K
u	fluid velocity	cm min^{-1}
u^i	value of u in zone i	cm min^{-1}
u_i	fluid velocity in the large pores of a perfusive medium	cm min^{-1}
u_{crit}	critical fluid velocity for compressible beds	cm min^{-1}
U_T	total interaction potential	J
V	volume of liquid	mL
$\overline{V}$	lumped volume of the particles	mL
V_0	extraparticle volume of liquid in the column	mL
$V_{acc,A}$	pore volume accessible to A	mL
V^{BSTR}	volume to reach a given conversion in a BSTR	mL
V_{col}	column volume	mL
VCF	volume concentration factor	
V^{CSTR}	volume to reach a given conversion in a CSTR	mL
V^F	injection volume in single-column process	mL
V^{PFR}	volume to reach a given conversion in a PFR	mL
V_{pore}	pore volume	nm^3
$V_{r,A}$	retention volume of A	mL
VR	volume ratio	
V_t	total volume of liquid in the column	mL
W	Fuchs stability ratio	
X_A	conversion of A	
Y_A^R	yield of product A with respect to reactant R	
z	longitudinal coordinate	cm
z_A	apparent number of charges of A	
Z_A^R	selectivity of product A with respect to reactant R	

Greek Symbols

Symbol	Description	Common Units
α	parameter in the conjugation model	mol g^{-1}
β	ratio between actual and maximal theoretical feed times	
γ_A	eluent consumption associated with the purification of A	L g^{-1}
δ	thickness of the stagnant fluid layer	nm
δ_i	Kronecker row vector	
δ_K	coefficient in the Kozeny–Kármán equation	bar min cm^{-2}
ΔP	pressure drop	bar
ΔP^{max}	maximal acceptable pressure drop	bar
Δt	switch time	min
Δt^{DC}	switch time of the disconnected phase	min
Δt^{IC}	switch time of the interconnected phase	min
Δt^F	feed time	min
$\Delta t^{F,DC}$	feed time of the disconnected phase	min
$\Delta t^{F,IC}$	feed time of the interconnected phase	min
Δt_g	duration of modifier gradient	min
Δt_p	production duration	min
Δt^R	recovery and regeneration time	min
ΔUV	difference in UV signal	
ε_e	external porosity	
ε_i	internal porosity	
ε_t	total porosity	
ε_e'	external porosity of the compressed bed	
η	fluid viscosity	Pa s
ϑ_A	recovery of A	
Θ	fraction of free sites at the chromatographic medium surface	
Θ^*	value of Θ at thermodynamic equilibrium	
κ	parameter in the conjugation model	$Å^2$
λ	bed compression factor	
λ_{crit}	critical bed compression factor	
$\mu_{0,A}$	zeroth order moment of the peak of the solute A	
$\mu_{1,A}$	first order moment of the peak of the solute A	min
$\mu_{2,A}$	second order moment of the peak of the solute A	min^2
ξ	initial molar ratio between the add-on molecule and the protein	
ρ	fluid density	kg L^{-1}
ς_A	steric parameter of A in the SMA model	
σ_A^2	variance of the peak of the solute A	min^2

$\sigma_{\overline{K}}$	standard deviation of the distribution of lumped Henry coefficients	min^2
τ	pore tortuosity	
τ_{CSTR}	residence time in a CSTR to reach a given conversion	min
τ_{PFR}	residence time in a PFR to reach a given conversion	min
ϕ_e	phase ratio	
φ_A	productivity associated with the purification of A	
Φ	occupied volume fraction	
χ_A	purity of A	
ψ	hindrance factor	
Ψ_j	steric factor for protein with j grafted add-on molecules	
ω_i	roots of the ω-transform equation ($i = 1, 2$)	

Acronyms

ADC	antibody drug conjugate
AUC	analytical ultracentrifugation
BSTR	batch stirred tank reactor
BT	breakthrough
CD	circular dichroism
CHO	Chinese hamster ovary
CSTR	continuous stirred tank reactor
CV	column volume
DF	diafiltration
DLS	dynamic light scattering
DLVO	Derjaguin Landau Verwey Overbeek
DMSO	dimethyl sulfoxide
DNA	deoxyribonucleic acid
DoC	degree of conjugation
DSC	differential scanning calorimetry
Fab	fragment antigen binding
Fc	fragment crystallizable
FDA	food and drug administration
FFF	flow field-flow fractionation
FMA	film model approximation
GSSR	gradient with steady state recycle
HCP	host cell protein
HETP	height equivalent to a theoretical plate
HIC	hydrophobic interaction chromatography
HPLC	high performance liquid chromatography
HX-MS	hydrogen deuterium exchange-mass spectroscopy
ICH	international conference on harmonization
IEC	ion exchange chromatography
Ig	immunoglubulin
IMAC	immobilized metal affinity chromatography
LDF	linear driving force
LGE	linear gradient elution
mAb	monoclonal antibody
MALS	multiangle light scattering
MC	mixing cells
MCSGP	multicolumn solvent gradient purification
MMAE	monomethyl auristatin E
MPC	model predictive control
NHS	N-hydroxysuccinimide
PBS	phosphate-buffered saline
PCC	periodic countercurrent chromatography

PD	plug flow with axial dispersion
PEG	polyethylene glycol
PF	plug flow
PFR	plug flow reactor
pI	isoelectric point
PTFE	polytetrafluoroethylene
RC	repetitive control
RMPC	repetitive model predictive control
RPC	reversed-phase chromatography
SEC	size exclusion chromatography
SG-SMB	step gradient SMB
SLS	static light scattering
SMA	steric mass action
SMB	simulated moving bed
SPA	succinimide propionic acid
TFF	tangential flow filtration
TMB	true moving bed
UF	ultrafiltration
UV	ultraviolet
3F-SMB	three fraction SMB

1 Introduction to Biopharmaceutical Processes

1.1 Context

Thanks to major progress in the understanding of the biological processes involved in human diseases, protein-based drugs have emerged as an important class of therapeutics in the 1980s [1, 2]. The main asset of proteins is rooted in their ability to perform highly specific and complex sets of biological functions that can hardly be mimicked by traditional chemical drugs [2, 3]. The development and commercialization of protein-based drugs have been fostered by significant advances in protein and genetic engineering to such an extent that they are nowadays mostly produced by living organisms [1, 4, 5]. To name just a few examples, therapeutic monoclonal antibodies, hormones, and blood factors are produced at the commercial scale using the so-called *recombinant technology*. Besides recombinant proteins, a few proteins are still extracted from their native source, such as pancreatic enzymes or plasma proteins [2]. A third strategy to produce protein-based drugs is chemical synthesis, which is nevertheless limited to rather small proteins and peptides [2, 6]. Therapeutic peptides represent a kind of hybrid class of drugs, which bridge the gap between small-molecule and protein drugs in terms of physical properties, clinical applications, and means of production [3]. In the following, they will be placed under the umbrella "therapeutic proteins" due to their amino acid content. In any production process, whether based on the recombinant technology, on the extraction of a protein from its native source, or on chemical synthesis, the target protein has to be isolated from a complex mixture, and efficient purification processes are crucial to reach the high purity grades required for medical use.

In this chapter, we provide a succinct description of typical biopharmaceutical processes in order to contextualize the three main aspects that will be addressed in this book: chromatography, protein conjugation, and protein aggregation. In a first part, we briefly present the main unit operations encountered in the production of therapeutic proteins, and then we discuss *if* and *how* the biopharmaceutical industry could benefit from the use of continuous technologies.

1.2 Single-Unit Operations

Biopharmaceutical production processes are commonly divided into two parts: *upstream processing*, where the protein is produced by living organisms, and *downstream processing*, where a series of purification steps are performed to meet certain purity specifications. Although the terms *upstream* and *downstream* are usually reserved for the production of recombinant proteins, they could in principle be extended to synthetic proteins, where the term *upstream* would designate the chemical synthesis step.

There exist a wide variety of biopharmaceutical production processes, which greatly differ both in their upstream and downstream parts. Over the past years, there were some attempts to develop a universal protein purification platform [7, 8]. However, the number of available protein sequences rises dramatically with time, and protein databases now contain more than 20 million different amino acid sequences [9]. The classification of these proteins into comprehensive domain families led to the identification of about 16,000 different families, among which about two-thirds are singletons (i.e., single-member families) [10, 11]. Because the choice of the purification process is strongly dictated by the protein structure, this undoubtedly annihilates the vision of a universal purification platform. It is thus impossible to describe here a general production process applicable to any therapeutic protein.

Nevertheless, for some families of proteins, it is possible to identify rather general production platforms, as it is the case for monoclonal antibodies (mAbs) [12]. This is related to the fact that mAbs all share a common structure, with one constant domain (Fc) that is identical within a class of immunoglobulin (Ig) and two antigen-binding domains (Fab) that are specific to the antibody of interest. Besides their industrial relevance [13], mAbs therefore represent a good didactic example to introduce general concepts about biopharmaceutical processes in general and about downstream processes in particular, which is at the core of this book. Accordingly, we will take mAbs as a reference in the following. A representative example of a mAb production process is represented in Figure 1.1, and we will largely follow this scheme throughout this section. Deviations from this general scheme will be pointed out in the text, and appropriate references for nonantibody proteins will be indicated when possible.

1.2.1 Cell Culture

As mentioned earlier, a great number of tailor-made therapeutic proteins are nowadays produced by living organisms. Recombinant proteins can be expressed in a wide variety of systems including bacteria, mammalian cells, yeasts, fungi, as well as transgenic plants and animals [5].

Among these systems, *Escherichia coli* (*E. coli*) bacteria are particularly attractive due to their well-characterized genetics, great versatility, rapid growth, low media cost, and high expression level [14]. However, *E. coli* bacteria suffer from the inability to perform complex posttranslational modifications, such as glycosylation, and from the difficulty to achieve proteolytic protein maturation and disulfide bond formation. Moreover, proteins expressed by *E. coli* are often produced as inclusion bodies, which are intracellular insoluble deposits containing the protein at high concentration in an inactive aggregated state. In this case, it is necessary to dissolve these inclusion bodies and to properly refold the protein of interest *in vitro*, which may turn out to be a very challenging operation [15].

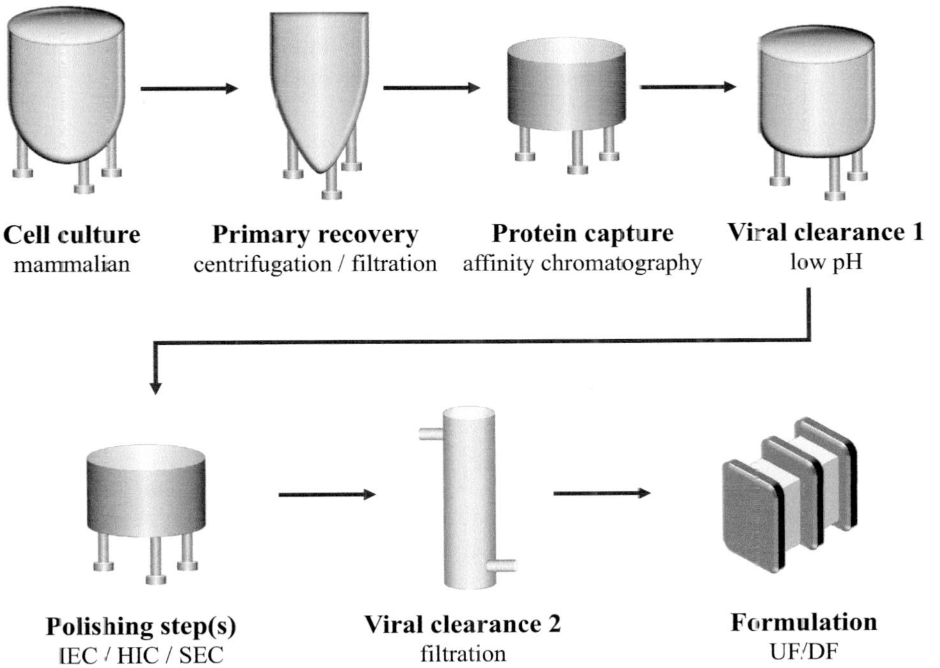

Figure 1.1 Schematic example of a process for the manufacturing of mAbs.

The aforementioned limitations prohibit bacteria from expressing some pharmaceutically relevant proteins, such as mAbs. The latter are largely produced by mammalian cell culture, and in particular with Chinese hamster ovary (CHO) cells [4]. Despite their high cost and complex genetics, mammalian cells are the system of choice for expressing large proteins (above 100 kDa) whose biological activity is strongly affected by both folding and posttranslational modifications [16].

Cell culture falls out of the scope of this book, and the reader interested in biochemical engineering is referred to the following references [17–19].

1.2.2 Primary Recovery

After the cell culture step, the protein of interest is retrieved either by collecting the culture broth supernatant (if the protein is expressed extracellularly) or after supernatant removal and cell disruption (if the protein is expressed intracellularly). In both cases, the first step consists in separating the supernatant from the cells and cell debris formed during cell autolysis, which always occurs to some extent during the cultivation step. This solid–liquid separation step is usually performed with centrifugation or filtration [20, 21] and is relatively simple due to the large size of cells and cell debris as compared to proteins.

In the case where the supernatant is collected and the cells are discarded, as it is typically the case for mammalian cell culture, this first purification step is termed *clarification*, and the recovered liquid is usually referred to as the *harvested cell culture fluid*.

1.2.3 Protein Capture

The fluid obtained after primary recovery contains a wide range of impurities, which include components secreted by the living cells and intracellular components released during cell lysis (e.g., proteins, nucleic acids, lipids, and endotoxins), unused components from the cell culture media (e.g., salts, sugars, amino acids, surfactants, growth factors), and viruses that potentially infected the cells [21]. Among all these species, endotoxins and viruses are extremely hazardous and thus need to be thoroughly eliminated.

At this level, the objective is to extract the maximum amount of target protein from the biological feedstock and to remove as much as possible of the aforementioned impurities (although not necessarily satisfying the final purity requirement). Appropriately, this step is commonly called the *protein capture step*. During this operation, it is also particularly important to recover the protein at a sufficiently high concentration in order to reduce the volume to be treated during the subsequent purification steps. One of the most powerful techniques to perform protein capture is certainly affinity chromatography, which relies on the highly specific reversible interaction between the protein of interest and a natural or synthetic ligand covalently linked on the solid support. The majority of antibody purification platforms employ Protein A as affinity ligand, which interacts specifically with the Fc domain of immunoglobulin G (IgG) [21–23]. Other specific ligands include antibodies, which are used in the industrial purification of recombinant factor VIII for instance [24], heparin that is widely employed in the purification of plasma proteins [25], as well as substrates, co-factors, and inhibitors that are typically selected for enzyme purification [26]. The high price of affinity chromatographic media promotes the development of alternative methods to perform the capture step [22, 27], and two notable exceptions of antibody purifications using ion exchange chromatography are Zenapax® (Daclizumab) and Humira® (Adalimumab) [22]. More details about affinity chromatography and other types of chromatography will be given in Chapter 2, while multicolumn chromatographic processes relevant for the capture step will be presented in Chapter 4.

Although protein chromatography is traditionally performed with packed beds, membrane chromatography has also been examined as an alternative [20, 28–31]. In addition, aqueous two-phase partitioning, which was popular in the 1980s for enzyme purification [32], has recently regained substantial interest in the biotechnology industry as a potential strategy to perform the capture step [20, 33–35]. Finally, precipitation, either of the impurities or of the protein of interest, has often been used as a first step in protein purification processes [20, 36–38]. A noteworthy example among mAbs is Orthoclone OKT®3 (Muromomab CD3), which is recovered from the clarified supernatant by ammonium sulfate induced precipitation [22].

1.2.4 Polishing Steps

After the capture step, a number of *polishing steps* aim at removing the residual impurities in order to obtain the product at the required purity. The expression "intermediate purification steps" is sometimes used to reserve the term *polishing* to the very last purification operation. In this case, the purification process is divided into three parts: a capture step, one or several intermediate purification steps, and a polishing step. There is nevertheless no conceptual difference between intermediate and polishing steps, and we will therefore use the term *polishing* to denote any purification step subsequent to the capture step.

Typical impurities that need to be removed after the capture step include the residual host cell proteins and nucleic acids that were not entirely eliminated during the capture step, potential affinity ligands that leached from the chromatographic medium, as well as fragments, aggregates, and variants of the protein of interest [20]. The latter correspond to proteins that vary slightly in their primary sequence as compared to the target protein. These variations are typically the result of posttranslational modifications, for example due to protein oxidation, deamidation, or different glycosylations. Protein variants are usually difficult to remove due to their high similarity with the target protein.

Ion-exchange chromatography (IEC) or hydrophobic interaction chromatography (HIC) are typically employed during the polishing steps [22], exploiting differences in charge and hydrophobicity, respectively, between the solutes to be separated [21]. The use of size exclusion chromatography (SEC) has also been reported for the large-scale purification of antibodies and notably for the removal of aggregates [22]. The fundamental principles underlying the separation mechanisms by chromatography will be described in Chapter 2, while the multicolumn processes relevant for the polishing steps will be presented in Chapter 5.

As for the capture step, nonchromatographic techniques can also be envisaged for the polishing steps. For instance, the purification process of recombinant insulin involves the precipitation of impurities by ethanol addition followed by insulin crystallization [39].

1.2.5 Viral Clearance

As mentioned earlier, the removal of viruses is a major concern for the biopharmaceutical industry, and conventional purification processes typically target a reduction of the level of retroviruses by 12–18 orders of magnitude [40]. Several methods are available for *viral clearance*, such as low-pH inactivation, heat inactivation, filtration, solvents/detergents treatment, and gamma irradiation [40, 41]. Chromatographic steps have also been shown to reduce the content of viruses by several orders of magnitude [40].

The European Medicines Agency recommends "to investigate the contribution of more than one production step for virus reduction and at least two orthogonal steps should be assessed" [42]. For mAbs, a low-pH hold after the capture step and a viral filtration operation after the polishing steps are typically sufficient to meet the specifications in terms of virus clearance [22]. Specific process development is needed to extend these methods to continuous manufacturing [43].

1.2.6 Formulation

After the purification process, a *formulation step* is necessary to obtain the therapeutic protein at the desired concentration and in the selected formulation solution, which usually contains a buffering agent to regulate the pH as well as various excipients (e.g., salts, polyols, amino acids, surfactants). A proper formulation solution is essential to ensure proper drug delivery into the organism and to guarantee a sufficiently long shelf life of the biopharmaceutical product [44–47]. With this respect, the strong tendency of proteins to form aggregates represents a major issue because the presence of aggregates may compromise drug efficacy and drug safety. The impact of formulation conditions on protein stability will be discussed in Chapter 7.

Diafiltration (DF) and ultrafiltration (UF) are commonly used to perform buffer exchange and to increase protein concentration, respectively [22, 48, 49]. UF and DF processes are typically performed with the tangential flow filtration (TFF) technology, where the feed stream flows parallel to the filter surface in order to prevent fouling or clogging of the system [48, 49].

Finally, the manufacturing process ends with a sterile filtration step and fill/finish operations, which are sometimes performed by specialized contractors [22, 50].

To date, the development of antibody-based drugs has focused primarily on injectable routes of administration, thus favoring the choice of liquid formulations. These are in general cheaper and faster to develop than alternative formulation types. However, proteins in solution are prone to physical and chemical modifications (e.g., unfolding, aggregation, fragmentation, oxidation, deamidation), which prompted the use of lyophilized formulations in several cases including some blood factors, growth hormones, antibodies, and PEGylated interferon [44, 51, 52].

1.2.7 Additional steps

Protein Refolding

As mentioned in Section 1.2.1, high expression levels of proteins in bacteria often result in the formation of inclusion bodies containing the protein of interest in an inactive aggregated state. In this case, an additional step is required in the production process so as to solubilize and properly refold the therapeutic protein [15]. Protein refolding is performed by exchanging the buffer used to solubilize the protein with a buffer favoring native protein conformations. It has been shown that this step is facilitated by the use of chromatographic columns, although the underlying mechanisms of matrix-assisted refolding has not been fully elucidated yet [21, 53].

Protein Conjugation

Protein-based drugs have emerged as a major class of pharmaceuticals due to their outstanding targeting properties. However, therapeutic proteins often suffer from a rapid *in vivo* clearance, a low solubility, and a limited stability. A possibility to overcome these limitations is to attach a suitable chemical molecule to the protein. The idea is to benefit from the targeting properties of the protein, while further enhancing its therapeutic action by a controlled chemical modification [54]. For example, the attachment of polyethylene glycol (PEG) chains to small proteins has been shown to improve the *in vivo* circulation half-life of the drug [55]. Moreover, the conjugation of antibodies targeting cancer cells to cytotoxic drugs, the so-called antibody drug conjugates (ADC), has led to significant improvements in the field of oncology [56]. Protein conjugation is thus highly important for the pharmaceutical industry, and Chapter 6 will be dedicated to this topic, addressing both reaction and purification aspects.

1.3 Overview of the Impurities to Be Removed

1.3.1 Process-Related and Product-Related Impurities

In Section 1.2, we have enumerated a number of impurities that need to be removed during the downstream processing of recombinant proteins. Figure 1.2 presents an overview of these impurities for a typical mAb production process. Following the nomenclature of the international council for harmonization (ICH), impurities may be classified as *product-related impurities* and *process-related impurities* [57].

Product-related impurities are similar to the target protein and include protein aggregates, protein fragments, and various protein variants such as glycoforms, deamidated proteins, and oxidized proteins. Among these impurities, protein aggregates are of

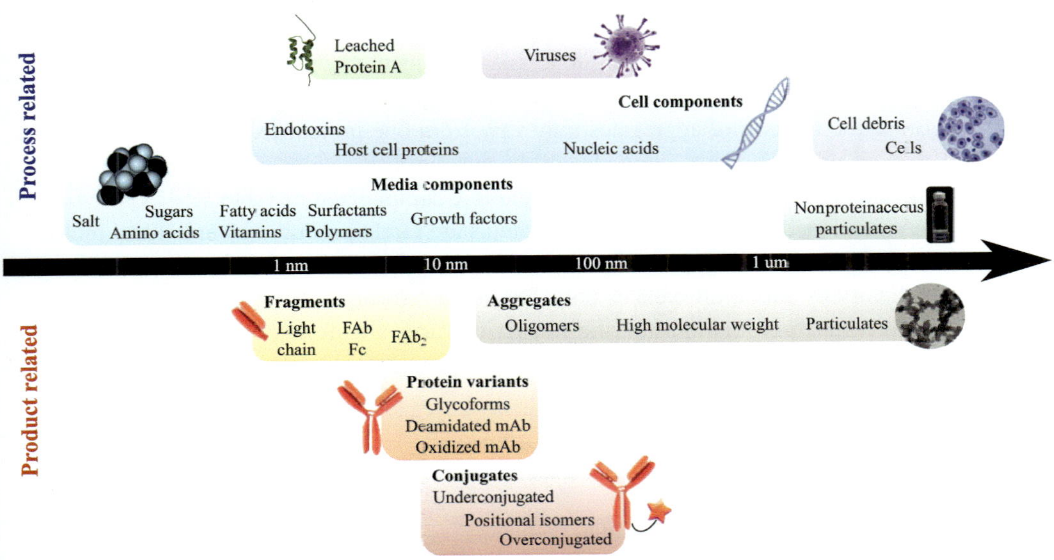

Figure 1.2 Summary of the main process-related impurities and product-related impurities encountered in the manufacturing of mAbs.

particular concern because they may affect drug efficacy and safety [58, 59]. The formation of protein aggregates and their removal are topics covered in Chapter 7.

When dealing with conjugated proteins, the situation is possibly more complex because the conjugation reaction typically leads to a mixture of conjugates, which differ in terms of both number and position of the grafted molecules. This implies that additional impurities need to be removed, such as unconverted proteins, residual chemical reactants, by-products of the reaction, and under-, over- and misconjugated proteins. Strategies to purify conjugated proteins will be addressed in Chapter 6.

On the other hand, process-related impurities do not share common features with the target protein. They derive from the manufacturing process and include host cell proteins, nucleic acids, viruses, and leached affinity ligands.

Overall, when considering all the aforementioned impurities, two main requirements of the purification process can be identified: *(i)* it should remove numerous and diverse impurities in a limited number of steps, *(ii)* it should separate molecules that differ only by slight variations in their size, charge, or hydrophobicity.

Chromatography is currently the technique of choice for the purification of biopharmaceuticals as it is a versatile, selective, and efficient separation technique satisfying these two points. Chromatography is extensively used to purify mAbs [20, 38], a wide range of other recombinant proteins [21, 60], nonrecombinant proteins (e.g., purification of plasma proteins) [25, 61, 62] as well as synthetic peptides [63].

1.3.2 Purity Specifications

The purity requirements for biopharmaceuticals are specific to each product as they depend on various criteria, such as the therapeutic indication or the dose administrated. With regard to viruses, ICH guidelines recommend to review viral clearance studies on a case-by-case basis according to a risk assessment analysis [64, 65]. There is also no general rule on the maximum limit for the aggregate content in biopharmaceuticals because the toxicity of aggregates may vary from one protein to another [66]. Nevertheless, in order to give a rough idea of the purity specifications of biopharmaceuticals, some indications for few relevant impurities are summarized in Table 1.1 [21].

As mentioned earlier, recombinant proteins are rarely produced as individual molecular entities, but rather as a large number of similar variants, also called isoforms, which vary by subtle differences only. Despite a growing understanding of the protein structure–function relationship, the impact of various isoforms on the biological and pharmaceutical activity of a drug remains largely unclear and can only be assessed

Table 1.1. Examples of indications for purity specifications.

Aggregates	<1.0%
HCP	<100 ng/mg
Ligand leakage	<1 ppm
Viruses	<1 particle per million doses
DNA	<10 pg per dose

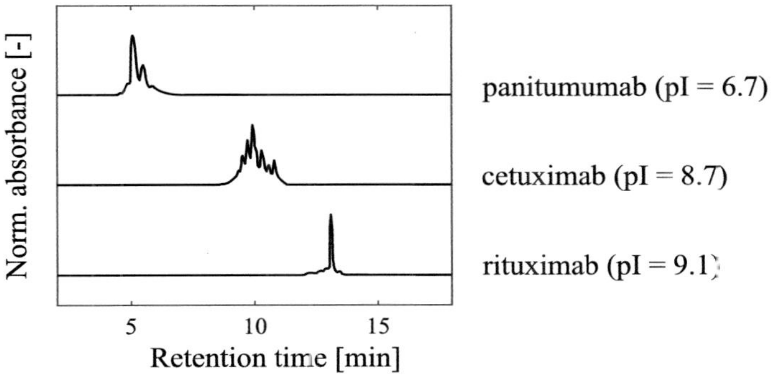

Figure 1.3 Chromatograms of three commercial mAbs obtained by analytical cation exchange chromatography using a pH gradient, illustrating the heterogeneity of biopharmaceutical products in terms of charge variants. The three selected mAbs elute at different times due differences in their isoelectric points (pI). Adapted from [67], copyright 2015, with permission from Elsevier.

by experimentation [16]. In order to guarantee drug quality, efficacy, and safety, it is paramount to maintain the distribution of these isoforms within established acceptable ranges in agreement with the ICH guidelines [57]: "An inherent degree of structural heterogeneity occurs in proteins due to the biosynthetic processes used by living organisms to produce them; therefore, the desired product can be a mixture of anticipated post-translationally modified forms (e.g., glycoforms). [...] The manufacturer should define the pattern of heterogeneity of the desired product and demonstrate consistency with that of the lots used in preclinical and clinical studies." For illustrative purposes, the degree of heterogeneity of some marketed biopharmaceutical products can be appreciated in Figure 1.3, which shows the chromatograms of three mAbs obtained by analytical cation exchange chromatography [67]. Each peak can be associated with at least one mAb variant.

Although it is challenging to control precisely the distribution of product variants, it is possible to influence the type and amount of these variants by acting both on the upstream and downstream parts of the production process. These considerations become particularly relevant in the case of biosimilars, where it is crucial to demonstrate that there are "no clinically meaningful differences" with respect to the reference product [68, 69].

1.4 Continuous Production Processes

In the previous sections, we described briefly the unit operations that are commonly used in the production of protein-based drugs. These considerations are sufficiently general to hold true for both batch and continuous manufacturing processes. The goal of this section is to compare these two modes of production, first in general terms and then with specific reference to biopharmaceutical processes.

1.4.1 Definition of Batch and Continuous Processes

A discussion on batch and continuous processes requires first to give clear definitions of these two terms, which is probably less obvious than one would expect. Generally, a process is termed *continuous* if it is constituted by open units with uninterrupted inlet and outlet material flows. *Batch processes* are instead constituted by units that are prevalently closed, and the feed materials/final products are loaded/withdrawn according to some time schedule. A domestic dishwasher represents a common example of a batch installation, where a given quantity of dishes is introduced in the device, undergoes successive operations (such as washing, rinsing, and drying) and is then taken out before another washing program is started. If the operations are repeated over regular periods (for example, the washing program is started every morning), the process is termed *periodic*. On the other hand, the continuous equivalent of the domestic dishwasher would be a conveyor washing machine used in large dining facilities. Importantly, while in the batch process, cleaning operations are performed one after the other in a single unit; they are performed simultaneously but at different locations along the conveyor belt in the continuous process, allowing constant inlet/outlet of dirty/clean dishes.

It is worth noting that the notion of constant inlet/outlet depends on the scale of observation and on the level of averaging. For example, the number of dishes cleaned per minute with the conveyor washing machine is constant during the working hours of the restaurant. However, if one looks at the number of dishes cleaned per minute over a week, one observes a discontinuous process, with peaks at lunch and dinner times. On the other hand, even though the overall cleaning process in a domestic dishwasher is discontinuous, single operations may be regarded as continuous at the time scale of the considered operation and for some material flows. This is, for example, the case for water (but of course not for the dishes) during the rinsing step if water is injected and evacuated at constant flow rates. It is thus clear that the distinction between "batch" and "continuous" processes is necessarily associated with a relevant time scale and an observed variable.

Moreover, it is important to avoid confusion between the concepts of being continuous, which as mentioned earlier characterizes a system with uninterrupted inlet and outlet flows, and being at *steady state*, where all the internal variables of the units (e.g., temperature, pressure, reactant concentration) are constant in time.

Considering the particular case of cell culture, several reactor types can be envisaged [19]. *Batch bioreactors* are rather rare because the depletion of nutrients eventually causes a decrease in cell viability, which is defined as the ratio of the number of viable cells over the total number of cells. Therefore, *fed-batch bioreactors*, where some feed solution is introduced intermittently, are usually preferred. On the other hand, two types of continuous bioreactors can be considered, namely *chemostats* and *perfusion bioreactors*. In both cases, the cells are introduced in the bioreactor before starting to operate the process, so that the continuous inlet flow contains only the culture medium and no cells. However, the two bioreactors differ regarding their outlet flows. Chemostats are characterized by a single continuous outlet flow containing the supernatant and the cells in the same proportions as inside the bioreactor. On the other hand, perfusion bioreactors

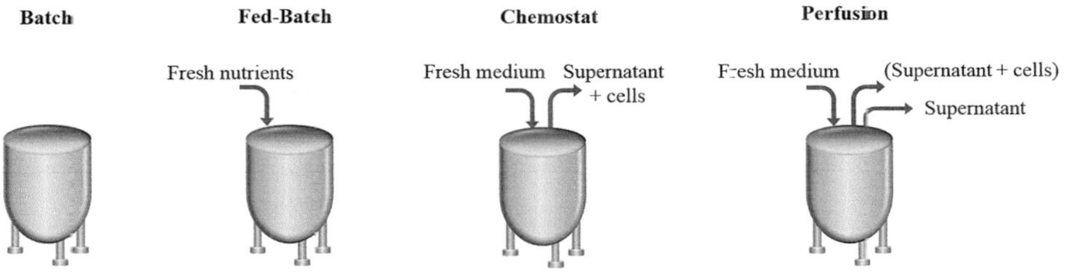

Figure 1.4 The four most common bioreactor types.

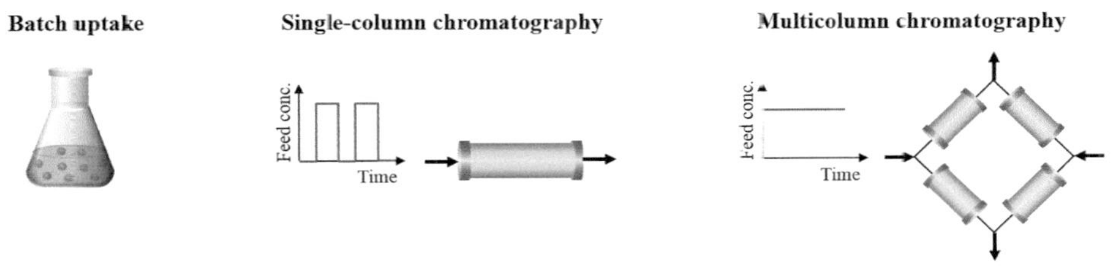

Figure 1.5 Schematic representations of batch uptake, single-column, and multicolumn chromatography.

are equipped with a cell-retention device that allows removing the supernatant, while keeping the cells inside the bioreactor [70, 71]. If one wishes to reach steady state, it is necessary to maintain a constant cell density in the bioreactor. To compensate the increase in cell density due to cell growth, a second outlet stream referred to as "bleed" is required. This outlet flow contains the supernatant and the cells in the same proportions as inside the bioreactor. The four types of reactors mentioned earlier are schematically represented in Figure 1.4.

Let us now analyze the case of chromatography, for which there is often confusion with the terms *batch* and *continuous*. By strictly applying the definitions given earlier, batch chromatography corresponds to a process in which the chromatographic medium, solutes, and eluent are mixed together in a closed vessel until thermodynamic equilibrium is reached. This process is also known as *batch uptake chromatography* and is commonly used to determine solute partitioning between the liquid and solid phases at equilibrium (see Chapter 2). However, in the literature, the term batch chromatography also designates a process in which a defined volume of the mixture to be separated is injected into a chromatographic column and eluted before the next injection starts, while the flow of eluent through the column is maintained constant. To avoid confusion with batch uptake chromatography, we prefer the term *single-column chromatography* in this book. As we will see in Chapter 2, single-column chromatographic processes are not capable of separating feed mixtures in a continuous manner. On the other hand, *multicolumn chromatography* makes it possible. Figure 1.5 illustrates schematically the three types of chromatography aforementioned: batch uptake, single-column, and multicolumn chromatography. In particular, it is seen that the feed mixture is introduced

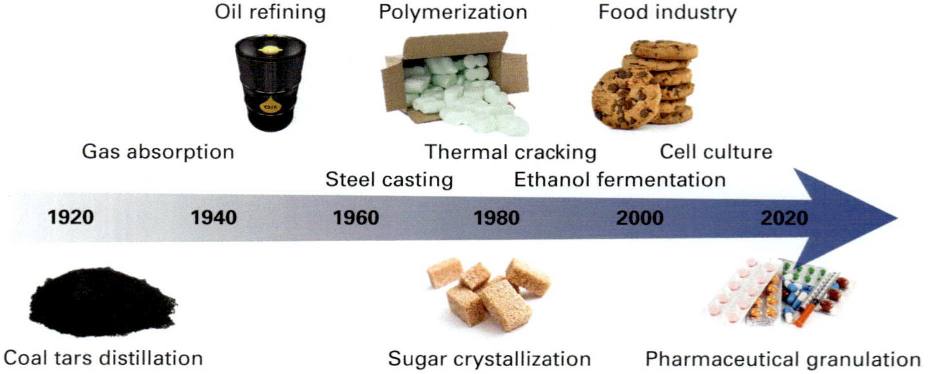

Figure 1.6 Timeline of the establishment of continuous production technologies in various industrial sectors.

by periodic injections in single-column chromatographic processes, while it can be introduced at a constant flow rate in multicolumn chromatographic processes. Fundamental concepts of multicolumn chromatography will be introduced in Chapter 3, while practical applications for the capture and polishing steps will be presented in Chapters 4 and 5, respectively.

1.4.2 Industrial Context

Continuous manufacturing has been successfully used for decades in a number of industries [72, 73], from the petrochemical to the polymer and the food industries. For illustrative purposes, Figure 1.6 shows the timeline for the establishment of continuous production processes in various industrial sectors.

These successes have driven a lot of expectations in the (bio)pharmaceutical industry for the production of both small-molecule [74–77] and protein-based drugs [78–93]. Among those expectations, one can mention a decrease of the capital expenditure (CAPEX) and a decrease of the operational expenditure (OPEX). Such prospects are often attributed to an expected reduction of the turnaround times, equipment size, solvent consumption, and manpower.

However, one needs to realize that the economic challenges faced by various industries may differ greatly, and it would thus be illusive to believe that continuous manufacturing is a one-size-fits-all solution. A careful analysis of the objectives and constraints of the particular process under consideration should be carried out in order to evaluate the pros and cons of the batch and continuous production modes, which may actually differ depending on several factors, such as the scale considered (i.e., laboratory, clinical, or commercial). Although capital and operational costs are critical for the massive production of low-value products such as chemical commodities or sugars, one can wonder what is the contribution of production costs in the manufacturing of biopharmaceuticals, where the costs associated with research and clinical tests are significant. The economic pressure may increase in the near future due to the loss of patent protection for several commercialized drugs and the subsequent emergence of biosimilars and biobetters [68]. Another relevant aspect in this context is the growing importance of alternative industrial approaches, such as the Indian pharmaceutical industry that strives to provide therapeutics at the lowest possible cost [94]. This new economic situation may motivate (bio)pharmaceutical companies to reconsider their current production strategies, and the transition from batch to continuous processes could therefore represent a "paradigm shift" in the words of Thomas Kuhn [95], meaning a necessary transformation to adapt to an ever-changing environment.

Finally, the regulatory requirements with respect to product quality control require some comments. The uncertainty regarding the definition of a "lot" is sometimes put forth as an obstacle for the implementation of continuous biopharmaceutical processes. However, the definition of *lot* given in section 210.3 of the code of federal regulations refers to an uniform amount of product, applicable to both batch and continuous production modes: "Lot means a batch, or a specific identified portion of a batch, having uniform character and quality within specified limits; or, in the case of a drug product produced by continuous process, it is a specific identified amount produced in a unit of time or quantity in a manner that assures its having uniform character and quality within specified limits." Therefore, agencies such as the FDA are definitely not an obstacle to the transition from batch to continuous processes and rather recommend "*cleaner, flexible, more efficient continuous manufacturing*" [96, 97].

1.4.3 Some Engineering Considerations

Let us now address the question of continuous pharmaceutical processes from an engineer point of view. For this, we first analyze separately the upstream and downstream parts of the production process. Then, we discuss their possible integration in a unique continuous process.

Upstream Processes

Before considering the specific and complex case of cell culture, we compare the intrinsic performances of batch and continuous reactors. This is a classic theme in chemical reaction engineering [98, 99] that we review here for the case of a simple homogeneous reaction involving only one reactant A and characterized by the reaction rate $\mathcal{R}_A = k_A C_A^{n_A}$, where C_A is the reactant concentration, k_A the reaction rate constant, and n_A the reaction order. We consider first a perfectly mixed reactor and analyze whether it would be more advantageous to operate it in batch or in continuous mode. These two types of reactors are commonly referred to as *batch stirred tank reactor* (BSTR) and *continuous stirred tank reactor* (CSTR), respectively, and are illustrated in Figure 1.7(a).

It can be easily shown with mass balance equations that a BSTR performs better than a CSTR for any positive value of n_A, which corresponds to the most common case where an increase in the reactant concentration leads to an increase in the reaction rate. This is illustrated in Figure 1.7(b) for $n_A = 1$, where the conversion is reported as a

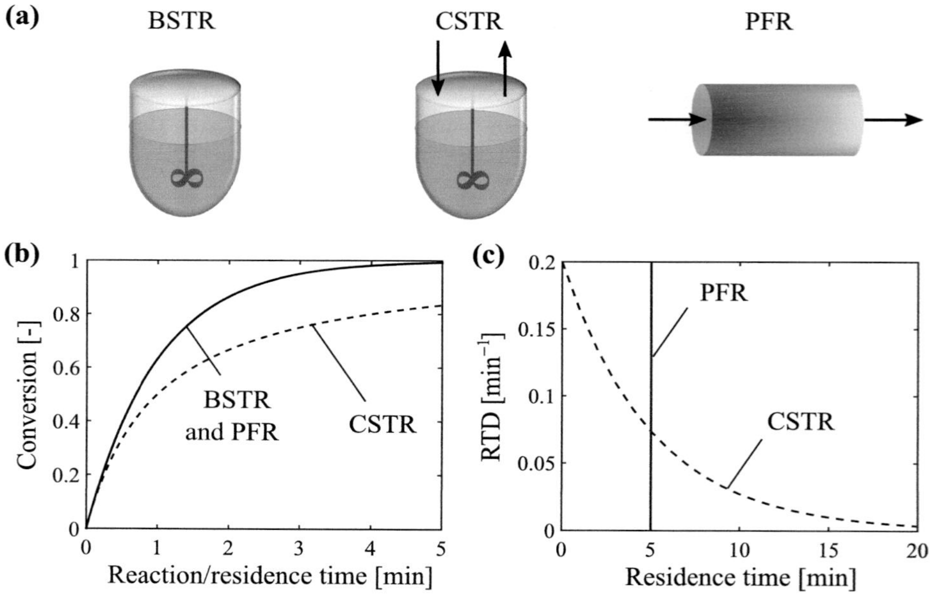

Figure 1.7 (a) Schematic representation of the most common ideal reactors. (b) Time necessary to reach a given conversion in the three ideal reactors considering a first-order reaction kinetics with a reaction rate constant of 1 min^{-1}. The term *time* refers to the reaction time in the case of the batch reactor (BSTR) and to the average residence time in the case of the continuous reactors (CSTR and PFR). (c) Residence time distribution (RTD) for a CSTR and a PFR. The average residence time, defined as the ratio between the reactor volume and the volumetric flow rate, was set to 5 min in both cases.

function of the "time" spent by the reactant inside the reactor. This "time" corresponds to the reaction time in the case of the batch reactor. On the other hand, it corresponds to the average residence time inside the reactor in the case of the continuous reactor, which is defined by the ratio between the reactor volume and the volumetric flow rate. It appears clearly in Figure 1.7(b) that the CSTR is less efficient than the BSTR at any conversion, meaning that more time is required to obtain a certain conversion as compared to the BSTR.

However, we can now consider another ideal continuous reactor called the *plug flow reactor* (PFR). It is represented as a tube, where the fluid is flowing as a series of infinitely thin plugs. It is assumed that each plug is perfectly mixed in the radial direction, while no mixing occurs in the axial direction. In such a reactor, all the molecules spend exactly the same time in the unit, which is given by the ratio between the reactor volume and volumetric flow rate. This is in contrast with the CSTR, where different molecules may follow different paths inside the reactor and thus exit at different times. The residence time distribution of a CSTR is given by a decreasing exponential function, as shown in Figure 1.7(c) for an average residence time of 5 min. As a comparison, the residence time distribution of a PFR is a Dirac function, which is also shown for an average residence time of 5 min in Figure 1.7(c).

It is important to realize that there is a strong analogy between the behavior of a PFR in space and the behavior of a BSTR in time. This is evidenced in Figure 1.7(b), where it is seen that the conversion profile in a PFR is identical to that obtained in a BSTR, provided that the reaction time is replaced by the residence time in the reactor. It results that the continuous plug flow reactor performs better than the continuous stirred tank reactor, at least for this example. This conclusion can in fact be generalized to any positive reaction order. However, it must be emphasized that the situation may change for more complex reacting systems. Overall, this simple example shows that moving from a batch to a continuous process does not necessarily bring any improvement and may actually even decrease process performances. It is thus crucial to consider the nature of the reactions and their kinetics to select the most appropriate reactor type [98, 99].

In Chapter 6, we will apply these elementary chemical reaction engineering considerations to the particular case of protein conjugation. In this context, we will show how to design both batch and continuous reactors for the production of conjugated proteins. We will then analyze the performances of these reactors in terms of conversion, yield, and selectivity.

Here, we focus on the case of cell culture bioreactors. In light of the preceding discussion, batch or fed-batch bioreactors would be expected to perform better than chemostat and perfusion bioreactors, which can somehow be regarded as CSTRs. In the following, we examine what still makes CSTRs attractive in the particular case of cell culture.

At first, it is essential to keep in mind that running a bioreactor is a rather delicate task because cultivating living organisms requires maintaining a well-defined and well-controlled environment within the whole reactor volume and all along the protein expression step. Small variations in the operational conditions may in fact be detrimental to cell viability and cell productivity. Such variations may also affect the quality of the product, such as the level of aggregates, the distribution of charge variants or the glycosylation pattern [100]. The objective to have a cell culture medium characterized by a composition that is constant in time and uniform in space guides the choice of the bioreactor toward a CSTR. In contrast, in batch or even fed-batch bioreactors, the nutrients from the culture media are depleted during time, and inhibitory or toxic species produced by the cells accumulate. This leads to a changing environment, which may impact cell viability and cell productivity and promote the expression of heterogeneous products [101, 102]. A simplistic theoretical model has been used to compare the kinetics of protein expression in batch and continuous stirred tank reactors [103, 104], illustrating that the latter is a good candidate to maintain the constant environment necessary for cell culture.

Two types of continuous bioreactors may be envisaged to obtain conditions uniform in space and constant in time, namely chemostat and perfusion bioreactors. The possibility to reach a high cell density in perfusion bioreactors thanks to the cell retention device makes the latter very attractive for upstream processing [101]. Perfusion bioreactors are not only appealing as protein production units, but also as seed train reactors to achieve a high inoculum cell density for a subsequent fed-batch production bioreactor [105, 106]. On the other hand, perfusion bioreactors bring new technological challenges

[70, 71, 82, 107]. For example, the fouling of the membrane of the cell retention device is especially troublesome and may cause the premature termination of the perfusion process. It is also necessary to control cell culture conditions in the external loop and to develop aeration strategies suitable to high cell densities, which may be challenging at the large scale. In addition, the cells should be genetically stable, resistant to shear stress and to aggregation. Finally, guaranteeing sterility during the entire production period may be complicated by the continuous inlet/outlet flows of material.

When these obstacles can be overcome, it is possible to cultivate cells at a high density for extended periods of time and to reach steady state, which implies that the composition of the bioreactor remains constant. To illustrate this, Figure 1.8 shows the time evolution of the percentage of glycoforms of a mAb expressed in a perfusion bioreactor (closed symbols). It is observed that, after around seven days, a constant distribution of mAb variants is produced. As a comparison, the percentage of glycoforms produced at a given time by the same cell line in a fed-batch bioreactor is shown with open symbols. It is seen that the distribution of glycoforms significantly changes during time in the fed-batch cell culture experiment.

At steady state, not only the cell density, protein concentration, and isoform distribution remain constant in time, but also the intracellular concentrations. This has been shown by analyzing the concentrations of amino acids, metabolites, and proteins both in the supernatant and in the cells as a function of time [102, 110–112]. These constant conditions are reached after a transient period during which the product specifications are rather unlikely to be met. The duration of the transient period is impacted by two distinct processes [108, 109]. The first one refers to the hydraulics of the system, which depends on the inlet/outlet flow rates, reactor volume, and mixing. Like in any other stirred vessel, the duration of this process is in the order of a few times the average residence time. The second process is more complex and refers to the cell metabolism, independent of the reactor scale [113]. This process seems to be slightly slower than the first one, as indicated by the data of Figure 1.8, where the glycoform distribution achieves steady state in about seven days against the two or three that are expected based on the reactor hydraulics [109, 114].

In the following, we further compare (fed)-batch and perfusion bioreactors in terms of protein concentration, medium consumption, protein residence time, and reactor size. For the sake of simplicity, we assume that steady state is reached in the perfusion process.

Protein Concentration

Let us start with the protein concentration reached in the harvested cell culture supernatant, which is a key parameter for the economic viability of both the upstream and downstream parts of the production process. Nowadays, mAb concentrations in the order of few g/L are typically reached in fed-batch bioreactors, while concentrations two to five times lower are often reported for perfusion bioreactors [82, 85, 88]. However, this difference in concentration is most likely due to a sub-optimal operation of perfusion processes. Extensive time and efforts have in fact been devoted to optimize cell lines and culture media for fed-batch processes, which do not necessarily correspond to the

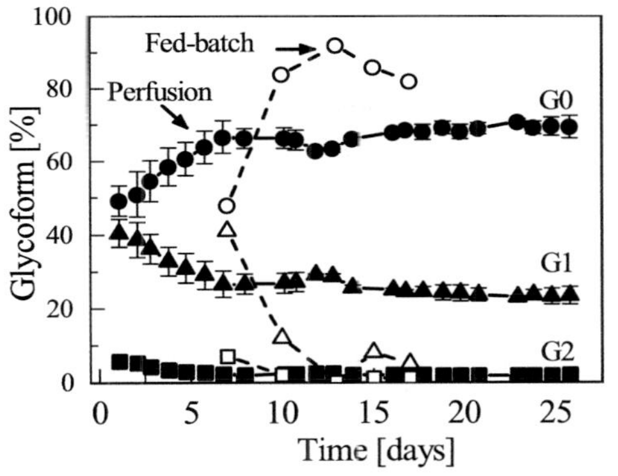

Figure 1.8 Percentage of mAb glycoforms produced by mammalian cell culture in a fed-batch bioreactor (open symbols) and in a perfusion bioreactor (closed symbols) as a function of time. Circles, triangles, and squares correspond to the G0, G1, and G2 glycoforms, respectively. In the case of the fed-batch experiment, the instantaneous distributions of glycoforms are reported (in contrast to cumulative distributions that would also account for the distributions produced at earlier times of the cell culture experiment). Adapted from [108], copyright 2017, with permission from Elsevier and from [109] by permission of John Wiley & Sons Inc.

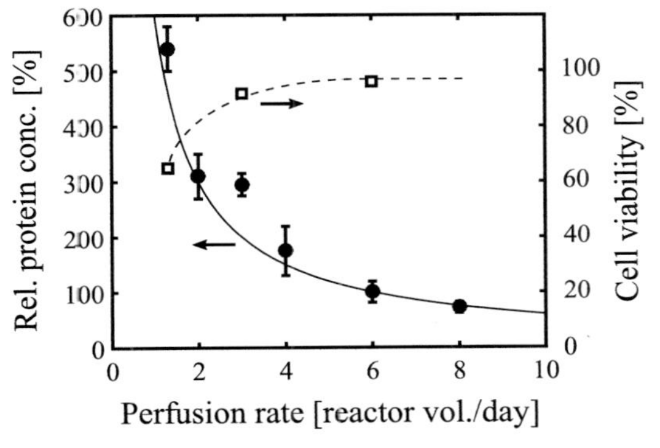

Figure 1.9 Influence of the perfusion rate on cell viability and protein concentration. The dashed line is a guide to the eyes, while the solid line represents the prediction of the decrease in protein concentration with the reciprocal of the perfusion rate. The values of the protein concentration are normalized with respect to the value corresponding to a perfusion rate of 6 reactor volume per day. Adapted from [115], with permission from Springer.

best candidates to run continuous operations. Minimizing cell growth at steady state (to reduce the cell discard rate) and improving the longtime stability of the cells against genetic mutations and clumping are examples of specific objectives to be reached in order to make perfusion bioreactors more competitive.

Figure 1.9 shows experimental results acquired during the production of a mAb by perfusion cell culture [115]. In these experiments, the volumetric flow rate of fresh medium (also called the perfusion rate) was varied between 1.3 and 8 volumes of bioreactor per day, while maintaining the cell density at 20×10^6 cells/mL by adjusting the bleed flow rate. It is observed that a high perfusion rate leads to the dilution of the protein inside the bioreactor and that the protein concentration is proportional to the reciprocal of the perfusion rate. In addition, it is seen that the cell viability is close to 100% for perfusion rates above 3 reactor volumes per day, but drops for lower perfusion rates. This indicates that the flow rate of fresh medium is not sufficient to ensure the minimum supply of nutriments and removal of inhibitory species necessary to maintain a good cell viability. The development and selection of proper cell lines and culture media specific to perfusion bioreactors is expected to allow reducing the perfusion rate while maintaining a high cell viability, thus alleviating the protein concentration difference between fed-batch and perfusion modes.

Medium Consumption

Another critical component in the economical evaluation of protein production by cell culture is medium consumption, often expressed in volume of cell culture medium per amount of produced protein. For a fed-batch bioreactor, the medium consumption is given by the reciprocal of the protein concentration at harvest. For a perfusion bioreactor, it is given by the reciprocal of the protein concentration at steady state if there is no bleed stream. Therefore, the lower protein concentrations currently reached in perfusion bioreactors result in higher medium consumption as compared to fed-batch processes. In addition, the loss of depleted medium in the bleed stream to compensate for the cell growth rate further contributes to the large medium consumption. Nevertheless, the development of dedicated cell lines and culture media mentioned earlier is expected to improve the protein titer and limit the growth rate, thus reducing the medium consumption in perfusion bioreactors.

Residence Time

A relevant process variable to be considered is the average residence time of the protein in the bioreactor. In a batch cell culture, proteins produced at the beginning of the cultivation step spend the whole production time inside the bioreactor, while proteins produced later on only spend a fraction of it. To give an order of magnitude, proteins spend on average roughly one week in a fed-batch bioreactor. In the case of perfusion bioreactors instead, the average residence time is given by the ratio between the volume of the bioreactor and the volumetric flow rate of fresh medium. For instance, proteins spend on average half a day in a bioreactor whose perfusion rate is two reactor volumes per day. This reduction in the residence time may be particularly useful for unstable proteins, which are, for example, prone to aggregation, fragmentation, deamidation, or oxidation. An emblematic example is the production of recombinant factor VIII, which is used in the treatment of hemophilia A. Indeed, the high aggregation propensity of this protein encouraged the development of the first worldwide licensing of a recombinant protein produced by perfusion mammalian cell culture (Kogenate®) [24]. The average residence time was reduced from 12 days in the case of the batch process to 2–3 h in the perfusion process. Nevertheless, it has to be noted that a short residence time is associated with a high perfusion rate, which means a high medium consumption and a low protein concentration. There is, therefore, a compromise to be found between protein degradation, medium consumption and protein concentration.

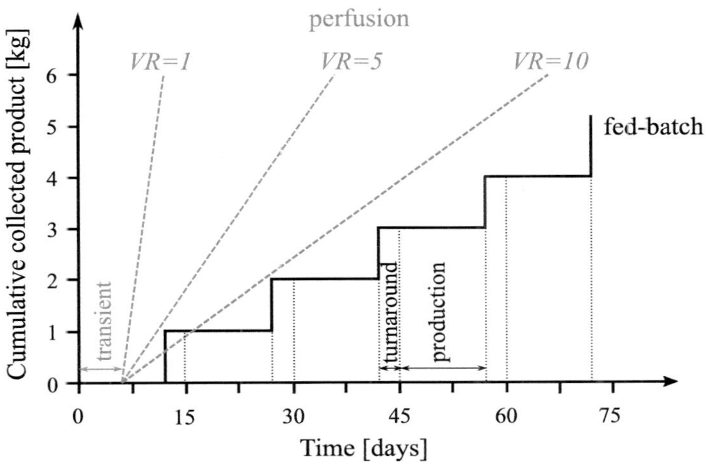

Figure 1.10 Comparison between fed-batch and perfusion cell culture in terms of reactor size. For the fed-batch bioreactor, a production time of 12 days and a turnaround time of 3 days are considered. For the perfusion bioreactor, it is assumed that 6 days are needed to reach steady state. The cell culture supernatant is collected at a protein concentration of 1 g/L in both cases. The perfusion rate is set to one reactor volume per day. The volume of the fed-batch reactor is set to 1000 L, while three different volumes are considered for the perfusion bioreactor, namely 1000 L, 200 L, and 100 L, corresponding to volume ratios (*VR*) equal to 1, 5, and 10, respectively.

Bioreactor Volume

Finally, let us compare fed-batch and perfusion cell cultures in terms of bioreactor size. To do so, we consider the illustrative example represented in Figure 1.10, which shows the cumulative quantity of product collected as a function of time for a typical process (see numerical values in the caption). In the case of the fed-batch bioreactor, no product is collected during the cell culture, and it is only at the end of the culture that the reactor is harvested and the product collected. Then, a turnaround time is necessary for cleaning and start-up operations before a new cell culture run can be started. This leads to the stepwise cumulative collected product profile represented by the black solid line in Figure 1.10. Regarding perfusion bioreactors, we assume that no product is collected during the transient period, i.e., the time before reaching steady state. Then, the quantity of collected product is linearly increasing during the whole time of operation. This is shown by the dotted straight lines in Figure 1.10, which correspond to different values of the volume ratio *VR* defined as the ratio of the volume of the fed-batch bioreactor over that of the perfusion bioreactor.

If the two reactors (fed-batch and perfusion) have the same volume and are operated at the same protein concentration, it is seen in Figure 1.10 that the quantity of product collected from the perfusion bioreactor is always much higher than the one collected from the fed-batch bioreactor, indicated as $VR = 1$ in Figure 1.10. The size of the perfusion bioreactor can therefore be significantly reduced while collecting similar quantities of product. This does not only have a positive impact on the investment costs, but also allows achieving a better mixing, i.e., a more uniform distribution of nutrients

Table 1.2. Examples of marketed proteins produced by perfusion cell culture. The dates correspond to the years of approval.

Name	Trade name	Company	Year
Abciximab	Reopro®	Janssen Biologics B.V.	1994
β-glucocerebrosidase	Cerezyme®	Genzyme	1994
^{111}In capromab pendetide	ProstaScint®	Cytogen	1996
Follicle-stimulating hormone	Gonal-f®	Merck-Serono	1997
^{99m}Tc votumumab	Humaspect®	Organon Teknika	1998
Interferon β-1a	Rebif®	Merck-Serono	1998
Basiliximab	Simulect®	Novartis	1998
Infliximab	Remicad®	Janssen Biotech / Merck & Co	1998
Factor VIII	Kogenate-FS®	Bayer	1998
Activated protein C	Xigris®	Eli Lilly	2001
Agalsidase β	Fabrazyme®	Genzyme	2003
Alglucosidase α	Myozyme®	Genzyme	2006
Golimumab	Simponi®	Johnson & Johnson / Merck & Co	2009
Ustekinumab	Stelara®	Janssen Biotech	2009
Siltuximab	Sylvant®	Janssen Biotech	2014

and oxygen in the bioreactor. For illustrative purposes, it is shown in Figure 1.10 that a reduction of the reactor volume by a factor 10 can be achieved to obtain a similar productivity in the perfusion process as compared to the fed-batch process (indicated as $VR = 10$) in our example. The actual volume reduction depends on the values of the process variables, such as the turnaround times and protein concentration. For instance, the volume reduction would be even more pronounced for a longer turnaround time between consecutive batches. On the other hand, if the perfusion bioreactor were operated at a protein concentration lower than in the fed-batch bioreactor, as it is still often the case, the volume reduction would be lower.

Overall, the choice of the bioreactor should be dictated by the constraints and objectives of the process, taking into account several factors such as cell stability, protein degradation, medium consumption and productivity. In the literature, the comparison between fed-batch and perfusion technologies has been widely discussed, also in economic terms, indicating the variety of scenarios that can be encountered at the industrial level [82, 85, 86, 88]. Several marketed biologics are currently produced by perfusion cell culture, including blood factors, enzymes, and mAbs [82]. Examples are listed in Table 1.2.

Downstream Processes

Let us now consider the downstream part of the manufacturing process. Chromatography is a crucial technique in the purification of therapeutic proteins, and it is thus important to understand if and how continuous multicolumn chromatography performs better than single-column chromatography. We have seen with simple reaction engineering considerations that moving from a batch to a continuous process does not necessarily lead to improved process performances. The same holds true for chromatography. Indeed, a continuous chromatographic process can be obtained by properly parallelizing multiple columns whose loading phases are simply shifted in time so as to allow a continuous feed of the material to be treated and a continuous recovery of the products. This situation is, for example, typical for wastewater treatment, where the feed stream comes from the outlet of a continuous production unit. A peculiar realization of this parallelization is annular chromatography, where the feed is introduced on top of a rotating chromatographic bed packed in between two concentric cylinders [116–119]. Due to the negligible mixing in the radial direction, the contact between the liquid and solid phases in annular chromatography is the same as in a single-column process [120, 121]. Therefore, even though continuous, parallelized and annular chromatographic processes do not bring any improvement in terms of separation performances as compared to single-column processes.

In fact, rather than their continuous nature, it is their ability to simulate a countercurrent movement between the fluid and solid phases that makes multicolumn chromatographic units particularly efficient. The fundamentals of countercurrent chromatography will be presented in Chapter 3, while here we simply show a schematic representation of countercurrent contact in Figure 1.11 for illustrative purposes. This example is inspired from the aquatic respiration of fishes, whose gills allow the uptake of oxygen dissolved in water. Briefly, fish gills are composed of lamellae densely irrigated with blood vessels that are permanently put in contact with water pumped through the fish mouth. The orientation of the gill lamellae is such that the water flows in the opposite direction to the blood [122]. In order to better understand how this countercurrent flow helps the aquatic respiration, it is easier to first take a look at the opposite case, i.e., when

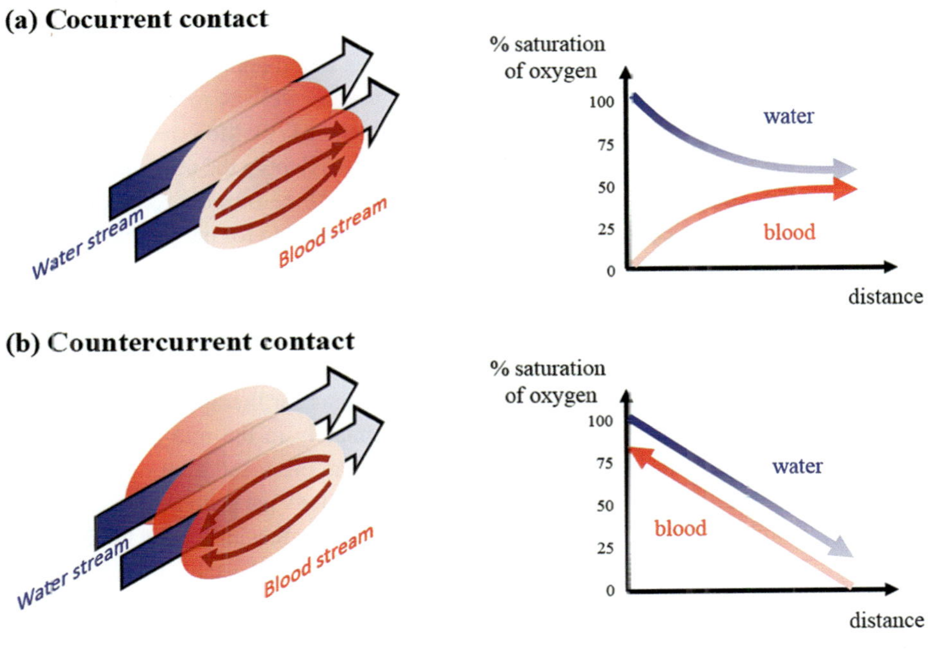

Figure 1.11 Illustration of the oxygen transfer from water to blood in fish gills. (a) Hypothetic cocurrent configuration. (b) Countercurrent configuration.

the water and the blood flow cocurrently. To do so, a hypothetic cocurrent gas exchange system is represented in Figure 1.11(a). In this cocurrent oxygen exchange system, the inlet water, which is saturated in oxygen (represented in dark blue) comes into contact with the bloodstream which contains very little oxygen (represented in light red). In this configuration, the oxygen transfer from the water to the bloodstream is very fast at the entrance of the gill due to the large difference in oxygen concentration. All along the lamella, the oxygen concentration in the water diminishes (light blue) and is transferred to the bloodstream (dark red). Eventually, the gas transfer stops once the concentrations in the two streams are at equilibrium, as schematically shown on the right-hand side of Figure 1.11(a). On the other hand, when the water flows in the opposite direction than the blood, i.e., the two streams are flowing countercurrently as shown in the left-hand side of Figure 1.11(b), gas transfer occurs all along the lamella. In this configuration, the oxygen-rich blood is brought in contact with the inlet water stream, which is saturated in oxygen. Oxygen uptake by the oxygen-rich blood is still possible because it is in contact with the water stream that contains oxygen at the highest possible concentration. The outlet water stream, which is already partly depleted in oxygen, still contains enough gas to ensure transfer with the oxygen-poor blood. This is schematically represented in the diagram on the right-hand side of Figure 1.11(b), where it is seen that the oxygen concentration in the blood can reach higher values as compared to the cocurrent case. This simple example illustrates how countercurrent exchangers outperform cocurrent ones, which will be particularly important to develop efficient separation processes, as discussed in the following chapters.

The evolution of multicolumn countercurrent chromatography over the years has been quite rich [79]. It started with the invention of the simulated moving bed (SMB) process by Broughton for the separation of xylene isomers in the middle of the twentieth century [123]. Since then, SMB processes found applications in the food industry, and in particular in the production of sugars, as well as in the pharmaceutical industry, especially for the separation of optical isomers [79]. More recently, several variations of the SMB process were developed in order to adapt to the needs of the biopharmaceutical industry [83]. The classical SMB process will be introduced in Chapter 3, while other countercurrent chromatographic processes suitable for the protein capture and polishing steps will be presented in Chapters 4 and 5, respectively.

Although various countercurrent multicolumn chromatographic systems are available for the continuous purification of therapeutic proteins at the laboratory and pilot scales, these processes are not well established at the commercial scale yet. The first applications of continuous chromatographic systems for therapeutic protein purification were reported in the patent literature approximately in the mid-2000s, which is more than ten years after the first marketed mAb was produced continuously by perfusion cell culture. In fact, Table 1.2 shows that continuous perfusion bioreactors are already well implemented in the biopharmaceutical industry. In addition, technologies for continuous centrifugation and continuous filtration are also well established [81]. Therefore, continuous chromatography appears to be the bottleneck for the implementation of integrated continuous processes in the biopharmaceutical industry, which is one of the motivations for writing this book.

Integrated Processes

After having analyzed the unit operations separately, considering now the complete process as a whole offers additional perspectives. Exploiting the unity of the process and the connections between the different units is referred to as *process integration* and is applicable to both batch and continuous modes [124–126].

The choice between batch or continuous mode for the individual unit operations is largely case-specific. However, it is important to realize that ensuring a proper continuity between the various units is key to an effective process integration. In particular, turning only one unit from batch to continuous mode implies to invest either in hold tanks to store inlet and outlet streams or in more units to parallelize the rest of the process and match the different residence times. Therefore, the transition from batch to continuous production technologies may necessitate a holistic redesign of the manufacturing process to achieve end-to-end integration.

The integration of reaction and purification steps will be further discussed in the context of protein conjugation in Chapter 6. A few examples of integrated continuous processes for the production of mAbs, integrating the upstream and downstream operations, have been published in the literature [127–129]. For illustrative purposes, we report two of such examples in the following. These two examples deal with the continuous production of mAbs with perfusion bioreactors equipped with an alternating tangential flow (ATF) system and mainly differ in their downstream part.

In the first example [129], the protein capture step was performed with affinity chromatography using a two-column process, namely the CaptureSMB, which is further described in Chapter 4. A surge bag between the bioreactor and the CaptureSMB was necessary because the feed flow rate of the selected chromatographic unit is periodically changing in time, while the outlet flow rate of the perfusion bioreactor is constant. It is also worth mentioning that sterile filters were added at the inlet and outlet of the surge tank in order to minimize the risk of contamination. The Protein A eluate was kept in a retention device at pH 3.2 for 30 min. Such low pH hold is a standard viral inactivation procedure, as described in Section 1.2.5. Then, two polishing steps were implemented: the first one using a two-column solvent gradient purification unit (MCSGP, further described in Chapter 5) packed with a cation exchange chromatographic medium; and the second one using a single-column unit packed with an anion exchange medium. Before loading the content of the viral inactivation device in the MCSGP, the pH of the solution was adjusted thanks to inline dilution to approximately 5.5 to ensure sufficient binding on the cation exchange medium. Due to the functioning mode of the MCSGP, the mAb was recovered only intermittently at the outlet of the first polishing step. This explains that operating conditions could be found to perform the second polishing step with only one single-column unit. In the case where a continuous stream would exit the preceding unit, several single-column processes in parallel would need to be implemented to avoid intermediate storage (or one should consider using another continuous unit).

The reactor was operated for 18 days continuously, and in the last 4 days it was connected to the downstream units. The concentration of the mAb and of three representative impurities (HCP, DNA and leached Protein A ligands) at steady state at different

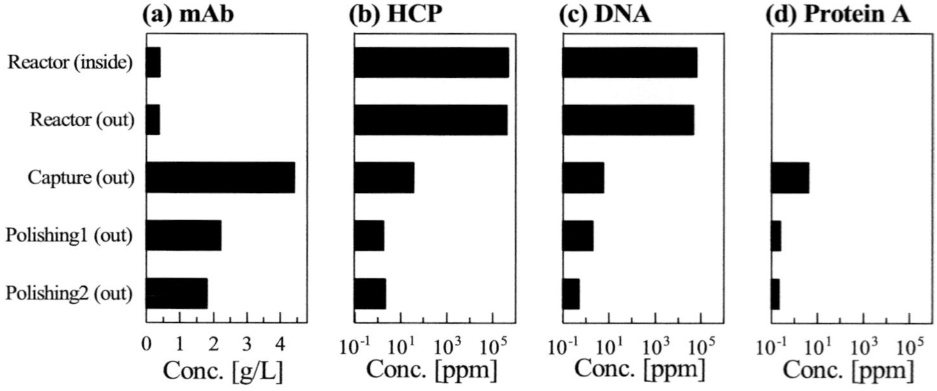

Figure 1.12 Illustrative example of an integrated process for mAb production. Concentrations of (a) mAb, (b) HCP, (c) DNA, (d) leached Protein A at steady state at different locations. Adapted from [129] by permission of John Wiley & Sons, Inc.

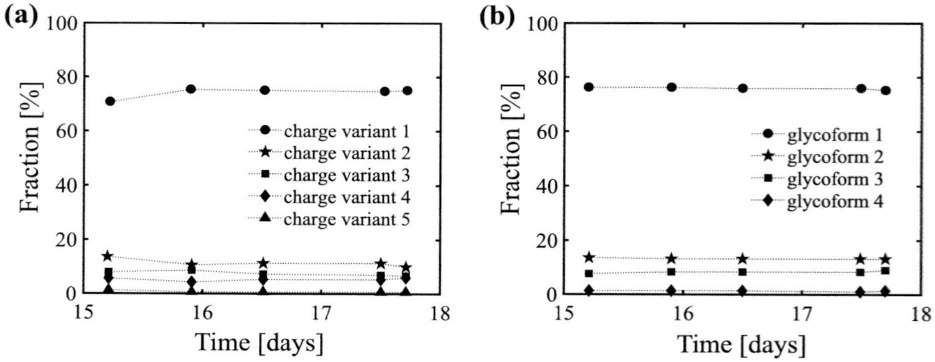

Figure 1.13 (a) Charge variant and (b) glycoform distributions at the outlet of the integrated process reported in Figure 1.12. Adapted from [129] by permission of John Wiley & Sons, Inc.

locations of the process are presented in Figure 1.12. It is seen that no differences in composition are observed between the reactor content and the harvested cell culture fluid, indicating that no undesired retention in the cell filtration device occurred. Then, it is observed that the protein capture step allowed reducing the concentration of both HCP and DNA by approximately 4 orders of magnitude, while achieving an increase in the mAb concentration by a factor of 12. The subsequent polishing steps further reduced the concentration of residual HCP and DNA as well as the concentration of affinity ligands leaching from the mAb capture step. This also resulted in a decrease in the mAb concentration by a factor of around 2. Overall, the polishing steps allowed depleting the content of HCP and DNA by a factor of 5 approximately with respect to the mAb. Figure 1.13 shows the charge variant and glycoform distributions during the last 3 days of steady state operation at the outlet of the integrated process, showing that a product of constant quality is produced over time.

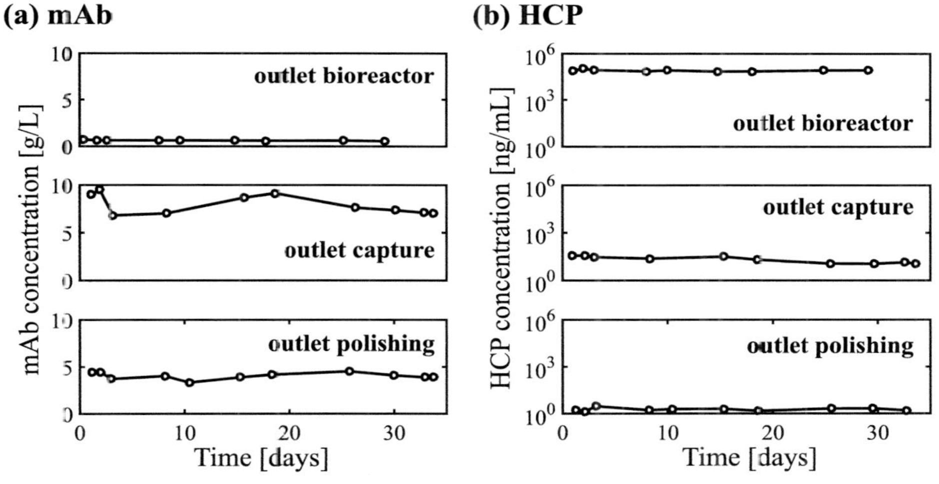

Figure 1.14 Illustrative example of an integrated continuous process for mAb production. (a) Profiles of the mAb concentration at the outlet of the bioreactor, capture and polishing steps during 1 month of operation. (b) Profiles of the HCP concentration at the outlet of the same three steps. Adapted from [128], copyright 2015, with permission from Elsevier.

In the second example [128], the protein capture step was performed with affinity chromatography using a three-column periodic countercurrent chromatographic process, which is abbreviated as 3C-PCC. As in the previous example, the clarified harvested fluid exiting the bioreactor was injected in a surge bag through a sterile filter before being loaded in the first chromatographic unit. The mAb concentration at the outlet of the bioreactor was maintained at 1 g/L during 1 month of operation, as seen in the top panel of Figure 1.14(a). As an example of impurities that need to be removed by the downstream process, the HCP concentration profile at the outlet of the bioreactor is shown in the top panel of Figure 1.14(b). It is observed that the HCP concentration is constant in time at the bioreactor outlet and in the order of 10^5 ng/mL.

The concentration profiles of mAb and HCP after the capture step are shown in the middle panels of Figure 1.14(a) and (b), respectively. It is observed that the capture step allowed increasing the protein concentration by almost an order of magnitude and decreasing the concentration of HCP by four orders of magnitude. The Protein A eluate was then pumped in a stirred vessel with a residence time of 1 h for viral inactivation at pH 3.75. The polishing step was performed with a second 3C-PCC system using a cation exchange chromatographic medium. Importantly, the pH and ionic strength of the solution had to be adjusted in-line before loading the cation exchange column, as in the previous example. The concentration profiles of mAb and HCP after the polishing step are shown in the bottom panels of Figure 1.14. It is seen that the mAb concentration decreased from a bit less than 10 g/L to around 5 g/L and that cation exchange chromatography further reduced the concentration of HCP by one order of magnitude. Overall, the polishing step allowed depleting the content of HCP by a factor of 5 approximately with respect to the mAb.

These two examples demonstrate the feasibility of integrated continuous bioprocesses. They also highlight some important aspects that need to be accounted for when designing such processes: *(i)* the flow rates and times associated with the different units should be properly adapted to minimize the use of surge tanks and parallel units, which increase the equipment footprint; *(ii)* the buffer solutions should be selected with a view to reducing the use of in-line dilution, which increases the volumes of solution to be treated; and *(iii)* particular care should be taken to avoid contamination during prolongated periods of time. Another important aspect to be considered in the future development of continuous bioprocesses is the implementation of online monitoring strategies and control systems. These are needed to fully exploit the potential of an integrated process, where measurements taken all along the process can be used to adapt the operating conditions of the different units in order to reject disturbances and maintain optimal process performance [130, 131]. Overall, envisioning continuous technologies in the biopharmaceutical industry needs to be part of a general reflection about the integration and the organization of the different unit operations.

2 Fundamentals of Protein Chromatography

2.1 Introduction

We saw in Chapter 1 that chromatographic processes are essential in the industrial-scale purification of biopharmaceutical products. Chromatography is indeed a versatile and powerful separation technique capable of accommodating the requirements of the downstream process, namely removing in an efficient manner numerous and diverse impurities among which some may be very similar to the protein of interest.

The goal of this chapter is to provide the fundamentals of protein chromatography. We start by describing the physical phenomena underlying solute separation in a chromatographic system and by introducing the various chromatographic modes. Then, we give the necessary tools to model protein chromatography and to estimate model parameters. Finally, we define process performance criteria and discuss how to design a chromatographic process.

2.2 Interactions with Chromatographic Media

Chromatography relies on the partitioning of solutes between a liquid phase and a solid phase, which is typically constituted of porous particles packed in a column. The separation of solutes is driven by the strength of the interactions with the solid phase: the stronger a solute interacts with the solid phase, the more it is retained in the column and the later it elutes. Several types of chromatographic media can be envisaged to separate a mixture of solutes, which exploit various physicochemical properties of the solutes such as their size, charge, and hydrophobicity [21]. In the following, we provide a brief overview of the main chromatographic techniques and classify them according to the type of interactions underlying the separation process.

2.2.1 Steric Interactions

Steric interactions set the basis of *size exclusion chromatography* (SEC), also known as *gel filtration chromatography* [21]. The underlying idea of SEC is to exploit the internal porosity of the chromatographic medium and use it as a molecular "sieve" to separate solutes according to their size: large molecules are sterically excluded from the pores of the chromatographic medium and thus elute faster than small molecules that can enter the pores.

Ideally, the separation is exclusively based on size differences, and surface interactions should not play any role. To minimize protein–surface interactions, SEC media are typically constituted of highly porous hydrophilic polymer gels such as agarose, dextran, and polyacrylamide [132]. These polymers are often cross-linked to improve their mechanical properties. The composition of the eluent can also be tuned to reduce the possible adsorption of the protein on the solid phase [133]. For instance, the addition of arginine has been shown to improve the recovery of protein aggregates [134].

Size exclusion chromatography is widely used for analytical purposes in a large range of applications. For preparative purposes, it is especially attractive to separate solutes of significantly different size. For instance, SEC is quite often used to perform buffer exchange, during which the small buffer components are easily separated from the biomacromolecules that are excluded from the pores of the chromatographic medium. Another typical application of SEC to be discussed in Chapter 7 is the separation of aggregates from the monomeric form of therapeutic proteins. Depending on the separation to be carried out, it is important to properly select the pore size of the chromatographic medium. Each chromatographic medium is in fact characterized by an exclusion limit in terms of solute size above which solutes are completely excluded from the pores, and thus all co-elute in a single peak [132].

An important advantage of SEC lies in the use of mild operating conditions that are not detrimental to the stability of the proteins. On the other hand, SEC suffers from its inability to separate solutes of similar size.

2.2.2 Hydrophobic Interactions

Hydrophobic interactions, which are driven by van der Waals forces between nonpolar regions of the solutes and a hydrophobic chromatographic medium, are at the core of *hydrophobic interaction chromatography* (HIC) and *reversed-phase chromatography* (RPC) [21]. The term *reversed-phase* has a historical background because it was introduced to contrast with what has been originally considered as "normal phase" chromatography. The latter involves hydrophilic chromatographic media containing silanols groups on silica backbones and utilizes apolar solvents.

Both in HIC and RPC, highly hydrophobic solutes interact more strongly with the chromatographic medium than with the polar solvent, and thus elute later than less-hydrophobic solutes. The difference between HIC and RPC lies in the strength of the interactions: a greater hydrophobicity characterizes RPC media compared to HIC media [21]. For RPC, alkyl chains of various length (e.g., C4, C8, or C18) are typically grafted on silica backbones, while for HIC, butyl, hexyl, and phenyl groups are often grafted on polymer particles [135, 136].

Interactions with HIC media can be modulated by the addition of a salt (e.g., NaCl, Na_2SO_4). Typically, the protein is loaded at a high salt concentration, which reduces its solubility and thus promotes solute–surface interactions. It is then eluted by decreasing the content of salt in the solvent [21]. On the other hand, interactions with RPC solid phases are modulated with the addition of water-soluble organic solvents (e.g., acetonitrile, methanol, ethanol, isopropanol). In this case, the protein is loaded in the column with an aqueous eluent and eluted by increasing the content of organic solvent. In both HIC and RPC, an aspect to be considered with caution is the protein solubility because a high salt concentration or a high organic solvent content may trigger protein precipitation.

Because the use of organic solvents often promotes protein unfolding and aggregation, RPC is mainly employed for the purification of small proteins and peptides [136]. The milder conditions used in HIC make this technique applicable to most proteins, and a number of commercial monoclonal antibodies are purified by HIC [22].

2.2.3 Electrostatic Interactions

Another strategy consists in exploiting electrostatic interactions between charged chromatographic media and charged regions of the solutes, which provides the basis for *ion exchange chromatography* (IEC) and *ion exclusion chromatography* [21, 137].

Depending on the charge of the chromatographic medium, ion exchange chromatography is referred to as *anion exchange chromatography* or *cation exchange chromatography*. Anion exchange chromatography relies on the interaction of negatively charged solutes with a positively charged chromatographic medium, while cation exchange chromatography relies on the interaction of positively charged solutes with a negatively charged chromatographic medium. The term *exchange* reflects the idea that the counter-ions of the chromatographic medium are replaced by the charged molecules to be separated. This exchange mechanism must be accounted for in the modeling of IEC, as further discussed in Section 2.4.2.

IEC is referred to as *strong* when the ionizable groups of the chromatographic medium are fully dissociated, as it is for example the case with quaternary amines and sulfonyl groups for anion and cation exchange, respectively [138]. On the other hand, IEC is referred to as *weak*, when the ionizable groups of the chromatographic medium are only partially dissociated, as it is for example the case with ternary amines and carboxyl groups for anion and cation exchange chromatography, respectively [138].

The proper selection of solution pH is crucial in ion exchange chromatography. Proteins are indeed composed of a large number of amino acids whose side chains can be neutral, positively, or negatively charged depending on the solution pH. The sum of all these individual contributions confers to the protein a net charge that is positive at low pH values and negative at high pH values. The pH value at which the protein net charge is null is known as the isoelectric point, pI. To illustrate the impact of solution pH on protein separation with IEC, let us consider the example shown in Figure 2.1. When the solution pH is below the pI of the two proteins to be separated, both proteins are positively charged. Provided that the two proteins interact differently with the chromatographic medium, a separation can be achieved by cation exchange chromatography. The protein carrying the highest charge usually elutes the latest. On the other hand, no separation can be obtained in this pH range with anion exchange chromatography because none of the two proteins is retained on the chromatographic medium. When the pH of the solution is above the pI of both proteins, the situation is reversed. Both proteins are negatively charged, and the separation can be obtained

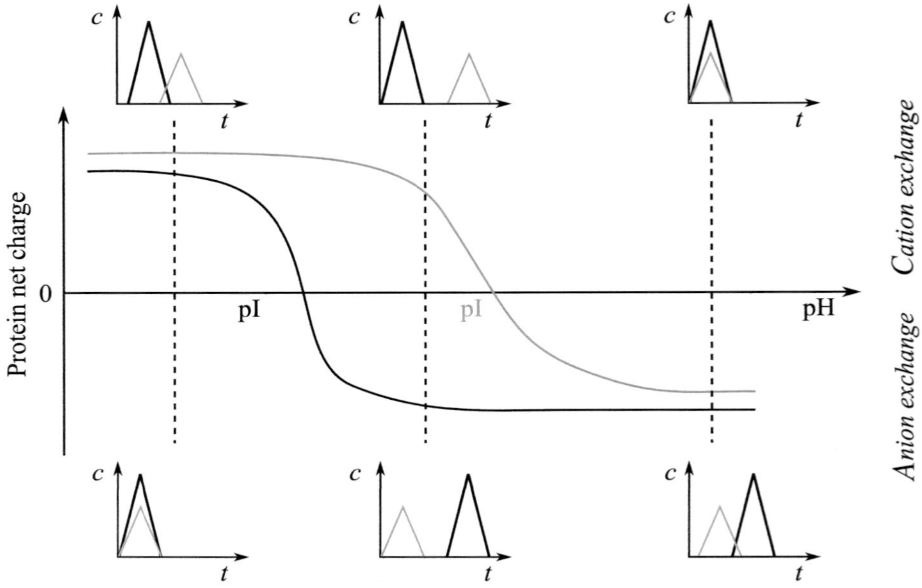

Figure 2.1 Principle of ion exchange chromatography as a function of the solution pH. The schematic chromatograms illustrate whether or not a separation can be achieved.

with anion exchange chromatography but not with cation exchange chromatography. At intermediate pH values, either cation or anion exchange chromatography can be used: The protein carrying the same charge as the chromatographic medium elutes without being retained, while the second protein interacts with the chromatographic medium and thus elutes at later times.

Ion exchange chromatography is a well-established technique for the purification of therapeutic proteins due to its versatility, high saturation capacity, and ability to preserve the biological activity of the purified product. Most of the downstream processes in the industrial production of monoclonal antibodies employ at least one polishing step based on ion exchange chromatography [22].

In contrast to ion exchange chromatography, repulsive electrostatic interactions are exploited in ion exclusion chromatography [139]. Strongly charged ions carrying the same charge as the surface of the chromatographic medium are repelled and elute rapidly. On the other hand, neutral and weakly charged ions access completely, or at least partially, the pores of the chromatographic medium, depending on their charge and size. Ion exclusion chromatographic applications include acid/sugar separations, amino acid purifications, and desalting processes [137, 140].

2.2.4 Biospecific Interactions

The biological activity of proteins involves specific interactions between protein active sites and other molecules as enzymes, inhibitors, substrates, etc. These interactions include complementary van der Waals forces, hydrophobic and electrostatic interactions, as well as hydrogen bonds. The high specificity of such interactions is at the

Table 2.1. Overview of proteins and associated affinity ligands.

Protein	Affinity ligand
IgG	Protein A
Antibody (IgG, IgY, etc.)	Antigen peptide
Antigen	Antibody
Fibronectin	Gelatin
Glycoprotein	Lectin
Growth factor	Heparin
Enzyme	Substrate

root of *affinity chromatography*. The idea is to attach covalently specific ligands to an inert chromatographic matrix. The mixture to be purified is injected under conditions that favor specific and reversible binding of the target protein to the ligands, so that the protein strongly binds on the chromatographic medium, while the other solutes flow through the column without being retained. Subsequently, the target protein is recovered by changing the elution conditions to weaken the protein/ligand binding.

Affinity chromatography was discovered in 1910 [141], but it was only in the late 1960s that its potential as a purification technique was really exploited for the purification of enzymes with immobilized substrates [142]. Since its discovery, this technique has been extended to various types of proteins. A brief overview of common ligand/protein couples often used in affinity chromatography is reported in Table 2.1 [21]. The best-known example of affinity chromatography at the industrial scale is certainly the purification of human immunoglobulins G (IgG) with *Protein A* affinity ligands. Protein A is a natural protein found in the cell wall of *Staphylococcus aureus*, whose biological function is to bind antibodies to camouflage the bacterium and protect it from the reaction of the immune system. Protein A contains five domains that can specifically bind the Fc regions of IgG1, IgG2 and IgG4. However, due to steric hindrance, only two to three IgGs usually bind to one Protein A ligand. Even though affinity chromatography is extremely attractive due to its very high selectivity toward the target protein, the availability of specific ligands restricts its use to a few classes of proteins. Nevertheless, the situation is constantly evolving with the development of new technologies to design custom affinity ligands, which may turn affinity chromatography into a more versatile purification technique [143].

The affinity ligands need not only to specifically bind the target protein, but they should also be stable under the operating conditions and easily retrieve their binding properties after the protein/ligand association has been disrupted to recover the target protein. Moreover, the linkage between the ligands and the chromatographic support should be chemically stable under the adopted conditions so as to avoid ligand leaching.

2.2.5 Complexation

Immobilized metal affinity chromatography (IMAC) relies on the formation of a complex between electron donor groups of the protein and metal ions immobilized on the chromatographic support [21]. Transition metals including Cu^{2+}, Zn^{2+}, Ni^{2+}, or Fe^{3+} are frequently employed and are attached to the chromatographic support by means of chelating ligands [144–146]. It is thought that proteins interact with the immobilized metal ions mainly via histidine groups exposed to the protein surface. A single histidine group is generally sufficient to bind an immobilized Cu^{2+} ion, and the presence of several histidine groups may lead to very strong binding. In a sense, IMAC can be regarded as an intermediate technique between affinity chromatography, which is highly specific, and other less specific techniques, such as IEC. Three methods can be considered to elute the protein: *(i)* decrease the eluent pH, which protonates the electron donor and thus disrupts the protein-ion complex; *(ii)* add a competitive ligand such as imidazole; or *(iii)* add a chelating agent such as EDTA to strip off the protein–ion complex [144, 145].

IMAC is readily applied to proteins containing surface-exposed histidine groups, as, for example, some growth hormones, some antibodies, or human interferon γ [145]. However, histidine is not found in abundance in natural proteins, and a potential strategy consists in grafting a histidine group to the protein prior to performing purification by IMAC [144, 145]. In the case where the His-tag domain is found to interfere with the biological function of the protein, enzymatic or chemical cleavage of the tag followed by an additional purification step are required [145]. The IMAC is a conventional technique at the laboratory scale, but it has to date only few industrial applications, presumably due to the potential risk of metal ion leakage. Notable successful examples are the purification of factor IX [147] and of aprotinin [148].

2.2.6 Multimodal Interactions

Finally, *mixed mode chromatography* exploits multiple types of interactions simultaneously [21]. Most often, hydrophobic and electrostatic interactions are combined, but other types of multimodal interactions can be envisaged such as SEC/IEC [149].

The presence of multimodal interactions is actually almost unavoidable in protein chromatography. This is due to the complex structure of biomolecules, which possess charged, polar, and apolar amino acid residues simultaneously interacting with the chromatographic medium. For example, hydrophobic interactions are always present to some extent in IEC, and the same goes for electrostatic interactions in HIC/RPC. Similarly, it is difficult to prevent completely proteins from interacting with the surface of a SEC medium, while a size exclusion effect is observed with any chromatographic material whose pore size is similar to the protein size. However, in all these examples, one type of interaction is strongly prevailing over the others. This is in contrast with mixed-mode chromatography, where the chromatographic medium is intentionally designed to exploit more than one type of interactions. Provided that the different interactions are properly combined, mixed-mode chromatographic media can lead to unique separation performances [150].

Two main strategies can be considered to generate bimodal interactions on the surface of a chromatographic medium [149, 151]. The first method consists in grafting two different types of functional molecules that differ in their physicochemical properties on the same chromatographic support. In this case, the ratio between the two molecules can be tuned to modulate the relative contributions of the two types of interactions. For example, the incorporation of few charged groups at the surface of reversed-phase media offers the possibility to combine an ion exclusion effect to classical reversed-phase chromatography [152]. The second method consists in grafting on the chromatographic support a single molecule that carries different functional groups as, for example, one ionizable group as well as one aromatic ring in the case of IEC/HIC.

In a sense, hydroxyapatite, which is an inorganic material identical in composition to calcium phosphate found in human bones and teeth, can also be regarded as a mixed-mode material. It combines two types of functional groups: positively charged calcium ions and negatively charged oxygen atoms [153, 154].

Due to its high selectivity, mixed-mode chromatography is currently gaining popularity in the biopharmaceutical industry [150]. In particular, it represents an alternative to affinity chromatography for protein capture applications [155, 156]. Hydroxyapatite chromatographic media are also playing an increasingly relevant role in the purification of biological products [157]. They are especially well suited for the purification of pharmaceutical plasmid DNA [157] and are also used for the polishing steps of two marketed monoclonal antibodies [22].

2.3 Modes of Operation

In Section 2.2 we reviewed the most common chromatographic media and the corresponding types of interactions, which drive the separation of the solutes. In this section, we describe the possible modes of operation of a chromatographic system once an appropriate chromatographic medium is selected.

2.3.1 Single-Column Systems

Elution and Frontal Chromatography

Single-column chromatographic systems are characterized by a single inlet and a single outlet stream. The feed mixture is periodically injected at the column inlet, while the product of interest is periodically collected at the column outlet.

This can be realized according to two main chromatographic modes that differ in terms of the volume of feed mixture injected into the column. In *elution chromatography*, the injection volume is small compared to the retention volumes of the solutes, whereas it is relatively "large," possibly leading to the complete saturation of the chromatographic medium, in the case of *frontal chromatography* [21]. Figure 2.2 schematically illustrates how the solutes are distributed inside a chromatographic column in these two operating modes considering a simple binary mixture. It also shows schematic chromatograms corresponding to each mode of operation. The choice between elution and frontal chromatography is often driven by the "easiness" of the separation. When a solute is significantly more retained than the others, frontal chromatography is usually selected. This is typically the case for the protein capture

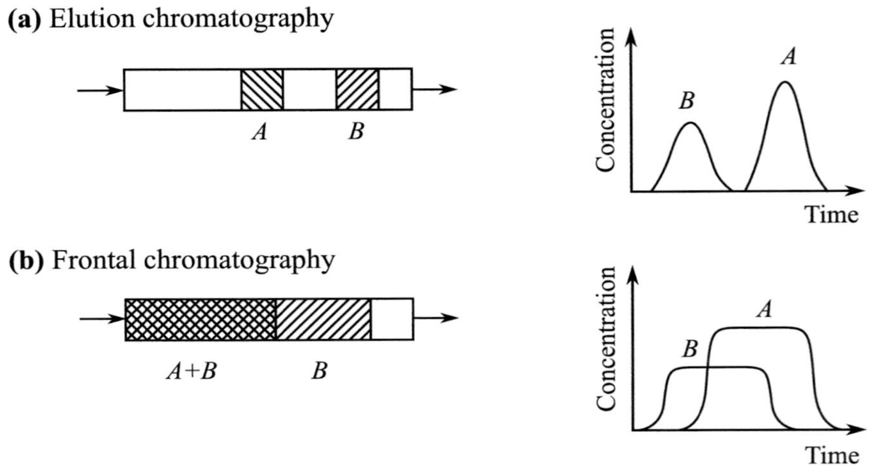

Figure 2.2 Schematic representation of (a) elution chromatography and (b) frontal chromatography for a binary mixture. The left-hand side of the figure shows how the solutes are distributed in the column, after the injection in panel (a) and during the injection in panel (b). The right-hand side of the figure shows illustrative chromatograms recorded at the column outlet.

step, where the protein of interest binds to the chromatographic medium while the impurities flow through the column. A similar situation is encountered in the polishing steps that are performed in the so-called flow-through mode where the impurities bind to the chromatographic medium while the protein of interest flows through the column. On the other hand, elution chromatography is preferred when the different solutes have similar retention behaviors.

Note that both in elution and frontal chromatography, the purified fractions can be collected only during a certain time interval, and not continuously. The time intervals during which the purified fractions can be collected in elution or frontal chromatography depend on the process constraints (e.g., purity, recovery) as well as on the position and shape of the chromatograms.

Chemical Modulation

The protein solutions purified in biopharmaceutical processes often contain solutes spanning a very broad range of interaction strengths with the chromatographic medium, from solutes that are only weakly retained to solutes that are strongly retained. In this case, a constant chemical composition of the eluent, which is referred to as *isocratic* conditions, would lead to extensively long elution times to achieve a satisfactory separation. In order to minimize the elution time while maintaining a sufficiently good separation, the chemical composition of the eluent can be modulated in time [21]. Chemical modulation can, for example, be obtained by modifying the solution pH or by introducing an additional chemical species in the eluent, such as electrolytes or organic molecules. Most often, the chemical composition of the eluent is varied according to either a step gradient or a linear gradient. The former is typically implemented for the recovery of strongly bound solutes and the regeneration of the chromatographic medium, as it is, for example, the case in protein capture applications (see Chapter 4). On the other hand, linear gradient elution (LGE) conditions are quite popular to rapidly obtain acceptable separations at the laboratory scale, especially when the feed mixture contains a wide variety of unknown impurities. However, LGE conditions are currently seldom used at the large scale.

In the particular case where the chemical modifier binds more strongly than any species composing the feed mixture, it is referred to as a displacer, and this chromatographic mode is sometimes called *displacement chromatography* [21].

Finally, it is worth mentioning that, in principle, physical modulation could also be used, for example, by modifying the temperature in time [158]. However, physical modulation is rather rare in the case of proteins.

2.3.2 Multicolumn Systems

We have seen in the previous section that it is not possible to collect purified fractions in a continuous manner by simply loading the feed mixture continuously in a single-column system. As a matter of fact, frontal chromatography allows collecting only one purified fraction during a limited time interval. On the other hand, the use of multicolumn chromatographic systems makes continuous purification possible.

As mentioned in Chapter 1, a continuous chromatographic process can be implemented by using multiple columns in parallel and shifting their load phases in time. This simple idea sets the basis of annular chromatography, which can be seen as a process composed of an infinite number of columns connected in parallel [116–118]. It must be highlighted that, even though continuous, such parallel processes do not bring any improvement in terms of separation performances as compared to single-column systems.

On the other hand, in the case of multicolumn processes that simulate a *countercurrent contact* between the solid phase and the liquid phase, the chromatographic medium is better utilized than in single-column systems, and an increase in the separation performances is observed [79]. Countercurrent chromatography will be introduced and discussed in Chapter 3.

2.4 Mechanistic Models

Despite the great popularity of chromatographic processes in the biopharmaceutical industry, their implementation is still often based on heuristic knowledge and trial-and-error approaches, which are often costly and lengthy, especially at the large production scale. On the other hand, model-based approaches offer the possibility to gather solid knowledge and ultimately optimize chromatographic processes thanks to a better understanding of the physical phenomena underlying the separation mechanisms.

In this section, we provide key tools to model chromatographic processes. We focus on protein purification, while the reader can refer elsewhere for a more general treatment of the subject [158–162]. Numerous models, considering different physical phenomena and varying in terms of complexity, are available in the literature. In this frame, the chromatographic column can be modeled as a plug flow, with axial dispersion (or not), with radial concentration profiles (or not), isothermal (or not), at thermodynamic equilibrium (or not), subject to mass transfer limitations (… or not). The selection of the model to be used for a given separation process is then dictated by the specific needs. To quote the statistician George E. P. Box, it should be remembered that "all models are wrong but some are useful" [163]. One should always keep a critical eye when using any model, which is a "truncated" representation of reality, necessarily excluding some effects to access an intelligible description of a given phenomenon or process. Models of chromatographic processes are no exception to this general recommendation.

2.4.1 Overview

Before entering in any mathematical description of a chromatographic process, it is essential to identify the key physical phenomena to be considered. Figure 2.3 shows a possible classification of the phenomena occurring in a chromatographic column.

At the scale of the column, solutes are transported by *convection* due to the fluid motion. The chromatographic bed is packed with porous solid particles, and the presence of eddies between these particles leads to solute *dispersion* along the column longitudinal axis. To this process is superimposed the *molecular diffusion* of the solute molecules due to the presence of concentration gradients in the column axial direction. At the scale of the particles, solutes need to overcome several mass transfer resistances before

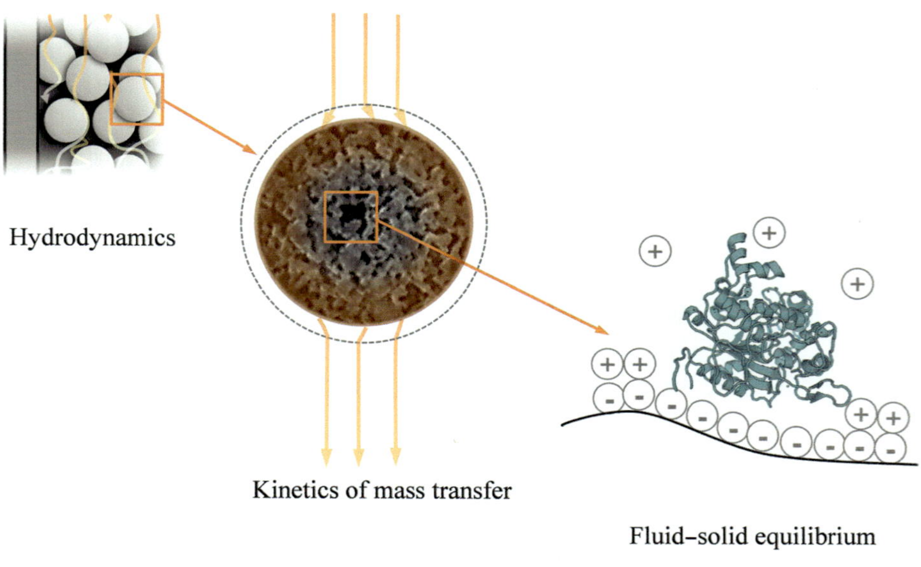

Figure 2.3 Relevant physical phenomena involved in chromatography.

reaching the surface of the chromatographic medium: diffusion through the stagnant fluid layer surrounding the particle (referred to as *external transport*) and diffusion in the pores of the particles (referred to as *internal transport*). Finally, solutes interact with the surface of the chromatographic medium, thus resulting in their partitioning between fluid and solid phases.

The physical phenomena described earlier can be classified into three types:

- the *hydrodynamics* in the chromatographic bed
- the *kinetics* of solute transfer from the liquid to the solid phase
- the *thermodynamics* of solute partitioning between fluid and solid phases

Thermodynamics is the engine driving the separation and characterizes the feasibility of the purification. On the other hand, hydrodynamic dispersion and mass transfer limitations are only reducing the efficacy of the separation process. Therefore, focusing only on the thermodynamic aspects and neglecting the other effects provides a simple reference case to start with, which is referred to as the *equilibrium theory* or sometimes also to as the *ideal model*. The mathematical description of the fluid–solid thermodynamic equilibrium will be presented in Section 2.4.2.

Besides thermodynamics, the description of a chromatographic process derives from the combination of appropriate hydrodynamic and kinetic submodels [158]. Table 2.2 summarizes the chromatographic models that we briefly introduce in the following, while more details about common hydrodynamic and kinetic submodels will be given in Sections 2.4.3 and 2.4.4, respectively.

Let us start with the models describing the hydrodynamics in a chromatographic bed. As a first approximation, the eluent convective flow can be described by a simple plug flow model. A more refined model would also account for dispersion along the column longitudinal axis. Two options are available to account for hydrodynamic dispersion: *(i)* consider the column as a plug flow with axial dispersion, or *(ii)* consider the column as a series of mixing cells in which the solute concentrations are assumed to be uniform. These two models are equivalent since a plug flow with axial dispersion can in general be modeled as a cascade of continuous stirred tanks [98]. Figure 2.4(a) illustrates schematically the three models just mentioned and shows how axial dispersion leads

Table 2.2. Summary of the most common chromatographic models indicating the submodels describing the hydrodynamics and kinetics of mass transfer. Common names given in the literature are also reported.

Hydrodynamics	Ext. transfer	Int. transfer	Binding	Common names
PF	Equil	Equil	Equil	Equilibrium/ideal model
PD	Equil	Equil	Equil	Dispersive model
MC	Equil	Equil	Equil	Plate model
PF	Equil	LDF	Equil	Transport model
PD	Equil	Equil	Ads	Reaction dispersive model
PD	Equil	LDF	Equil	Lumped kinetic model
PD	LDF	Fick	Equil	General rate model

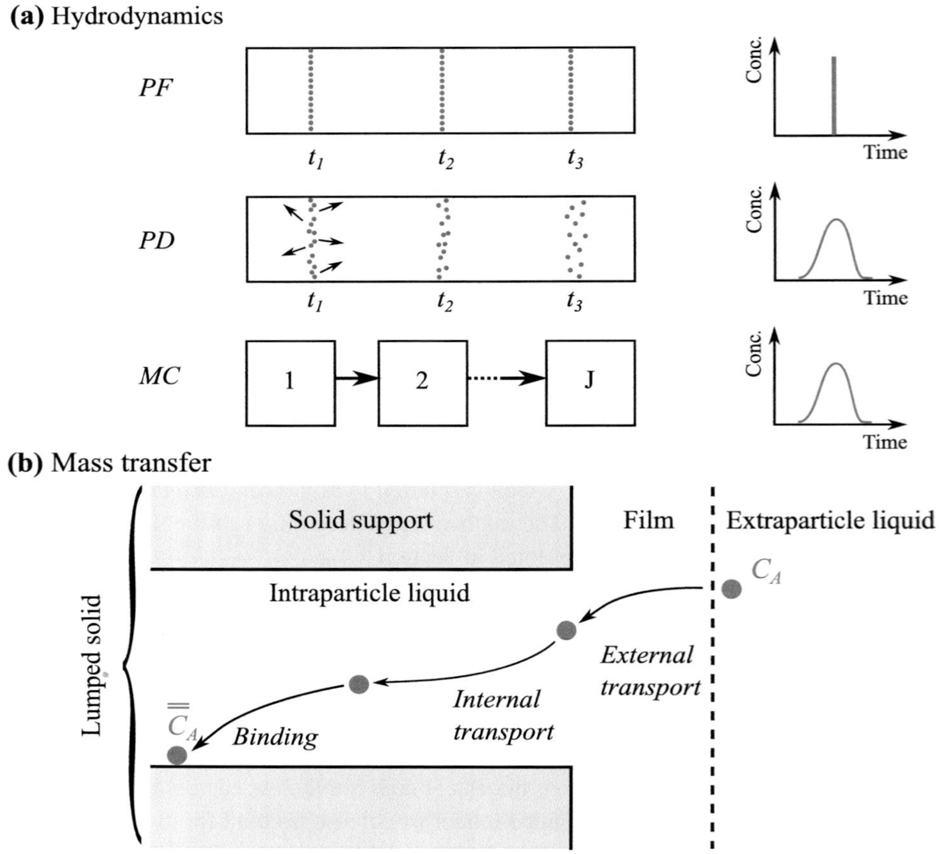

Figure 2.4 Transport phenomena in a chromatographic column illustrating both (a) hydrodynamic and (b) mass transfer aspects. The concentrations C_A and $\overline{\overline{C}}_A$ are the solute concentrations in the liquid and at the solid surface, respectively.

to peak broadening in the case where a narrow pulse is injected in the column. In the following, the basic plug flow model, the plug flow model with axial dispersion, and the mixing cell model will be denoted PF, PD, and MC, respectively.

Regarding the kinetics of mass transfer from the fluid to the solid phase, one can identify three consecutive steps, which are schematically represented in Figure 2.4(b):

1. the diffusion in the fluid layer surrounding the particles
2. the diffusion in the particles
3. the kinetics of adsorption of the solute on the solid surface

In the frame of the equilibrium theory, thermodynamic equilibrium between the fluid and solid phases is assumed to be reached instantaneously at all locations along the column. More realistic models relax the assumption of equilibrium, which is indicated as Equil in Table 2.2 for the different steps. To account for the kinetics of external mass transfer, the fluid layer surrounding the particles is generally modeled as a stagnant film of constant thickness through which solutes move by molecular diffusion. In this case, the external transport can be modeled using a simple linearized form of Fick's law, which states that the flux of solute diffusing through the film is proportional to the difference in concentration between the bulk fluid and the extraparticle fluid at the particle–fluid interface. The difference in concentration corresponds to the driving force for diffusion, and this approximation is therefore referred to as the *linear driving force* (LDF) *approximation*. Describing internal transport is more challenging than external transport due to the steric hindrance of solutes in the pores of the chromatographic medium and to the interactions with the solid surface. Internal mass transfer limitations result in solute concentration gradients along the particle radius, both in the intraparticle liquid and on the solid phase. At thermodynamic equilibrium, the solute concentrations become uniform along the particle radius. In principle, the diffusive flux of solutes inside the particles can be described by Fick's law, as indicated by Fick in Table 2.2. However, this approach is rather strenuous, and a reasonable simplification consists in averaging the solute concentration within the particles and using the LDF approximation. Overall, the LDF approximation can thus be applied both for external and internal diffusion as shown in Table 2.2. Finally, the binding process at the solid surface can be described either by assuming local instantaneous equilibrium or by introducing a kinetic mechanism, the latter being indicated as Ads in Table 2.2.

Further levels of details can then be included in these models, for example, by accounting for distribution profiles along the column radius [158], for the pore size distribution of the particles [164], or for the compressibility of the bed [165].

2.4.2 Thermodynamics of Fluid–Solid Equilibrium

Solute Partitioning

Separation by chromatography relies on the partitioning of solute molecules between a liquid phase and a solid phase. The solid phase is constituted of particles, which are most often porous in order to increase their specific surface area. It is thus essential to distinguish between two types of porosity: the *external porosity* ε_e, which corresponds to the void comprised between the particles packed in the column and is also referred to as the interparticle porosity; and the *intraparticle porosity* ε_i, which corresponds to the void due to the porous structure of the particles. The total porosity of the chromatographic bed is then given by:

$$\varepsilon_t = \varepsilon_e + (1 - \varepsilon_e)\varepsilon_i \tag{2.1}$$

Overall, each solute is partitioned between three phases: *(i)* in the liquid in between the particles, *(ii)* in the liquid in the pores of the particles, *(iii)* or bound to the surface of the chromatographic medium.

The concentration of a solute A in the extraparticle liquid is denoted C_A, as shown in Figure 2.4(b). The concentration of A in the intraparticle liquid depends on the accessibility of this solute to the pores of the particles. If the solute can access all the pores, its concentration in the intraparticle liquid is equal to its concentration in the extraparticle liquid C_A at equilibrium. However, if the solute is partially excluded from some pores due to its large size (steric exclusion) or to its charge (ion exclusion), the solute concentration in such pores may drop below C_A, while it remains equal to C_A in the accessible pores. This situation is treated in the case of linear thermodynamic equilibrium in the following. Finally, we denote $\overline{\overline{C}}_A$ the concentration of solute A bound to the surface of the chromatographic medium, as represented in Figure 2.4(b).

It is then convenient to define the lumped solute concentration in the particles as the volume average between the concentration of solute in the intraparticle liquid and bound to the solid surface. Mathematically, the lumped concentration $\overline{C}_A$ is defined as follows in the absence of exclusion phenomena:

$$\overline{C}_A = \varepsilon_i C_A + (1 - \varepsilon_i)\overline{\overline{C}}_A \tag{2.2}$$

The concentration C_A is defined with reference to a volume of liquid, whereas $\overline{\overline{C}}_A$ is defined with reference to the volume of solid and $\overline{C}_A$ with reference to the particle volume.

Linear Equilibrium

The Lumped Henry Coefficient

In the case of a linear equilibrium relation and in the absence of solute exclusion, the concentration of solute bound to the surface of the chromatographic medium $\overline{\overline{C}}_A$ is related to the concentration of solute in the liquid phase C_A by:

$$\overline{\overline{C}}_A = \overline{\overline{K}}_A C_A \tag{2.3}$$

where $\overline{\overline{K}}_A$ is the Henry coefficient.

By introducing Equation 2.3 in Equation 2.2 it follows that the lumped concentration of A in the particles, i.e., both in the liquid contained in the pores and on the solid surface, is:

$$\overline{C}_A = \overline{K}_A C_A \tag{2.4}$$

where the lumped Henry coefficient $\overline{K}_A$ is given by:

$$\overline{K}_A = \varepsilon_i + (1 - \varepsilon_i)\overline{\overline{K}}_A \tag{2.5}$$

One can then define the selectivity of a binary separation as the ratio between the lumped Henry coefficients of the two solutes of interest, i.e., $\overline{K}_A/\overline{K}_B$ with A more retained than B. The closer the selectivity to one and the more difficult the separation.

Let us now consider the case where the solutes cannot access all the pores, which is typically the case for size exclusion chromatography. In this situation, it is convenient to introduce the partition coefficient $\grave{K}_A$, which represents the volume fraction of the pores accessible to the solute A. The partition coefficient is therefore ranging from 0 to 1: 0 for solutes that are completely excluded from the pores, and thus only access the extraparticle volume, and 1 for solutes accessing all the pores. In this frame, the product $\varepsilon_i \grave{K}_A$ can be regarded as an apparent internal porosity specific to the solute A. This accessible internal porosity depends on the solute, the structure of the pores and the concentration of solute already bound to the solid surface [164]. The product $\grave{K}_A C_A$

instead represents the solute concentration in the intraparticle liquid phase averaged over the particle volume, while $\overline{\overline{C}}_A = \overline{\overline{K}}_A \grave{K}_A C_A$ corresponds to the solute concentration bound to the solid surface. This leads to $\overline{C}_A = \varepsilon_i \grave{K}_A C_A + (1 - \varepsilon_i)\grave{K}_A \overline{\overline{K}}_A C_A$ which, combined with Equation 2.4, gives the following expression of the lumped Henry coefficient:

$$\overline{K}_A = \varepsilon_i \grave{K}_A + (1 - \varepsilon_i)\overline{\overline{K}}_A \grave{K}_A \tag{2.6}$$

The lumped Henry coefficient is related to the retention time of the solute A, $t_{r,A}$, as follows [158]:

$$t_{r,A} = t_0\left(1 + \phi_e \overline{K}_A\right) \tag{2.7}$$

where $t_{r,A}$ is defined as the average elution time measured from pulse injection experiments under dilute conditions. The phase ratio ϕ_e is equal to the ratio between the volume of the particles (including the chromatographic medium and the intraparticle fluid) and the volume of extraparticle fluid:

$$\phi_e = \frac{1 - \varepsilon_e}{\varepsilon_e} \tag{2.8}$$

In addition, t_0 is the zero-retention time, i.e., the retention time of a completely excluded tracer computed as:

$$t_0 = \varepsilon_e \frac{V_{col}}{Q} \tag{2.9}$$

where V_{col} is the volume of the column and Q is the volumetric flow rate. One can also define the retention volume of a completely excluded and inert tracer as $V_0 = Qt_0 = \varepsilon_e V_{col}$. Similarly, the retention volume of the solute A is defined as $V_{r,A} = Qt_{r,A}$.

We will see later that it is useful to think in terms of *column volumes* (CV), which represent the ratio between a given volume (e.g., retention volume, feed volume, eluent volume) and the volume of the column V_{col}. In particular, the CV associated with the retention of the solute A is defined as follows:

$$CV_A = \varepsilon_e + (1 - \varepsilon_e)\overline{K}_A \tag{2.10}$$

The Case of Size Exclusion Chromatography

In this paragraph, we consider the particular case of size exclusion chromatography, which is a simple subcase of the preceding section. Indeed, it is reasonable to assume that there is no solute bound to the surface of the SEC chromatographic medium, which translates to $\overline{\overline{K}}_A = 0$. Therefore, the lumped Henry coefficient is restricted to its steric part, and Equation 2.6 reduces to $\overline{K}_A = \varepsilon_i \grave{K}_A$. It results from Equation 2.7 that:

$$\grave{K}_A = \frac{V_{r,A} - V_0}{V_t - V_0} \tag{2.11}$$

where V_t is the total volume of liquid in the SEC column computed as:

$$V_t = \varepsilon_t V_{col} \tag{2.12}$$

In other words, V_t is the retention volume of a tracer accessing all the pores.

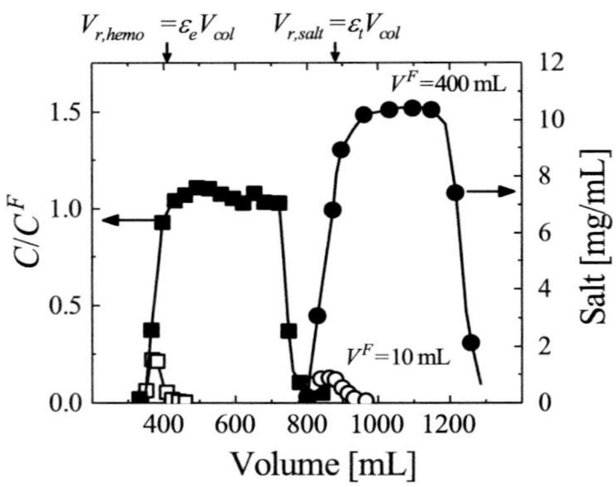

Figure 2.5 SEC chromatograms corresponding to the desalting of hemoglobin samples using a 4×85 cm column (1.07 L) packed with a dextran gel and using a flow rate of 4 mL/min. The peaks of hemoglobin and salt are represented with squares and circles, respectively. Open symbols correspond to an injection of 10 mL ($C^F_{hemo} = 10$ mg/mL, $C^F_{salt} = 10$ mg/mL), while closed symbols correspond to an injection of 400 mL ($C^F_{hemo} = 1$ mg/mL, $C^F_{salt} = 10$ mg/mL). Adapted from [166], copyright 1961, with permission from Elsevier.

As an example of separation by SEC, let us consider the case of hemoglobin desalting, which is shown in Figure 2.5 [166]. Sodium chloride is separated from the protein using a column packed with a dextran gel ($V_{col} = 1.07$ L). Two experiments were performed with different injection volumes, namely 10 mL and 400 mL, which are represented with open and closed symbols, respectively. It can be seen in Figure 2.5 that for the pulse injection experiment (10 mL injection), the retention volumes of hemoglobin and salt are $V_{r,hemo} = 370$ mL and $V_{r,salt} = 880$ mL, respectively. In this example, the chromatographic medium was selected so that hemoglobin is completely excluded from the porous particles (i.e., $\grave{K}_{hemo} = 0$) while sodium chloride accesses all the pores (i.e., $\grave{K}_{salt} = 1$). Therefore, the values of the retention volumes can be used to estimate the total and external porosities by using Equations 2.9, 2.11, and 2.12. We obtain $\varepsilon_e = 0.370/1.07 = 0.35$ and $\varepsilon_t = 0.88/1.07 = 0.82$, which leads to $\varepsilon_i = 0.73$ by using Equation 2.1.

It is also interesting to comment on the shape of the peaks observed in Figure 2.5. It is observed that the peaks are almost Gaussian for the small injection volume, which is characteristic of linear fluid–solid equilibria. When the injection volume is increased, the position of the front does not change, and the peaks become rectangular, which also indicates a linear equilibrium. In this case, the maximum of the peak corresponds to the concentration of the feed, and its width to the injection volume. This implies that the dilution of the feed mixture in SEC is much less pronounced at large injection volumes than at small injection volumes. In the example under consideration, the protein solution was diluted 10 times in the 10 mL injection experiment but only 1.25 times in the 400 mL injection experiment [166].

In the situation illustrated in Figure 2.5, the solutes could access either all the pores, or none of the pores. In the following, we investigate intermediate cases where solutes can access only a fraction of the intraparticle volume depending on their size. For this,

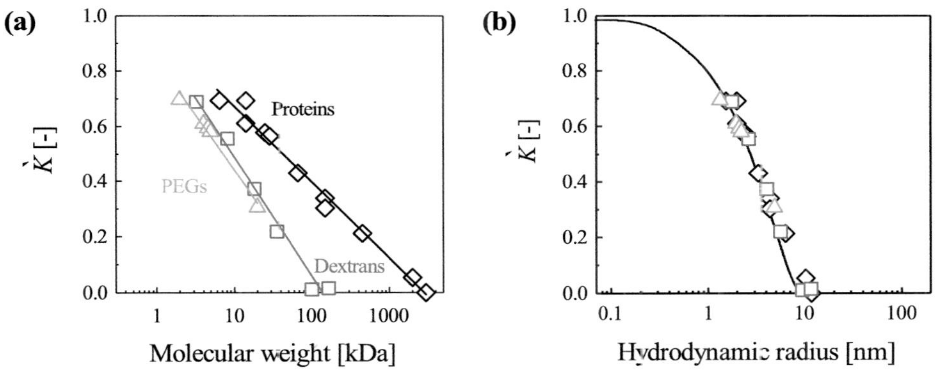

Figure 2.6 Partition coefficients in SEC (a) as a function of the molecular weight and (b) as a function of the hydrodynamic radius. Experimental data are represented by symbols and were obtained with protein standards (from insulin B chain to apoferritin), dextrans and PEG polymers on Superdex 200 (GE Healthcare) at a flow rate of 0.5 mL/min. In panel (a), lines correspond to fitting with Equation 2.13. In panel (b), the partition coefficient was estimated with Equation 2.15, assuming uniform cylindrical pores and using $r_{pore} = 10.4$ nm. Adapted from [55], copyright 2014, with permission from Elsevier.

it is convenient to relate the partition coefficient to the solute molecular weight with an empirical correlation of the form:

$$\grave{K} = a - b \times \log(MW) \tag{2.13}$$

where a and b are fitting parameters that depend on the type of solutes considered and on the pore size distribution of the chromatographic medium, while MW is the molecular weight of the solute in Daltons.

Figure 2.6(a) shows retention factors for three different kinds of solutes, namely proteins, dextrans, and PEGs, of various molecular weights. The values of the retention factors were computed with Equation 2.11 using retention volumes measured in a SEC column. The solid lines correspond to the regression given by Equation 2.13, and it is seen that the latter provides an acceptable description of the experimental data. It is observed that three different curves are obtained, namely one for each kind of solute. Indeed, the correlation given by Equation 2.13 is only valid within a given solute class (e.g., proteins, dextrans and PEGs).

Considering the nature of the SEC partitioning, it is more rigorous to use the hydrodynamic radius of the solute rather than its molecular weight to determine the partition coefficient. In practice, the hydrodynamic radius is often measured by dynamic light scattering and related to the solute molecular weight by a power law. Typical correlations are [167–169]:

$$\begin{aligned} &\text{Globular proteins:} && R = 0.082(MW)^{1/3} \\ &\text{Dextran polymers:} && R = 0.027(MW)^{0.498} \\ &\text{PEG polymers:} && R = 0.019(MW)^{0.559} \end{aligned} \tag{2.14}$$

where the hydrodynamic radius R is in nm.

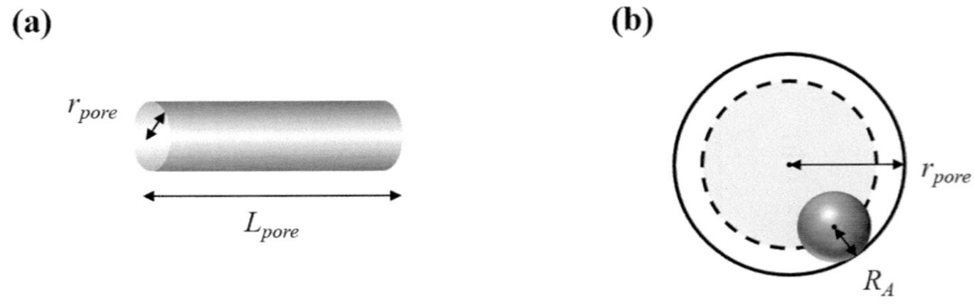

Figure 2.7 (a) Schematic representation of a cylindrical pore. (b) Schematic representation of the volume accessible to the center of mass of a solute of hydrodynamic radius R_A.

The experimental values of the retention coefficient presented in Figure 2.6(a) are plotted as a function of the solute hydrodynamic radii in Figure 2.6(b). Interestingly, it is observed that all the experimental data (represented by symbols) now follow the same trend.

Let us propose a reasonable fit for the experimental data shown in Figure 2.6(b). The retention coefficient of a solute A can be estimated from $\grave{K}_A = V_{acc,A}/V_{pore}$, where V_{pore} is the volume of one pore and $V_{acc,A}$ is the volume of this pore that is accessible to the center of mass of the solute A. For the sake of simplicity, it can be assumed that all the pores are identical cylinders of radius r_{pore} and length L_{pore} so that $V_{pore} = L_{pore} \times 4\pi r_{pore}^2$. Figure 2.7(a) shows a schematic representation of such pore, while Figure 2.7(b) illustrates what is the pore volume accessible to the center of mass of a solute. Assuming that the solute A can be described as a hard sphere of hydrodynamic radius R_A, we obtain $V_{acc,A} = L_{pore} \times 4\pi(r_{pore} - R_A)^2$ [170]. It follows that $\grave{K}_A = (r_{pore} - R_A)^2/r_{pore}^2$ for the solute A. Therefore, the following dependency of the retention coefficient $\grave{K}_A$ on the solute hydrodynamic radius R_A is found for $R_A < r_{pore}$:

$$\grave{K}_A = \left(1 - \frac{R_A}{r_{pore}}\right)^2 \tag{2.15}$$

It is worth recalling that $\grave{K}_A = 0$ for $R_A \geq r_{pore}$, which implies that solutes bigger than r_{pore} are not separated and co-elute in the void peak.

From Equation 2.15, it is seen that the partition coefficient $\grave{K}_A$ can be estimated a priori knowing the hydrodynamic radius of the solute (or at least a correlation relating the molecular weight to the radius) and the average pore radius. The dependency of the partition coefficient with the hydrodynamic radius according to Equation 2.15 is shown as a solid line in Figure 2.6(b) using r_{pore} as a fitting parameter. It is seen that this curve, which was obtained under the assumption of uniform and cylindrical pores, describes well the experimental data. For more detailed models accounting for the pore size distribution, the reader is referred to the following works [164, 168, 171].

Nonlinear Equilibrium

The Langmuir Adsorption Model

In the previous section, we dealt with linear thermodynamic equilibrium relations to describe solute partitioning at the fluid–solid interface. In the following, we introduce nonlinearities by considering first the simple case of a single solute A adsorbing on one adsorption site S on the surface of the chromatographic medium in the absence of exclusion effects. The equilibrium law between A and S can be represented as follows:

$$A + \bar{\bar{S}} \rightleftarrows \overline{\overline{AS}} \tag{2.16}$$

where the overlines refer to the adsorbed phase.

The total concentration of adsorption sites $\bar{\bar{S}}$ is equal to the saturation capacity $\bar{\bar{N}}$. The concentration of occupied sites $\overline{\overline{AS}}$ corresponds to the concentration of adsorbed solute, $\bar{\bar{C}}_A$. The concentration of free sites $\bar{\bar{S}}$ is simply equal to $\bar{\bar{N}} - \bar{\bar{C}}_A$.

At thermodynamic equilibrium, we can define the equilibrium constant $\tilde{K}_A$ according to the mass action law:

$$\tilde{K}_A = \frac{\bar{\bar{C}}_A}{C_A\left(\bar{\bar{N}} - \bar{\bar{C}}_A\right)} \tag{2.17}$$

This leads to the well-known Langmuir adsorption isotherm:

$$\bar{\bar{C}}_A = \frac{\bar{\bar{N}}\tilde{K}_A C_A}{1 + \tilde{K}_A C_A} = \frac{\bar{\bar{K}}_A C_A}{1 + C_A \bar{\bar{K}}_A / \bar{\bar{N}}} \tag{2.18}$$

where the Henry coefficient is defined as $\bar{\bar{K}}_A = \bar{\bar{N}}\tilde{K}_A$ and corresponds to the slope of the linear part of the adsorption isotherm. The term *isotherm* reflects the fact that adsorption equilibria are temperature dependent and are thus defined at a given temperature. A graphical representation of a typical Langmuir adsorption isotherm is given in Figure 2.8, where it is seen that the linear relation can be used at low solute concentration, while the nonlinearity arises at higher concentrations. A plateau is observed

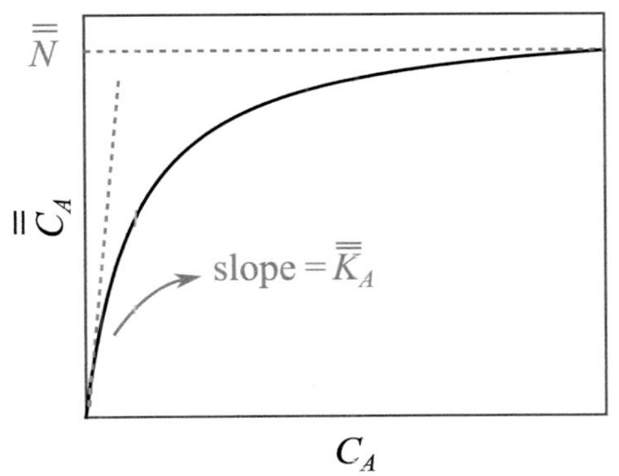

Figure 2.8 Illustrative representation of the Langmuir adsorption isotherm linking the concentration of a solute A in the liquid C_A to the concentration of solute adsorbed on the solid surface $\bar{\bar{C}}_A$.

when increasing further the solute concentration in the liquid phase, indicating that the surface of the chromatographic medium gets saturated. Langmuir adsorption isotherms belong to the so-called class of favorable adsorption isotherms that are characterized by a concave shape.

As mentioned earlier in the case of linear fluid–solid equilibria, it is convenient to describe the solute concentration inside the particles as the volume average between the intraparticle liquid phase and the solid phase. By using Equations 2.2 and 2.18 it follows that:

$$\bar{C}_A = \varepsilon_i C_A + \frac{\bar{N}\tilde{K}_A C_A}{1 + \tilde{K}_A C_A} \tag{2.19}$$

where the lumped particle saturation capacity is defined by:

$$\bar{N} = (1 - \varepsilon_i)\bar{\bar{N}} \tag{2.20}$$

For a multicomponent system, which is necessarily the case in a separation operation, one needs to account for the fact that different solutes may compete for the same adsorption sites. Accordingly, the Langmuir adsorption isotherm of a component A in a mixture of solutes competing for the same adsorption sites becomes:

$$\bar{C}_A = \varepsilon_i C_A + \frac{\bar{N}\tilde{K}_A C_A}{1 + \sum_j \tilde{K}_j C_j} \tag{2.21}$$

where the sum in the denominator is accounting for all the species that can adsorb on the surface of the chromatographic medium. The preceding derivation of the Langmuir adsorption isotherm holds for molar concentrations, so care should be taken when using mass concentrations. In the classical derivation of the Langmuir adsorption isotherm, $\bar{N}$ is unique so that all the solutes have access to the same number of adsorption sites. It may sometimes be convenient to relax this assumption by using a parameter $\bar{N}_j$ per solute to bring more flexibility when estimating the different parameters of the adsorption isotherm. However, relaxing the assumption of a unique number of sites $\bar{N}$ for all the solutes does not lead to thermodynamically consistent adsorption isotherms as they do not satisfy the Gibbs–Duhem equation [172].

It is also important to recall that the derivation of the Langmuir adsorption isotherm proposed here is based on the assumption that each solute interacts with a single adsorption site. Although this assumption may be valid in certain cases, like for affinity chromatography for example, there are a number of situations where this assumption is not met. This is for example the case of HIC or RPC, where the hydrophobic molecules grafted on the chromatographic medium cannot be reconducted to the notion of discrete adsorption sites. Nevertheless, even in such cases, Equation 2.21 still gives a good description for many protein systems provided that the model parameters (saturation capacity and Henry coefficient) are fitted directly on experimental results (see Section 2.5) [21]. Other forms of isotherms, such as the anti-Langmuir, Toth, Freundlich, and Moreau adsorption isotherms, might be relevant in some applications and can be found in several books [21, 158–161].

Stoichiometric Displacement and Steric Mass Action Models

As already mentioned in Section 2.2.3, ion exchange chromatography conceptually differs from the other chromatographic techniques due to the exchange mechanism of charged solutes. Notably, the electroneutrality condition has to be fulfilled at the surface of the chromatographic medium and in the liquid phase. In this section, we show how to link the concentration of A in the liquid phase, C_A, to the concentration of solute bound to the surface, $\overline{\overline{C}}_A$. We consider the case where there is no solute exclusion and where the modifier ion is in excess with respect to other solutes.

The ion exchange process between a solute A interacting with z_A charges at the surface of the chromatographic medium and a monovalent modifier ion M can be represented as follows:

$$A^{z_A+} + z_A\overline{\overline{M}}^{+} \rightleftarrows \overline{\overline{A}}^{z_A+} + z_A M^{+} \tag{2.22}$$

The parameter z_A is referred to as the protein apparent charge. We arbitrarily selected a positively charged protein here, but the same reasoning could be applied to a negatively charged protein.

The equilibrium constant of the exchange process given by Equation 2.22 is defined as:

$$\tilde{K}_A = \frac{\overline{\overline{C}}_A}{C_A}\left(\frac{C_M}{\overline{\overline{C}}_M}\right)^{z_A} \tag{2.23}$$

The condition of electroneutrality at the surface of the chromatographic medium imposes that:

$$\overline{\overline{C}}_M + z_A\overline{\overline{C}}_A = \overline{\overline{N}} \tag{2.24}$$

where $\overline{\overline{N}}$ is the ionic capacity of the chromatographic medium.

By replacing $\overline{\overline{C}}_M$ in Equation 2.23, we obtain after rearrangement:

$$\overline{\overline{C}}_A = \overline{\overline{K}}_A C_A\left(1 - z_A\overline{\overline{C}}_A/\overline{\overline{N}}\right)^{z_A} \tag{2.25}$$

where the Henry coefficient $\overline{\overline{K}}_A = \tilde{K}_A\overline{\overline{N}}^{z_A}C_M^{-z_A}$ describes the linear part of the equilibrium relation when the solute concentration tends to zero. This model corresponds to the classical representation of IEC and is referred to as the stoichiometric displacement model [173]. In this model, the concentration of modifier C_M in the liquid phase is assumed to be unaffected by the exchange mechanism given by Equation 2.22, which is a reasonable approximation when the molar concentration of modifier is large with respect to that of the protein. Relaxing this assumption requires writing additional equations describing the solution chemistry (e.g., acid–base equilibria, complexation) and ensuring the solution electroneutrality.

Proteins can simultaneously interact with several charges at the surface of the chromatographic medium, which is reflected by a value of z_A greater than 1. Moreover, proteins are heterogeneous molecules, containing both charged and neutral residues. Consequently, a certain number of the charges at the surface of the chromatographic medium can be covered by the protein without interacting electrostatically with any of its residues. In this case, the condition of electroneutrality is given by:

$$\overline{\overline{C}}_M + (z_A + \varsigma_A)\overline{\overline{C}}_A = \overline{\overline{N}} \qquad (2.26)$$

where ς_A represents the number of charges at the surface of the ion exchange medium shielded by one protein molecule. Therefore, Equation 2.25 becomes:

$$\overline{\overline{C}}_A = \overline{\overline{K}}_A C_A \left(1 - (z_A + \varsigma_A)\overline{\overline{C}}_A / \overline{\overline{N}}\right)^{z_A} \qquad (2.27)$$

This model is referred to as the steric mass action model (SMA model) [174]. Both in the stoichiometric displacement model and in the steric mass action model, the Henry coefficient $\overline{\overline{K}}_A$ decreases when the modifier concentration C_M increases according to:

$$\overline{\overline{K}}_A = \tilde{K}_A \overline{\overline{N}}^{z_A} C_M^{-z_A} \qquad (2.28)$$

Therefore, one should pay particular attention when using Equations 2.25 or 2.27 to properly account for the variations of the modifier concentration, especially under gradient elution conditions.

For a multicomponent system, the SMA model leads to:

$$\overline{\overline{C}}_A = \overline{\overline{K}}_A C_A \left(1 - \sum (z_j + \varsigma_j)\overline{\overline{C}}_j / \overline{\overline{N}}\right)^{z_A} \qquad (2.29)$$

where the sum is accounting for all the charged solutes that are bound to the surface of the chromatographic medium excluding the modifier.

Figure 2.9 gives a pictorial summary of the three nonlinear models presented in this section: the Langmuir adsorption model, the stoichiometric displacement model, and the steric mass action model.

It is worth noticing that the steric mass action model (Equation 2.27) reduces to the stoichiometric displacement model (Equation 2.25) when $\varsigma_A = 0$. In addition, it is easily shown after rearrangement of Equation 2.27 that the steric mass action model reduces to a Langmuir adsorption isotherm when $z_A = 1$ and $\varsigma_A = 0$.

With illustrative purposes, Figure 2.10 shows the experimental concentration $\overline{\overline{C}}_A$ of α-chymotrypsinogen as a function of the concentration of protein in the liquid phase (open symbols) measured on a strong cation exchange resin for different concentrations

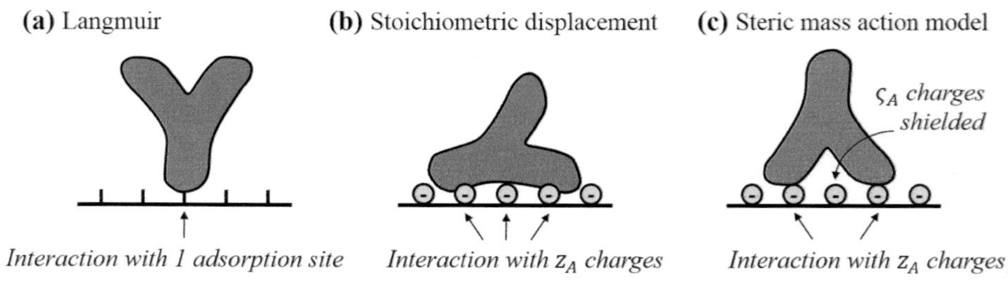

Figure 2.9 Common models used to describe nonlinear fluid–solid equilibrium: (a) Langmuir adsorption model, (b) stoichiometric displacement model, and (c) steric mass action model.

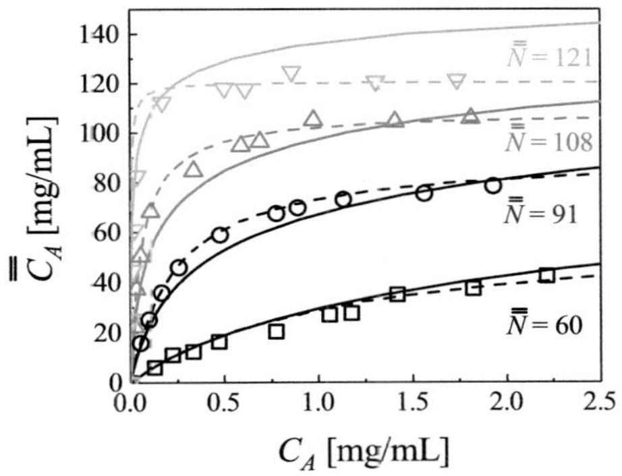

Figure 2.10 Adsorption isotherm of α-chymotrypsinogen on a strong cation exchange medium in 10 mM sodium phosphate buffer pH 6.5 measured at various NaCl concentrations: 0 mM (downward triangles), 25 mM (upward triangles), 50 mM (circles), and 100 mM (squares). Results of the SMA model (Equation 2.27, solid lines) with $\tilde{K}_A \overline{\overline{N}}^{z_A} = 0.034$ (mol/L)^{z_A}, $z_A = 3.5$, $\varsigma_A = 27$, $\overline{\overline{N}} = 173.5$ mg/mL, and of the Langmuir model (Equation 2.18, dashed lines) using the same values of $\overline{\overline{K}}_A$ but refitting $\overline{\overline{N}}$ for each salt concentration. Experimental data reprinted from [175], copyright 2000, with permission from Elsevier.

of NaCl (C_M from 0 to 100 mM) [175]. The results of the fitting using the SMA model (Equation 2.27) with the same parameters $\tilde{K}_A$, z_A, ς_A and $\overline{\overline{N}}$ for all the experiments are presented as solid lines. Note that Equation 2.27 is implicit so that it requires numerical resolution. As a comparison, the dashed lines in Figure 2.10 correspond to the predictions of the explicit Langmuir model (Equation 2.18). The parameters $\overline{\overline{K}}_A$ were kept identical to those used in the SMA model at each modifier concentration, while the parameter $\overline{\overline{N}}$ was refitted for each experimental condition. It can be seen that both models represent fairly well the experimental data, and that the good fitting obtained with the Langmuir expression is made possible by using $\overline{\overline{N}}$ as an adjustable parameter varying with the salt concentration.

Chemically Modulated Henry Coefficient

As described in Section 2.3.1, gradient elution conditions are often used in protein chromatography in order to reduce the total elution time while ensuring a sufficient separation. In the following, we quantify the impact of a change in the chemical composition of the eluent on the solute retention time, which is reflected by the value of the corresponding Henry coefficient.

Hydrophobic Interactions

Based on the solvophobic theory, an empirical model describing the variation of the Henry coefficient with the salt concentration in HIC has been proposed by analogy with the expression of the protein solubility in solution [176, 177]:

$$\ln\overline{\overline{K}}_A = a_0 - \frac{a_1\sqrt{C_M}}{1 + a_2\sqrt{C_M}} + s_A C_M \tag{2.30}$$

where a_0 is the value of the logarithm of the Henry coefficient in the absence of salt, C_M is the salt concentration, and a_1 and a_2 are two adjustable parameters. Here, we assume that the variation of $\overline{\overline{K}}_A = \tilde{K}_A\overline{\overline{N}}$ is primarily due to the alteration of the fluid–solid equilibrium constant $\tilde{K}_A$, while the saturation capacity $\overline{\overline{N}}$ remains unchanged. The parameter s_A is sometimes called the sensitivity coefficient and depends on the protein, salt, and chromatographic medium under investigation. At sufficiently high salt concentration, the previous equation reduces to:

$$\ln\overline{\overline{K}}_A = \ln\overline{\overline{K}}_{0,A} + s_A C_M \tag{2.31}$$

so that the Henry coefficient can be written as $\overline{\overline{K}}_A = \overline{\overline{K}}_{0,A}\exp(s_A C_M)$ with $\ln\overline{\overline{K}}_{0,A} = a_0 - a_1/a_2$.

A similar expression is used in the case of reversed-phase chromatography, where C_M refers to the concentration of organic solvent. Nevertheless, the sensitivity coefficient s_A is positive for HIC because the retention time increases with increasing salt concentration, while it is negative for RPC, where the retention time decreases with the concentration of the organic solvent.

Electrostatic Interactions

In the case of ion exchange chromatography, we have seen when deriving the steric mass action model that $\overline{\overline{K}}_A = \overline{\overline{K}}_{0,A}C_M^{-z_A}$, where $\overline{\overline{K}}_{0,A} = \tilde{K}_A\overline{\overline{N}}^{z_A}$. Therefore, we obtain:

$$\ln\overline{\overline{K}}_A = \ln\overline{\overline{K}}_{0,A} - z_A\ln C_M \tag{2.32}$$

In the case where electrostatic interactions are modulated by the addition of a salt, the saturation capacity $\overline{\overline{N}}$, the protein charge z_A, and the steric parameter ς_A are often assumed to be unaffected by the presence of salt. However, these assumptions do not hold any longer when the solution pH is varied. For instance, it was shown in Figure 2.1 that the protein net charge is pH-dependent, which implies that the parameter z_A is expected to vary with the pH of the solution. In addition, $\overline{\overline{N}}$ can also vary, as it is typically the case for weak ion exchange chromatographic media whose ionization state depends on the value of the pH with respect to the pKa of the charged groups. As a rough estimation, the fraction of ionized groups can be assessed from the Henderson–Hasselbalch equation as $(1 + 10^{\mathrm{p}K_a - \mathrm{pH}})^{-1}$, which is valid in the absence of electrolytes [178]. Multiplied by $\overline{\overline{N}}$, this gives the effective saturation capacity at a given pH value.

Mixed Hydrophobic–Hydrophilic Interactions

By using Equations 2.31 and 2.32, a general expression of the Henry coefficient valid for mixed hydrophobic–electrostatic interactions can be proposed [179]:

$$\ln\overline{\overline{K}}_A = A_0 + A_1 C_M + A_2\ln(C_M) \tag{2.33}$$

where A_0, A_1, and A_2 are three adjustable parameters.

Biospecific Interactions
In affinity chromatography, the strength of adsorption is commonly modulated by the solution pH. In most cases, step gradients are employed, so that the adsorption is very strong during the load phase at neutral pH and very weak during the recovery of the protein of interest at acidic pH. There was therefore only little effort to describe quantitatively the variations of the Henry coefficient with the solution pH in intermediate situations. We propose here a simple phenomenological model by considering the following equilibria between the protein and the affinity ligand:

$$A + \bar{\bar{S}} \rightleftarrows \overline{\overline{AS}} \qquad \tilde{K}_{0,A} = \frac{\bar{\bar{C}}_A}{C_A \bar{\bar{S}}} \tag{2.34}$$

$$\overline{\overline{AS}} + H \rightleftarrows A + \overline{\overline{SH}} \qquad \frac{1}{\tilde{K}_H} = \frac{C_A \bar{\bar{C}}_H}{C_H \bar{\bar{C}}_A} \tag{2.35}$$

where $\bar{\bar{S}}$ denotes the concentration of free affinity ligands. In Equation 2.34, it is assumed that the equilibrium between the protein and the affinity ligands is independent of the ionization state of A. The total concentration of adsorption sites $\bar{\bar{N}}$ is given by:

$$\bar{\bar{N}} = \bar{\bar{S}} + \bar{\bar{C}}_A + \bar{\bar{C}}_H = \bar{\bar{C}}_A \left(1 + \frac{1}{C_A \tilde{K}_{0,A}} + \frac{C_H}{C_A \tilde{K}_H}\right) \tag{2.36}$$

By introducing $C_H = 10^{-\mathrm{pH}}$, it is possible to show that the adsorption isotherm describing the previous equilibria simplifies to a Langmuir adsorption isotherm given by Equation 2.18 with an effective equilibrium constant $\tilde{K}_A$ given by:

$$\tilde{K}_A = \tilde{K}_{0,A} \left(1 + \frac{\tilde{K}_{0,A}}{\tilde{K}_H} 10^{-\mathrm{pH}}\right)^{-1} \tag{2.37}$$

2.4.3 Hydrodynamics

Hydrodynamic Dispersion

As mentioned in Section 2.4.1, two main models are available to describe the hydrodynamics in a chromatographic bed: the plug flow with axial dispersion model (PD) or the mixing-cells model (MC). These two models are based on mass conservation laws, and the general form of a mass balance equation is given by:

$$\text{inlet} + \text{production} = \text{outlet} + \text{accumulation} \tag{2.38}$$

Let us consider first the PD model and apply Equation 2.38 to a solute A on the column section comprised between the coordinates z and $z + dz$ shown in Figure 2.11:

$$\text{inlet} = C_A(z) \frac{u}{\varepsilon_e} \varepsilon_e S_{col} + D_{ax,A} \frac{\partial C_A(z)}{\partial z} \varepsilon_e S_{col} \tag{2.39}$$

$$\text{production} = 0 \tag{2.40}$$

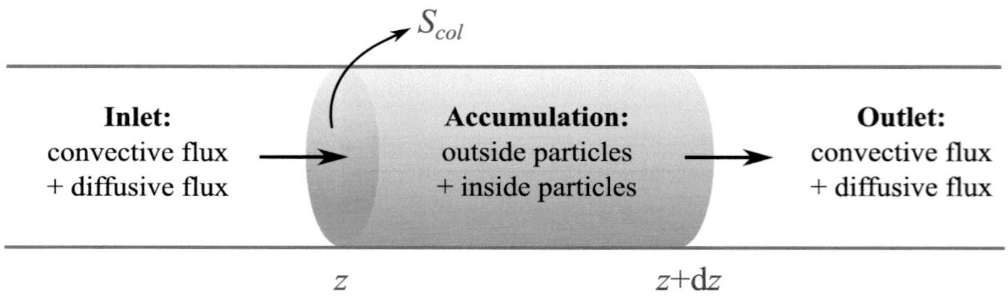

Figure 2.11 Mass balance on a slice of a chromatographic column.

$$\text{outlet} = C_A(z+\mathrm{d}z)\frac{u}{\varepsilon_e}\varepsilon_e S_{col} + D_{ax,A}\frac{\partial C_A(z+\mathrm{d}z)}{\partial z}\varepsilon_e S_{col} \tag{2.41}$$

$$\text{accumulation} = \left[\varepsilon_e\frac{\partial C_A}{\partial t} + (1-\varepsilon_e)\frac{\partial \bar{C}_A}{\partial t}\right] S_{col}\mathrm{d}z \tag{2.42}$$

where $D_{ax,A}$ is the axial dispersion coefficient of A, and u is the eluent superficial velocity (i.e., the volumetric flow rate divided by the column section). The interstitial velocity, representing the actual average velocity of the eluent through the packed bed, is given by u/ε_e. In addition, S_{col} is the column cross section and is computed as $S_{col} = \pi D_{col}^2/4$, where D_{col} is the column diameter. The inlet and outlet terms are constituted from a convective and a diffusive contribution, while the accumulation term accounts for the accumulation outside the particles and inside the particles (i.e., in the intraparticle fluid and bound onto the solid surface, respectively).

The combination of Equations 2.38–2.42 leads to:

$$\varepsilon_e\frac{\partial C_A}{\partial t} + u\frac{\partial C_A}{\partial z} + (1-\varepsilon_e)\frac{\partial \bar{C}_A}{\partial t} = \varepsilon_e D_{ax,A}\frac{\partial^2 C_A}{\partial z^2} \tag{2.43}$$

The boundary conditions of Equation 2.43 are given by [180]:

$$uC_A(z=0,t) - \varepsilon_e D_{ax,A}\frac{\partial C_A}{\partial z}(z=0,t) = uC_A^F(t) \tag{2.44}$$

$$\frac{\partial C_A}{\partial z}(z=L_{col},t) = 0 \tag{2.45}$$

where L_{col} is the column length and C_A^F is the feed concentration of solute A during the injection time Δt^F. The concentration at the inlet of the chromatographic column is described as follows:

$$C_A^F(t) = \begin{cases} C_A^F & 0 < t \le \Delta t^F \\ 0 & t > \Delta t^F \end{cases} \tag{2.46}$$

The initial conditions inside the column are given by the value of $C_A(z, t=0)$ for $0 < z < L_{col}$.

In the frame of the PD model, axial dispersion is quantified by the column Péclet number Pe_c defined as follows:

$$Pe_c = \frac{u\,L_{col}}{\varepsilon_e\,D_{ax}} \tag{2.47}$$

The Péclet number is a dimensionless number that characterizes the ratio between the rate of solute transport by convection over the rate of solute transport by dispersion in the chromatographic column. The index c refers to the column in order to avoid confusion with other Péclet numbers that will be introduced in Section 2.4.5.

Let us now consider the MC model, initially proposed by Martin and Synge [181]. Applying Equation 2.38 to a generic cell k, with k comprised between 1 and J, leads to:

$$QC_A^{k-1} = QC_A^k + \frac{V_{col}}{J}\left(\varepsilon_e\frac{\mathrm{d}C_A^k}{\mathrm{d}t} + (1-\varepsilon_e)\frac{\mathrm{d}\bar{C}_A^k}{\mathrm{d}t}\right) \tag{2.48}$$

where C_A^k represents the concentration of solute A in the cell k, and J is the total number of mixing cells. V_{col} represents the column volume, i.e., the total volume of all cells. The concentration at the inlet of the column is given by $C_A^0 = C_A^F$ for $t \leq \Delta t^F$ (0 otherwise). The initial conditions inside the column are given by the value of $C_A^k(t=0)$, for $k = 1$ to J.

Both the PD and MC models are one-dimensional models, i.e., axial dispersion is accounted for but radial heterogeneities are neglected. The two models provide substantially equivalent results for $Pe_c \geq 25$, and the choice to use one model or the other is to a large extent a matter of personal preference. They are connected by a relation between the two parameters describing the extent of axial dispersion: the Péclet number Pe_c in the PD model and the number of mixing cells J in the MC model:

$$Pe_c = 2J \tag{2.49}$$

Pressure Drop

The flow rate in a chromatographic column is limited by the maximal acceptable pressure drop, which ranges from few bars for preparative protein capture applications to a few hundred bars for high-performance analytical columns. A good estimation of the pressure drop ΔP in an incompressible chromatographic bed can be obtained from the Kozeny–Kármán model, which relies on the description of the porous bed as a bundle of unconnected cylindrical and tortuous pores with identical diameters [182, 183]:

$$\frac{\Delta P}{L_{col}} = \frac{h_K \eta \phi_e^2}{d_p^2 \varepsilon_e} u \tag{2.50}$$

where d_p is the particle diameter, η is the fluid viscosity, and h_K is a constant in the order of 150–180. We also define the parameter δ_K as $\delta_K = h_K \eta \phi_e^2/(d_p^2 \varepsilon_e)$. Equation 2.50 is valid under laminar flow conditions, i.e., when the fluid travels "smoothly" or "regularly" without turbulence. These conditions are in practice always met in the case of packed bed chromatography due to the relatively low fluid velocities typically employed.

Although rather simple, Equation 2.50 is very instructive because it shows that the pressure drop increases linearly with the fluid velocity, with the column length and with the fluid viscosity. Importantly, the pressure drop decreases with the square of the particle size. For analytical HPLC, particle sizes in the range 10–100 μm, and fluid velocities around 1 mm/s are typically used, which leads to pressure drops in the order of 50–200 bar/m for aqueous solutions. In large-scale low-pressure chromatography, larger particles in the range 100–500 μm are commonly used, leading to pressure drops in the order of few bar/m or few tens of bar/m for similar fluid velocities.

As mentioned earlier, the Kozeny–Kármán relation is only valid for uncompressible particles. However, soft particles are also quite popular in protein chromatography, and in particular hydrogel particles made of crosslinked agarose or dextran polymers are often used for bio-related applications. With such materials, the chromatographic bed compresses with increasing fluid flow rate until a critical velocity, u_{crit} beyond which the bed collapses and the pressure increases exponentially. The friction between the wall of the column and the particles helps sustaining the chromatographic bed so that narrow columns can stand higher flow rates. Therefore, the geometry of the column, represented by the ratio L_{col}/D_{col}, plays an important role in the hydrodynamics. The effect of the ratio L_{col}/D_{col} on the critical velocity u_{crit} can be accounted through the empirical relation developed by Stickel and Fotopolous [165]:

$$u_{crit}L_{col} = a\frac{L_{col}}{D_{col}} + b \tag{2.51}$$

where a and b are parameters characteristic of the chromatographic medium (e.g., for Sepharose 4FF, $a = 420$ cm^2/h and $b = 3600$ cm^2/h at 22 °C in PBS buffer solution).

In order to quantify the compressibility of the bed, let us introduce the bed compression factor, λ defined as $\lambda = (L_{col} - L'_{col})/L_{col}$, where L_{col} is the initial bed length while L'_{col} is the length of the compressed bed. This compression factor is related to the fluid velocity by the following empirical relation:

$$\lambda u_{crit} = \lambda_{crit} u \tag{2.52}$$

where λ_{crit} is a parameter characteristic of the chromatographic medium (e.g., for Sepharose 4FF, $\lambda_{crit} = 0.21$ at 22 °C in PBS buffer solution). Therefore, knowing the characteristic values λ_{crit} and u_{crit} for a particular chromatographic system, this equation allows computing the bed compressibility as a function of the fluid velocity. Using the so-obtained value of the compression factor λ, the porosity of the compressed bed ε'_e can be estimated as follows:

$$\varepsilon'_e = \frac{\varepsilon_e - \lambda}{1 - \lambda} \tag{2.53}$$

These reduced porosity and column length values (ε'_e and L'_{col}) can be used into the Kozeny–Kármán equation instead of ε_e and L_{col} to predict the evolution of the pressure drop with increasing fluid velocity for different column geometries.

Figure 2.12 shows experimental pressure drop data obtained at various fluid velocities for columns of various geometries packed with Sepharose 4FF [165]. It can be seen that

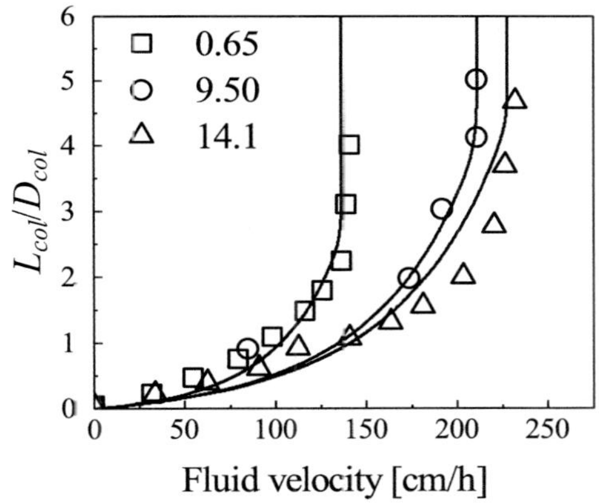

Figure 2.12 Experimental and predicted pressure drop profiles as a function of the fluid velocity for chromatographic columns of different geometries packed with Sepharose 4 FF. The following dimensions L_{col}/D_{col} have been considered: 19.6/30 (squares), 30.3/3.2 (circles), and 36.7/2.6 (triangles). The model predictions were obtained by estimating the porosity and the length of the compressed bed using Equations 2.51–2.53 and computing the pressure drop from Equation 2.50. Experimental data taken from [165] by permission of John Wiley & Sons, Inc.

large and short packed beds collapse at low velocity while thinner columns can be used at higher velocities and stand higher pressure drops. Moreover, it can be seen in Figure 2.12 that the model reported earlier represents fairly well the experimental data. Besides the empirical approach proposed by Stickel and Fotopolcus, more rigorous models are also available in the literature [158].

2.4.4 Kinetics of Mass Transfer

Approaches to Model Diffusion

A common approach to describe the kinetics of mass transfer consists in using Fick's law for diffusion. For a detailed treatment of the application of Fick's law in porous particles, the reader is referred to [158, 184]. Here, we simply intend to recall the commonly used equation to model the radial concentration profiles inside the particles:

$$\frac{\partial \bar{c}_A(r,z,t)}{\partial t} = \frac{1}{r^2}\frac{\partial}{\partial r}\left(r^2\overline{D}_A\frac{\partial \bar{c}_A(r,z,t)}{\partial r}\right) \tag{2.54}$$

where $\bar{c}_A$ denotes the local concentration of A lumping together the concentrations of solute in the intraparticle fluid and bound to the solid surface. It is assumed that the particles are spherical and of homogeneous porosity, so that $\bar{c}_A$ is defined at a given distance r from the center of the particle and does not depend on the angular coordinates. The boundary conditions are given by:

$$\bar{c}_A(r = d_p/2, z, t) = \overline{C}_A^*(z,t) \quad \text{and} \quad \frac{\partial \bar{c}_A(r=0,z,t)}{\partial r} = 0 \tag{2.55}$$

The average lumped concentration of solute A introduced in Section 2.4.2 ($\overline{C}_A$) can be computed by integrating the local concentration profile ($\bar{c}_A$) over the particle radius as follows:

$$\overline{C}_A(z,t) = \frac{24}{d_p^3}\int_0^{d_p/2} r^2 \bar{c}_A(r,z,t)dr \tag{2.56}$$

It is worth highlighting that Fick's law is strictly valid only for binary mixtures (i.e., one solute in one eluent in the case of chromatography) or in the limit of infinite dilution for more complex mixtures. Applying Fick's law to mixtures containing more than two solutes results in composition-dependent diffusion coefficients, which are difficult to estimate. As a first approximation, Fick's diffusion coefficients are often assumed constant, which may lead to significant discrepancies between model predictions and experimental chromatograms in the presence of multiple solutes, competitive effects, and strong interactions with the chromatographic medium.

A more general and physically sound approach consists in modeling the diffusion of the different solutes inside the particles using Stefan–Maxwell equations for diffusion [158, 184, 185]. Briefly, Stefan–Maxwell equations derive from a force balance written on each solute expressing the equality between the frictional force and the force deriving from the chemical potential. Stefan–Maxwell equations allow to better describe the coupling between solute diffusion and adsorption and provide diffusion coefficients that are relatively independent from the solutes concentration. Although certainly more precise, this approach substantially complicates the mathematical treatment of solute diffusion within the particles.

The Linear Driving Force Approximation

In this book, we will mainly use the *linear driving force* (LDF) approximation, which derives from the linearization of Fick's law given in Equation 2.54. In this frame, the kinetics of external mass transfer is described by:

$$\frac{d\overline{C}_A}{dt} = \frac{1}{t_{e,A}}(C_A - C_A^*) \tag{2.57}$$

where C_A^* is the solute concentration in the extraparticle fluid at the fluid–solid interface, and $t_{e,A}$ is the characteristic time for external mass transfer of the solute A.

On the other hand, internal mass transfer is described by:

$$\frac{d\overline{C}_A}{dt} = \frac{1}{t_{i,A}}(\overline{C}_A^* - \overline{C}_A) \tag{2.58}$$

where $\overline{C}_A^*$ is the lumped solute concentration at the outer surface of the particles at thermodynamic equilibrium with the extraparticle fluid (i.e., $\overline{C}_A^* = \overline{K}_A C_A^*$ for linear adsorption isotherms), and $t_{i,A}$ is the characteristic time for internal mass transfer for the solute A. Figure 2.13 shows a schematic representation of the concentration profiles in the frame of the LDF approximation.

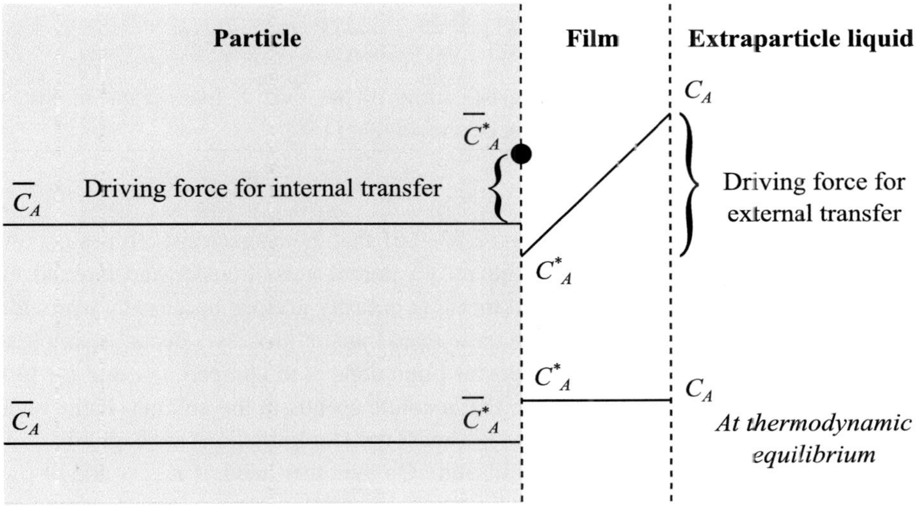

Figure 2.13 Schematic representation of the concentration profiles outside and inside the particles when the LDF approximation is used to describe mass transport. The top part of the graph highlights the driving forces for external and internal mass transfer, while the bottom part represents the situation at equilibrium.

Because there is no accumulation at the fluid–solid interface, the mass transfer rates of external and internal diffusion must be identical, and we can link the concentrations outside and inside the particles as follows:

$$\frac{\mathrm{d}\bar{C}_A}{\mathrm{d}t} = \frac{1}{t_{e,A}}(C_A - C_A^*) = \frac{1}{t_{i,A}}(\bar{C}_A^* - \bar{C}_A) \tag{2.59}$$

The concentrations $\bar{C}_A^*$ and C_A^* are related through the fluid–solid equilibrium relation, while the lumped concentration $\bar{C}_A$ is defined by Equation 2.2.

Let us now analyze two limiting cases:

- When the kinetics of mass transfer is limited by external transfer (i.e., internal diffusion is much faster than external diffusion), the internal concentration profile is flat, so that $\bar{C}_A \approx \bar{C}_A^*$.
- When the kinetics of mass transfer is limited by internal transfer (i.e., external diffusion is much faster than internal diffusion), the external concentration profile is flat, so that $C_A \approx C_A^*$.

With biomacromolecules the second case usually prevails, and external mass transfer limitations are often neglected.

Under linear equilibrium conditions, that is typically the case at low solute concentration, an overall LDF approximation describing a process for which both internal and external mass transfer limitations play a significant role can be written as follows:

$$\frac{\mathrm{d}\overline{C}_A}{\mathrm{d}t} = \frac{1}{t_{m,A}}(\overline{K}_A C_A - \overline{C}_A) \tag{2.60}$$

where $t_{m,A}$ is the characteristic time of the overall mass transfer. Still under linear adsorption conditions it can be shown that [158]:

$$t_{m,A} = t_{i,A} + t_{e,A}\overline{K}_A \tag{2.61}$$

It appears from Equations 2.57–2.61 that the characteristic times $t_{e,A}$ and $t_{i,A}$ are the key model parameters to quantify external mass transfer and internal mass transfer, respectively. Their physical meaning is easily grasped because they provide an estimate of the time scales on which these mass transfer processes occur. A simple way to assess the importance of mass transfer limitations is to compare $t_{e,A}$ and $t_{i,A}$ to the retention time, i.e., the average time that a solute spends in the column. If the retention time is much larger than $\overline{K}_A t_{e,A}$ and $t_{i,A}$, mass transfer limitations are negligible, and the system behaves as if it was at equilibrium. On the other hand, if $\overline{K}_A t_{e,A}$ and/or $t_{i,A}$ are close to or larger than the retention time, external and/or internal mass transfer limitations play a significant role in the shape of the chromatograms.

The kinetics of mass transfer is sometimes described in terms of mass transfer coefficients instead of characteristic times. The correspondence between these two approaches is straightforward because the mass transport coefficients correspond to the inverse of the characteristic times: $k_{i,A} = t_{i,A}^{-1}$, $k_{e,A} = t_{e,A}^{-1}$, and $k_{m,A} = t_{m,A}^{-1}$, for internal, external and overall mass transfer, respectively. It follows that $k_{m,A}^{-1} = k_{i,A}^{-1} + \left(k_{e,A}/\overline{K}_A\right)^{-1}$, which corresponds to the sum of resistances in series.

Expression of the Characteristic Time for Mass Transfer

The characteristic times for internal and external mass transfer can be expressed as follows [158]:

$$\begin{cases} \text{external} & t_{e,A} = \dfrac{d_p}{6}\dfrac{\delta}{D_A} \\ \text{internal} & t_{i,A} = \dfrac{d_p^2}{60\overline{D}_A} \end{cases} \tag{2.62}$$

where δ is the thickness of the stagnant fluid layer surrounding the particles, D_A is the molecular diffusion coefficient of A, and $\overline{D}_A$ is the lumped diffusion coefficient inside the particles. The latter is given by a suitable average between the diffusion coefficient of the solute in the liquid inside the pores $\grave{D}_A$ and the diffusion coefficient of the solute bound to the chromatographic medium $\overline{\overline{D}}_A$, which is generally much smaller than $\grave{D}_A$. Under linear conditions, $\overline{D}_A$ is given by:

$$\overline{D}_A = \frac{\varepsilon_i \grave{D}_A + (\overline{K}_A - \varepsilon_i)\overline{\overline{D}}_A}{\overline{K}_A} \tag{2.63}$$

The thickness of the stagnant fluid layer is often estimated from expressions that correlate the Sherwood (Sh), Reynolds (Re), and Schmidt (Sc) dimensionless numbers, which are defined as follows:

$$Sh = \frac{d_p}{\delta} \qquad Re = \frac{\rho u d_p}{\eta} \qquad Sc = \frac{\eta}{\rho D_A} \tag{2.64}$$

where ρ and η denote the density and the dynamic viscosity of the fluid, respectively, so that η/ρ is the kinematic viscosity. The Sherwood number quantifies the ratio between the particle size and the thickness of the boundary layer, the Reynolds number quantifies the ratio of inertial forces to viscous forces in the fluid flow, and the Schmidt number quantifies the ratio between the kinematic viscosity and the molecular diffusion coefficient.

Several correlations have been proposed in the literature and to give examples we report here the ones by Wakao and Funazkri [186], Carberry [187], and Wilson and Geankopolis [188]:

$$\text{Wakao \& Funazkri} \quad Sh = 2 + 1.1Re^{0.6}Sc^{1/3} \quad 3 < Re < 10^4 \tag{2.65}$$

$$\text{Carberry} \quad Sh = \left(\frac{1.32}{\varepsilon_e}\right)^{1/2} Re^{1/2}Sc^{1/3} \quad Re < 400 \tag{2.66}$$

$$\text{Wilson \& Geankopolis} \quad Sh = \left(\frac{1.09}{\varepsilon_e}\right) Re^{1/3}Sc^{1/3} \quad 1.5 \times 10^{-3} < Re < 55 \tag{2.67}$$

At low *Re* numbers, the *Sh* number computed by Equation 2.65 tends to the theoretical value of 2 expected under stagnant conditions. On the other hand, Equations 2.66 and 2.67 both lead to *Sh* values of 0 at low fluid velocity. Although physically unrealistic because it would imply that there is no mass transfer under stagnant conditions, the latter two correlations lead to satisfactory results under most conditions used in liquid chromatography.

Because different correlations typically lead to different estimates of the thickness of the stagnant fluid layer, the obtained value of δ should only be regarded as a rough indication. There is no general recommendation regarding the choice of the correlation, except that the range of validity, in terms of *Re* and *Sc* numbers, should be respected.

Considering Equations 2.62–2.67, interesting comments can be made regarding the impact of the fluid velocity and particle size on the kinetics of mass transfer. First, it is seen that the characteristic time for external mass transfer decreases with the fluid velocity (through a reduction of δ), while the characteristic time for internal mass transfer does not depend on the fluid velocity.

Depending on the correlation selected for computing the Sherwood number, the characteristic time for external mass transfer increases with the particle diameter with a power exponent varying between 1.5 and 2. On the other hand, the characteristic time of internal mass transfer $t_{i,A}$ always increases with d_p^2. Overall, an increase of the particle size has a detrimental impact on both external and internal mass transfer.

Equation 2.62 shows that the characteristic time for external mass transfer is inversely proportional to the molecular diffusion coefficient D_A to the power of 2/3, while the characteristic time of internal mass transfer is inversely proportional to the lumped diffusion coefficient $\overline{D}_A$. These two diffusion coefficients may be significantly different, with $\overline{D}_A$ being approximately up to two orders of magnitude lower than D_A in the

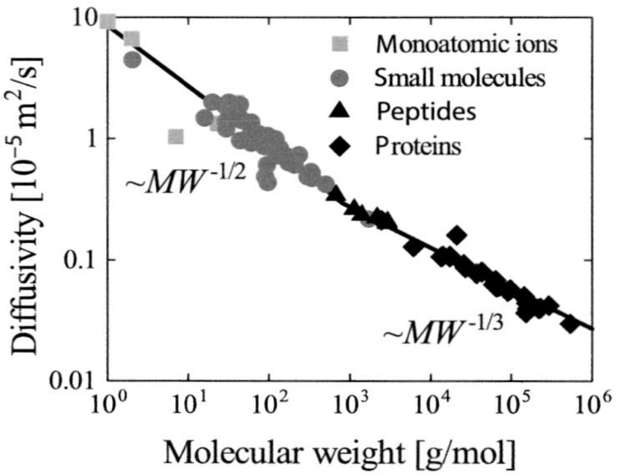

Figure 2.14 Diffusion coefficients of different solutes in water at 25 °C spanning a wide range of molecular weights [190–194]. Lines correspond to power law fits, as indicated in the figure.

case of proteins [21, 158]. The lumped diffusion coefficient averages resistances due to pore diffusion in the intraparticle fluid (characterized by $\grave{D}_A$) and to solid diffusion (characterized by $\overline{\overline{D}}_A$). In the following, we analyze briefly what are the main factors affecting D_A, $\grave{D}_A$, and $\overline{\overline{D}}_A$.

The molecular diffusion coefficient D_A primarily depends on the solute size, the temperature T, and the fluid viscosity. It is often estimated from the Stokes–Einstein equation as follows [189]:

$$D_A = \frac{k_B T}{6\pi \eta R_A} \tag{2.68}$$

where k_B is the Boltzmann constant.

Figure 2.14 shows the molecular diffusion in water for species of vastly different molecular weights, ranging from monoatomic ions to small molecules, peptides, and large proteins. It appears that the molecular diffusion coefficient decreases with the solute molecular weight to the power of 1/3 for peptides and proteins (see Equation 2.14), while the power law exponent is closer to 1/2 for smaller chemical species.

Based on a simplistic description of a porous particle, the following expression can be proposed to relate the diffusion coefficient in the liquid inside the particle pores $\grave{D}_A$ to the molecular diffusion coefficient D_A:

$$\grave{D}_A = \frac{D_A}{\tau} \psi \tag{2.69}$$

where the tortuosity factor τ characterizes the fact that the pores are not straight, and $\psi = \psi(R_A/r_{pore})$ is a hindrance factor that depends on the ratio between the solute hydrodynamic radius and the average pore radius. Correlations to assess this hindrance factor have been proposed in the literature [21]. To give rough estimates, tortuosity values between 2 and 5 are often reported, while the hindrance factor is in the order of 0.5 for a solute 10 times smaller than the average pore size. Nevertheless, obtaining an estimation of $\grave{D}_A$ from D_A is more challenging than it may seem. The main information to be remembered from Equation 2.69 is that solute diffusion in the pores of the particles may be reduced up to approximately one order of magnitude as compared to solute diffusion in solution. Such a reduction is particularly pronounced in pores that are tortuous and whose size is in the same order of magnitude as the solute size.

In porous chromatographic media, the relative contribution of $\grave{D}_A$ and $\overline{\overline{D}}_A$ to $\overline{D}_A$ depends on the value of the Henry coefficient. According to Equation 2.63, the lumped diffusion coefficient $\overline{D}_A$ decreases when increasing the adsorption strength from $\grave{D}_A$ in the absence of interactions, i.e., when $\overline{K}_A$ reduces to ε_i, to $\overline{\overline{D}}_A$ for large Henry coefficients. This brief analysis shows that the characteristic time for internal mass transfer is coupled with the adsorption strength: $t_{i,A}$ increases with $\overline{K}_A$.

In addition, $\overline{D}_A$ is often observed to be concentration dependent, which implies that $t_{i,A}$ may also vary with the solute concentration. By using Fick's law to describe the kinetics of internal mass transfer along the particle radius, it was shown that the lumped diffusion coefficient can be written as the value under linear conditions times a corrective factor [158]:

$$\overline{D}_A(C_A) = \overline{D}_A(C_A = 0) \times f(C_A) \tag{2.70}$$

where the function $f(C_A)$ depends on the derivative of the adsorption isotherm. This equation reflects that the presence of proteins already adsorbed on the surface reduces the possibility for the diffusing protein to interact with the chromatographic medium. In the case of favorable adsorption isotherms, this results in an increase of $\overline{D}_A$ and a decrease of $t_{i,A}$ with the protein concentration. Equation 2.70 implies that the LDF approximation can still be used under nonlinear conditions provided that a proper dependence of $\overline{D}_A$ (and thus of $t_{i,A}$) with the protein concentration is accounted for. An example of an LDF model with $t_{i,A}$ varying with the protein concentration will be presented in Section 4.2.3.

Kinetics of Binding

The kinetics of solute binding at the surface of a chromatographic medium is generally considered to be sufficiently fast compared to the kinetics of external and internal mass transfer so that it can be assumed instantaneous. Although this assumption is very reasonable for chromatographic processes where solutes binding occurs via physical interactions, the case of affinity chromatography deserves special considerations. Indeed, the binding of a protein to an affinity ligand might resemble a chemisorption process in the sense that it requires the formation of specific hydrogen bonds and that biospecific interactions may necessitate molecular rearrangement. These mechanisms may significantly slow down the binding process, so as to make its kinetics relevant.

This has motivated several works aiming at quantifying the kinetics of antibody binding to Protein A domains. In these studies, the kinetics of binding was described by:

$$\frac{d\overline{\overline{C}}_A}{dt} = k_{ads,A}\left[C_A(\overline{\overline{N}} - \overline{\overline{C}}_A) - \frac{\overline{\overline{C}}_A}{\tilde{K}_A}\right] \tag{2.71}$$

where $k_{ads,A}$ corresponds to the binding rate constant. Values of the characteristic time for binding, defined as $1/(k_{ads,A}C_A)$, ranging from approximately 1 to 20 s were reported [21]. By comparison, the characteristic time for internal mass transfer typically ranges from several minutes to one hour. As a result, the kinetics of binding is usually negligible in most industrial applications where relatively large particles are used. This might not be the case only in some special situations, for example where particles constituted of a nonporous core, referred to as pellicular particles, are used. A simple experiment to verify that the kinetics of solute partitioning is rate-limited by the binding step is to increase the particle size and check that no changes in the shape of the peaks is observed. Indeed, the characteristic time for binding is not affected by the particle size in contrast to the characteristic times of external and internal mass transfer that both increase with the particle size (see Equation 2.62).

An important aspect to consider when dealing with biomacromolecules is the potential reorganization of the protein structure on the surface of the chromatographic medium, which may slow down the apparent kinetics of binding. Such changes in the protein conformation have been reported to promote the formation of aggregates on various media, as further discussed in Chapter 7.

2.4.5 Column Efficiency

Number of Theoretical Plates

The efficiency of a chromatographic column is commonly characterized in terms of a finite number of equilibrium stages N, also called *theoretical plates*. Equivalently, one can define the *height of a theoretical plate* (HETP) as the ratio of the column length over the number of equilibrium stages: $HETP = L_{col}/N$. The higher is the number of plates, or the lower is the value of the HETP for a given column length, and the more efficient is the chromatographic column.

The HETP can be estimated experimentally from a chromatogram obtained by a single-pulse injection under linear conditions according to:

$$HETP = \frac{L_{col}}{N} = L_{col}\left(\frac{\sigma_A}{t_{r,A}}\right)^2 \tag{2.72}$$

where $t_{r,A}$ is the average retention time of the solute A and σ_A^2 is the variance of the peak. This mathematical expression simply means that sharp chromatographic peaks indicate a large number of theoretical stages. On the other hand, broad peaks indicate pronounced dispersive effects, due to either axial dispersion or mass transfer limitations.

The numerical estimation of the parameters $t_{r,A}$ and σ_A^2 necessitates to compute the normalized first- and second-order moments of the chromatographic peak, which are defined as follows:

$$\mu_{1,A} = \int_0^{\infty} \frac{C_A(L_{col}, t)t}{C_A^F \Delta t^F} \mathrm{d}t \tag{2.73}$$

$$\mu_{2,A} = \int_0^{\infty} \frac{C_A(L_{col}, t)t^2}{C_A^F \Delta t^F} \mathrm{d}t \tag{2.74}$$

where $C_A(L_{col}, t)$ is the concentration of A measured at the outlet of the column at time t. The first-order moment ($\mu_{1,A}$) and the second-order moment ($\mu_{2,A}$) describe the average position and the broadness of the peak, respectively. Higher-order moments can also be defined, where the order of the moment corresponds to the power exponent on the time in Equation 2.74. Notably, the third-order moment can be related to the peak symmetry. The following relations hold for the first- and second-order moments:

$$t_{r,A} = \mu_{1,A} \tag{2.75}$$

$$\sigma_A^2 = \mu_{2,A} - (\mu_{1,A})^2 \tag{2.76}$$

In the case of Gaussian peaks, it is possible to estimate the HETP directly from the chromatogram without computing the moments. Indeed, in this case, $t_{r,A}$ can be approximated to the position of the maximum of the peak, whereas σ_A^2 can be approximated by the width at mid-height divided by 5.54.

It must be stressed here that this analysis is valid only for a single-pulse injection performed under linear conditions, which means that a "sufficiently small" volume of a "sufficiently dilute" solution should be injected. In practice, these conditions are met as long as the average retention time does not depend on the injection volume nor on the feed concentration.

The van Deemter Equation

The broadening of a single solute chromatogram under linear equilibrium conditions is due to various dispersive mechanisms. The analysis developed by van Deemter allows estimating the contribution of the different mechanisms by reporting the HETP as a function of the superficial fluid velocity [195]:

$$HETP = \frac{a_{VD}}{u} + b_{VD} + c_{VD}u \tag{2.77}$$

where a_{VD}, b_{VD} and c_{VD} are three positive coefficients that are related to the molecular axial diffusion, to the Eddy axial dispersion, and to mass transfer resistances, respectively. Equation 2.77 derives from the computation of the moments obtained in the frame of the MC model under the LDF approximation. Figure 2.15 illustrates the shape of such curve and indicates the ranges of velocity where the different dissipative mechanisms dominate the expression of the HETP.

Van Deemter Parameters

Let us now analyze the contributions of these three terms in the case of proteins:

Parameter a_{VD}. The term a_{VD}/u turns out to be negligible in most liquid chromatography applications dealing with proteins due to the low value of the molecular diffusion coefficient of biomacromolecules.

Parameter b_{VD}. The term b_{VD} essentially depends on the quality of the packing of the chromatographic column. It is equal to:

$$b_{VD} = \frac{2\varepsilon_e D_{ax,A}}{u} \tag{2.78}$$

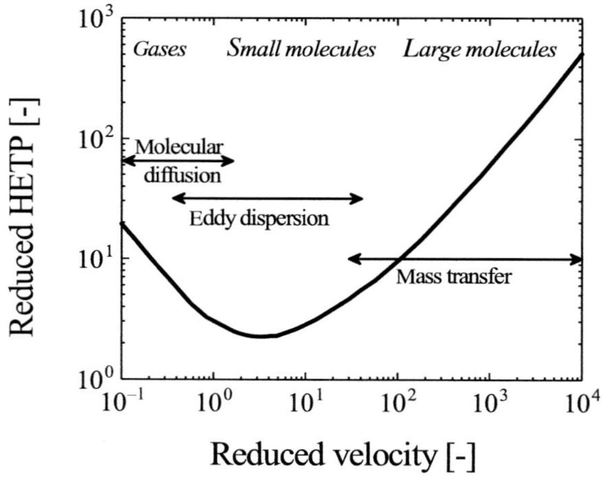

Figure 2.15 Van Deemter plot indicating the main contributions to the HETP at different fluid velocity values. The typical ranges of velocities used for gases, small molecules and large molecules are also specified. The reduced HETP is defined as $HETP/d_p$, and the reduced velocity is defined as $ud_p/(\varepsilon_e D_A)$. Adapted from [196] by permission of John Wiley & Sons, Inc.

In the range of fluid velocities commonly used for protein chromatography, it is a good approximation to consider b_{VD} values in the order of two to four times the particle diameter. The parameter b_{VD} can be related to the particle Péclet number Pe_p and to the number of mixing cells J, which both characterize the hydrodynamics of the column:

$$b_{VD} = \frac{2d_p}{Pe_p} = \frac{L_{col}}{J} \tag{2.79}$$

The particle Péclet number quantifies the ratio between the rate of solute transport by convection over the rate of solute transport by diffusion around the particles and is related to the column Péclet number by $Pe_p = Pe_c \times d_p/L_{col}$. When using $b_{VD} = 4d_p$, one obtains $Pe_p = 0.5$, which is a typical value for well-packed chromatographic columns.

Parameter c_{VD}. Under typical velocities used in protein chromatography, the contribution from mass transfer limitations dominates the expression of the HETP, as shown in Figure 2.15. The term c_{VD} can be computed from:

$$c_{VD} = \left(\frac{\phi_e \overline{K}_A}{1 + \phi_e \overline{K}_A}\right)^2 \frac{2t_{m,A}}{\varepsilon_e \phi_e \overline{K}_A} \tag{2.80}$$

where the overall characteristic time for mass transfer $t_{m,A}$ accounts for both internal and external mass transfer according to Equation 2.61. It is observed from Equation 2.80 that c_{VD} increases linearly with $t_{m,A}$. The impact of various factors on $t_{m,A}$ has been discussed in Section 2.4.4. For example, we have seen that $t_{m,A}$ roughly increases with d_p^2 and that the characteristic time for external mass transfer depends on the fluid velocity. Therefore, c_{VD} is independent of the fluid velocity only in the case where the kinetics of mass transfer is limited by internal diffusion. Moreover, we see in Equation 2.80 that the term c_{VD} depends on the lumped Henry coefficient (explicitly but also implicitly through $t_{m,A}$). Therefore, experiments carried out under different elution conditions will provide different van Deemter curves.

Thus summarizing, we have seen that in the case of biomacromolecules the expression of the van Deemter equation typically simplifies to a straight line given by $HETP = b_{VD} + c_{VD}u$, where b_{VD} is related to the axial dispersion while the term $c_{VD}u$ reflects the contribution of the mass transfer resistances. The term $c_{VD}u$ usually dominates the expression of the HETP. The estimation of c_{VD} allows assessing the overall characteristic time for mass transfer $t_{m,A}$, which is often close to $t_{i,A}$ due to strong internal mass transfer resistances.

HETP and the Mixing-Cell Model

When considering a mixing-cell model, the van Deemter equation can be written as follows:

$$\frac{1}{N} = \frac{1}{J} + \frac{c_{VD}}{L_{col}}u \tag{2.81}$$

where the number of theoretical plates $N = L_{col}/HETP$ should not be confused with the hydrodynamic parameter J. While the number of theoretical plates quantifies dispersive effects originating from both hydrodynamic and kinetic limitations, the parameter J reflects dispersive effects due to hydrodynamics only.

The question is then how many cells should be used in the MC model. Two options can be envisaged: either J mixing cells are considered and the kinetics of mass transfer is modeled independently, for example using the LDF approximation, or N mixing cells are considered and the fluid–solid equilibrium is assumed to be reached instantaneously. With the second approach, both hydrodynamic and mass transfer dispersive effects are accounted for simultaneously in a single parameter, namely the number of theoretical plates N.

Let us stress that HETP measurements should be performed with relevant solutes, at best those that are composing the feed mixture to be purified. Measuring the HETP with small and noninteracting molecules (e.g., salt or acetone) will not provide a reliable estimate of the number of plates. For such components, the value of the HETP gives only an indication of the hydrodynamic parameter J, which is larger than N, thus leading to an overestimation of the column efficacy.

Perfusive Chromatographic Media

The observation that the c_{VD} term dominates the expression of the HETP has encouraged the development of perfusive chromatographic media. The idea is to induce particle permeability by carefully creating sufficiently large pores to facilitate solute transport through the solid particles. In these large pores, solute transport is enhanced by a convection mechanism, in contrast to the smaller pores where solute transport is limited by the rate of diffusion. However, the surface area available for binding is decreasing with increasing pore size. Therefore, in order to benefit from the intraparticle convection mechanism while maintaining a sufficient surface area, it is important to ensure that the large pores are connected to smaller pores. In perfusive chromatographic media, the c_{VD} term depends on the fluid velocity according to [158, 197, 198]:

$$c_{VD}^{perf} = c_{VD} \times \frac{3}{Pe_i}\left(\frac{1}{\tanh\left(Pe_i\right)} - \frac{1}{Pe_i}\right) \approx c_{VD} \times \left(1 + \frac{Pe_i}{3}\right)^{-1} \tag{2.82}$$

where c_{VD} is given in Equation 2.80 and Pe_i is the intraparticle Péclet number, which quantifies the ratio between the rate of solute transport by internal convection within the particle over the rate of solute transport by diffusion in the pores. It is defined as:

$$Pe_{i,A} = \frac{u_i d_p}{6\grave{D}_A} \tag{2.83}$$

where u_i is the fluid velocity inside the large pores of the particles. The intraparticle velocity is often computed as $u_i = F_i u$ where the factor F_i depends upon the internal porosity. F_i values in the range 0.05% to 0.5% are reported in the literature [199, 200]. We underline here that we are dealing with three different Péclet numbers that should not be confused. These three Péclet numbers all quantify a ratio between the rate of solute transport by convection over the rate of solute transport by diffusion. However, they refer to different length scales: the chromatographic bed (Pe_c), the particle size (Pe_p) and the macropores (Pe_i).

Figure 2.16 shows experimental data of the HETP obtained with three chromatographic media: a commercial nonperfusive material (Fractogel EMD S03), a commercial perfusive material (POROS 50 HS), and a noncommercial perfusive material [201]. HETP measured at different flow rates with serum albumin under nonbinding conditions (i.e., the lumped Henry coefficient reduces to $\overline{K}_A = \varepsilon_i$ in Equation 2.80) are reported. It can be seen that the HETP obtained with the nonperfusive material (Fractogel

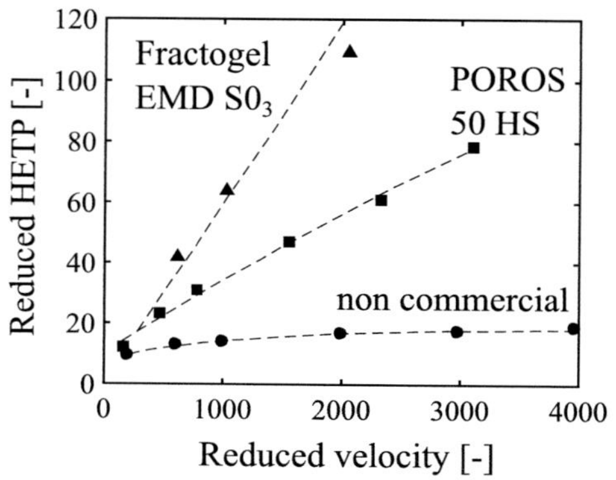

Figure 2.16 Van Deemter curves for a commercial nonperfusive material (Fractogel EMD S03), a commercial perfusive material (POROS 50 HS), and a noncommercial perfusive material. Symbols correspond to experimental data obtained with human serum albumin under nonbinding conditions. Lines correspond to fits performed with Equation 2.77, where the c_{VD} term is given by Equation 2.80 for the nonperfusive material, and by Equation 2.82 for the perfusive materials. The reduced HETP is defined as $HETP/d_p$ and the reduced velocity is defined as $ud_p/(\varepsilon_e D_A)$. Reprinted from [201], copyright 2014, with permission from Elsevier.

EMD SO3) increases linearly with the fluid velocity, which is consistent with the analysis carried out earlier. In addition, it is observed that the HETP values obtained with both perfusive chromatographic media are lower and tend to flatten out at high flow rates as predicted by Equation 2.82. This phenomenon is even more pronounced for the noncommercial material, which is characterized by a larger intraparticle Péclet number [201]. This highly perfusive behavior makes the HETP relatively independent of the fluid velocity. This has practical consequences because the chromatographic process can be operated at high flow rates (i.e., with a high throughput) without affecting the efficiency of the column. However, this can only be obtained at the expense of a lower saturation capacity due to a reduction of the surface area. To counteract this effect, it is possible to functionalize the pores with polyelectrolyte brushes or hydrogels so as to create a three-dimensional phase available for binding. This increases the saturation capacity of the chromatographic medium but reduces its perfusive behavior.

2.5 Model Parameter Estimation

2.5.1 Generalities

After having presented models to describe chromatographic systems, we provide here methods to estimate the corresponding model parameters. They can be classified into three categories: thermodynamic parameters, which describe fluid–solid equilibria, hydrodynamic parameters, which describe axial dispersion, and kinetic parameters, which describe external and internal mass transfer.

At first, the average position of the chromatogram is dictated by the thermodynamics of the fluid–solid equilibrium. The stronger a solute binds to the chromatographic medium, and the later it elutes. Then, the broadness and the asymmetry of the chromatogram depend on several phenomena: They both increase with the nonlinearity of the thermodynamic equilibrium, with hydrodynamic dispersion and with mass transfer limitations. In particular, Langmuir adsorption isotherms lead to elution peaks that tend to deviate from the symmetric Gaussian shape typical of linear adsorption isotherms and to exhibit a triangular shape characterized by a sharp front and a smooth tail.

Three main strategies can be envisaged to determine model parameters:

- Exploit some characteristic features of the chromatogram. In the case of elution chromatography, this corresponds to the various moments of the peak, which have been introduced in Section 2.4.5. In the case of frontal chromatography, this is, for example, the average position of the front.
- Exploit the entire chromatogram. The model parameters are regressed by comparing the experimental chromatograms with model simulations. This approach is referred to as the inverse method.
- Perform independent experiments that do not rely on the measurement of a chromatogram. This is, for example, the case of batch uptake experiments, to be discussed in Section 2.5.2.

Generally, it is advisable to design the set of experiments to be performed to decouple the different effects as much as possible and obtain model parameters independently. In the following, we discuss in more detail how to obtain relevant information on thermodynamic, hydrodynamic, and kinetic parameters.

2.5.2 Thermodynamics of Fluid–Solid Equilibrium

Isocratic Conditions

Our aim here is to determine the function $\overline{C}_A^* = \bar{f}_A(C_A^*)$, where the symbol * indicates equilibrium conditions. For the sake of simplicity, we present in this section the methods by considering noncompetitive equilibrium relations, i.e., assuming that each solute interacts with the chromatographic medium independent from all the others, while the case of competitive equilibrium relations will be discussed in Chapter 5.

There are a number of methods to determine thermodynamic equilibrium relations under isocratic conditions, and here we selected three of them for their relative ease of use and complementarity.

A common method consists in performing *batch uptake experiments*. The idea is to put in contact the protein solution with particles of the chromatographic medium, wait long enough to reach thermodynamic equilibrium, and measure the residual concentration in solution C_A^*. The lumped concentration of solute in the solid phase at equilibrium $\overline{C}_A^*$ can be computed with a mass balance:

$$VC_A^F + \overline{V}\overline{C}_A^F = VC_A^* + \overline{V}\overline{C}_A^* \tag{2.84}$$

where V and $\overline{V}$ are the volumes of the liquid phase and of the particles (accounting for both the solid and intraparticle liquid), respectively, while C_A^F and $\overline{C}_A^F$ are the concentrations of solute initially fed in the two phases. Figure 2.17(a) shows experimental results obtained with a monoclonal antibody on a Protein A chromatographic medium at several initial protein concentrations [202]. It is seen that, while the concentration in the liquid phase C_A^* increases quasi linearly with the initial protein concentration over the whole range of concentrations, the concentration in the solid phase $\overline{C}_A^*$ first increases linearly and then reaches a plateau. The fluid–solid equilibrium relation is obtained by plotting $\overline{C}_A^*$ as a function of C_A^*, and fitting the thermodynamic model parameters with a model, as, for example, a Langmuir adsorption isotherm given by Equation 2.18. In these experiments, it is necessary to investigate a sufficiently broad range of concentrations to access the nonlinear part of the equilibrium relation. In addition, batch-uptake experiments require a high accuracy in the measurement of the volumes of the protein solution and of the chromatographic medium brought into contact. This is why robotic methods are particularly well suited to screen a large number of chromatographic media [203, 204].

Frontal analysis is another popular technique to determine fluid–solid partitioning. It is based on the measurement of breakthrough curves obtained at various feed concentrations. As an example, Figure 2.17(b) shows breakthrough curves obtained with a monoclonal antibody on a Protein A chromatographic medium at three different feed concentrations [205]. The objective is to determine the solute concentration in the solid

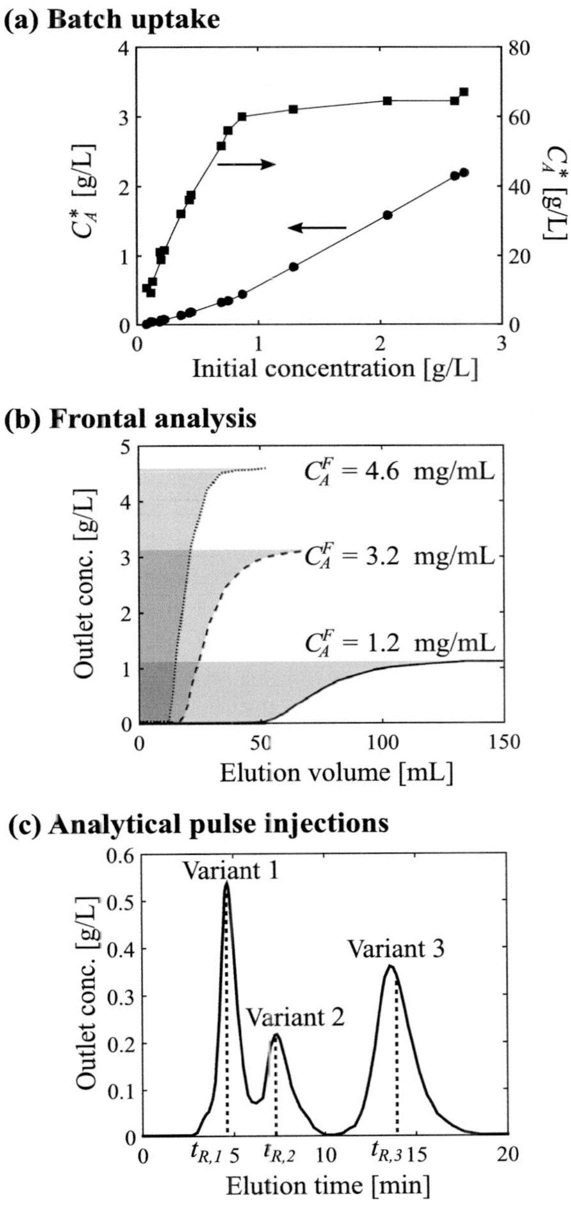

Figure 2.17 Common methods to estimate fluid–solid thermodynamic parameters. (a) Batch uptake experiments performed with a monoclonal antibody on a Protein A chromatographic medium. Experimental data taken from [202], copyright 2005, with permission from Elsevier. (b) Frontal analysis experiments performed with a monoclonal antibody on a Protein A chromatographic medium at three different feed concentrations. Adapted from [205] by permission of John Wiley & Sons, Inc. (c) Analytical pulse injections performed with three antibody variants on a weak cation exchange chromatographic medium. Reprinted from [206], copyright 2007, with permission from Elsevier.

phase in equilibrium with the solute concentration in the liquid phase, that is to say the feed concentration C_A^F. To do so, one has to realize that the area comprised between the breakthrough curve and the feed concentration (shaded in Figure 2.17(b)) is proportional to the amount of solute that is retained in the chromatographic column. This quantity is given by $Q \times \int_0^\infty \left(C_A^F - C_A\left(L_{col}, t\right)\right) dt$ and can be divided into two portions: one corresponding to the extraparticle liquid phase (equal to $\varepsilon_e C_A^F$) and the other one to the particles (equal to $(1 - \varepsilon_e)\overline{C}_A^*$). Therefore, the lumped concentration of solute in the solid phase at equilibrium is computed from:

$$\overline{C}_A^* = \frac{1}{1-\varepsilon_e}\left(\frac{u}{L_{col}}\int_0^\infty \left(C_A^F - C_A\left(L_{col}, t\right)\right) \mathrm{d}t - \varepsilon_e C_A^F\right) \tag{2.85}$$

By performing experiments at various feed concentrations, a plot of $\overline{C}_A^*$ as a function of $C_A^F = C_A^*$ can be constructed and fitted with a mathematical model. Equation 2.85 holds for a column initially free of solute, which implies that the chromatographic medium should be properly regenerated before starting a new experiment. As in the case of batch uptake experiments, a sufficiently broad range of concentrations should be selected to assess both the Henry coefficient and the saturation capacity.

Finally, performing *analytical pulse injections* under dilute conditions is a simple method to estimate the Henry coefficient from the measure of the retention time. Indeed, $t_{r,A}$ and $\overline{K}_A$ are related according to $t_{r,A}/t_0 = 1 + \phi_e \overline{K}_A$ (see Equation 2.7), where $t_{r,A}$ is computed from the first-order moment of the peak according to Equation 2.75. As an example, a chromatogram obtained by injecting three mAb variants in an IEC column is shown in Figure 2.17(c) [206]. An advantage of this method is that it allows determining the Henry coefficient of all the species contained in the feed mixture at once as long as the equilibrium relation is linear (and provided that the $\overline{K}$ values are sufficiently different). Moreover, under nonbinding conditions, as for example at high salt concentration for IEC, this method provides the steric exclusion coefficient because the lumped Henry coefficient reduces to $\overline{K}_A = \varepsilon_i \grave{K}_A$ (see Equation 2.6).

Besides these three techniques, the inverse method can be used to determine the fluid–solid equilibrium relation by fitting model parameters to chromatograms obtained at various feed concentrations and/or injection volumes. However, this method is accurate only for peaks whose shape is predominantly dictated by the fluid–solid equilibrium and not by dispersive effects such as axial dispersion and mass transfer resistances.

Chemical Modulation

In this section, we focus on the determination of the dependence of the Henry coefficient on the modifier concentration. Common methods to estimate the impact of C_M on $\overline{K}_A$ rely on the measurement of retention times from dilute pulse injection experiments. Two main strategies can be envisaged to perform such experiments: isocratic elution conditions or linear gradient elution conditions.

The former method is straightforward because the retention time is related to the lumped Henry coefficient by Equation 2.7. Once the lumped Henry coefficient is estimated at various modifier concentrations, a nonlinear regression can be used to assess

Table 2.3. Lumped Henry coefficients and average retention times for isocratic and LGE operations, where g is the gradient slope and t_0 is the retention time of a completely excluded tracer. For the derivation of the IEC expression it is assumed that $C_M^0 \approx 0$.

IEC	$\overline{K}_A = \varepsilon_i + (1-\varepsilon_i)\overline{\overline{K}}_{0,A} C_M^{-z_A}$	Isocratic / LGE
	$\frac{t_{r,A}}{t_0} = 1 + \phi_e \overline{K}_A$	Isocratic
	$\frac{t_{r,A}}{t_0} = 1 + \phi_e \varepsilon_i + \frac{\left((1-\varepsilon_i)\overline{\overline{K}}_{0,A}\phi_e(1+z_A)g t_0\right)^{\frac{1}{1+z_A}}}{g t_0}$	LGE
HIC/RPC	$\overline{K}_A = \varepsilon_i + (1-\varepsilon_i)\overline{\overline{K}}_{0,A}\exp(-s_A C_M)$	Isocratic / LGE
	$\frac{t_{r,A}}{t_0} = 1 + \phi_e \overline{K}_A$	Isocratic
	$\frac{t_{r,A}}{t_0} = 1 + \phi_e \varepsilon_i + \frac{1}{g t_0 s_A}\ln\left(1 + g t_0 s_A \phi_e \left((1-\varepsilon_i)\overline{\overline{K}}_{0,A}\right)\right)$	LGE

the different model parameters, i.e., $\overline{\overline{K}}_{0,A}$ and s_A for HIC/RPC and $\overline{\overline{K}}_{0,A}$ and z_A for IEC. Typical expressions of the lumped Henry coefficient as a function of the modifier concentration for HIC/RPC and IEC, which derive from Equations 2.31 and 2.32, respectively, are summarized in Table 2.3.

Another possible strategy consists in performing a series of analytical pulse injections under linear gradient elution (LGE) conditions. Although the mathematical treatment of LGE is substantially more complex than that of isocratic elution, LGE experiments are particularly useful to screen rapidly a wide range of modifier concentrations when no preliminary information is available on the system of interest. The implementation of gradient elution conditions implies that the concentration of modifier varies in time. In particular, for linear gradients, the concentration of modifier at the column inlet is given by:

$$C_M^F = C_M^0 + gt \qquad 0 \leq t \leq \Delta t_g \tag{2.86}$$

where C_M^0 is the initial modifier concentration, g is the gradient slope defined as $\Delta C_M/\Delta t_g$, where Δt_g is the gradient duration, and ΔC_M is the difference between the final and the initial modifier concentrations. The determination of the equilibrium parameters necessitates to measure retention times at various modifier gradient slopes. Analytical expressions of $t_{r,A}$ as a function of the gradient slope, g are given in Table 2.3 for both IEC and HIC/RPC. These expressions derive from the equilibrium theory using Equation 2.86 to describe the evolution of the inlet modifier concentration and assuming a negligible solute exclusion [21, 207]. They were, for instance, applied to rapidly scale up an ion exchange chromatographic process from experimental data [208]. In the presence of solute exclusion, i.e., when the different solutes access a fraction of the pore volume significantly lower than the modifier, similar equations can be found in the literature [209]. Finally, just like for the isocratic case, nonlinear regression methods are used to estimate the different model parameters by comparing experimentally measured retention times and model predictions from the expressions shown in Table 2.3.

To do so, it is recommended to determine the parameters t_0 and ε_i beforehand through independent experiments using suitable tracers.

For illustrative purposes, we apply the two methods given earlier (isocratic and LGE) to estimate the parameters z_A and $\overline{\overline{K}}_{0,A}$ for lysozyme on a strong cation exchange medium. On the one hand, six pulse injection experiments under isocratic conditions using various concentrations of salt were performed, and an overlay of the obtained chromatograms is shown in Figure 2.18(a). On the other hand, six pulse injection experiments under LGE conditions with various gradient slopes, i.e., increasing the salt concentration in the eluent from 0 to 1 M over different times (from 5 to 100 min), have been performed, and an overlay of the obtained chromatograms is shown in Figure 2.18(b). The retention times were then computed from the first-order moment of the chromatograms. The resulting values are plotted as a function of the modifier concentration in Figure 2.18(c) for the isocratic case and as a function of the gradient slope in Figure 2.18(d) for the LGE case. The model parameters $t_0 = 0.48$ min and

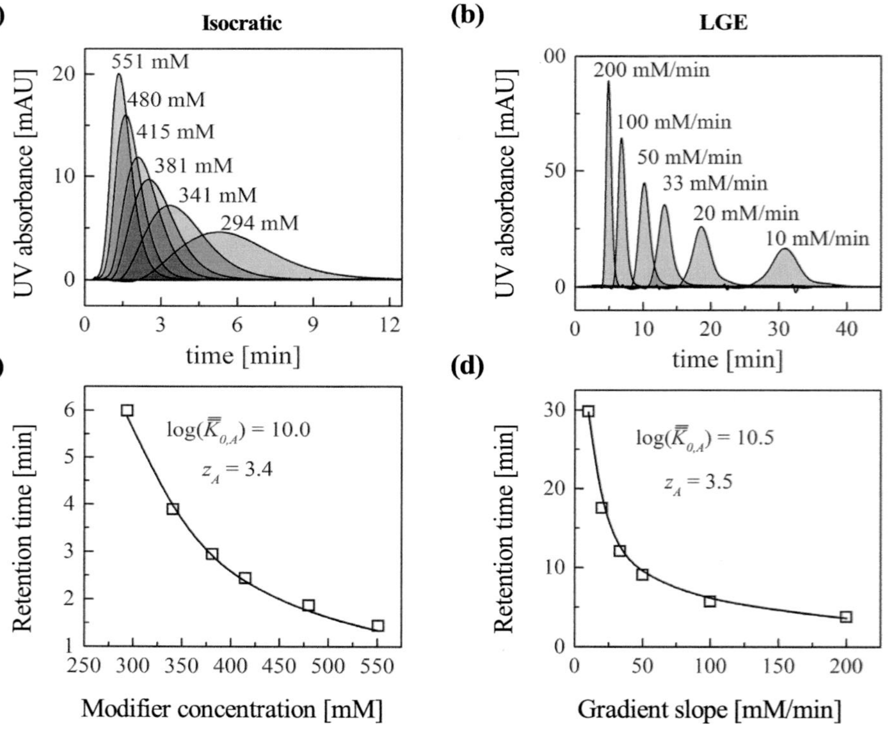

Figure 2.18 Estimation of the parameters z_A and $\overline{\overline{K}}_{0,A}$ for lysozyme on a strong cation exchange medium. (a) Overlay of the pulse injection chromatograms measured under isocratic elution conditions for various modifier concentrations. (b) Overlay of the pulse injection chromatograms measured under LGE conditions for various gradient slopes. (c) Retention times as a function of the modifier concentration in the eluent for the isocratic experiments. (d) Retention times as a function of the gradient slope for the LGE experiments. In panels (c) and (d), experimental data are represented as symbols and model results as solid lines.

$\varepsilon_i = 0.81$ were determined from independent experiments using blue dextran and acetone as tracers, respectively. Then, the model parameters z_A and $\bar{\bar{K}}_{0,A}$ were estimated from the equations given in Table 2.3 by nonlinear regression. The parameter values are shown in Figure 2.18(c) and (d) together with the model estimations. It can be seen that both approaches led to very similar parameters, 3.4 and 3.5 for z_A and 10.0 and 10.5 for $\ln\left(\bar{\bar{K}}_{0,A}\right)$, and provide a good description of the experimental data.

2.5.3 Hydrodynamics

Hydrodynamic dispersion in the chromatographic bed is quantified either by the column Péclet number Pe_c in the frame of the PD model, or by the number of mixing cells J in the frame of the MC model, as discussed in Section 2.4.3. As a rule of thumb, it is reasonable to assume that $Pe_p \approx 0.5$. From this, Pe_c or J can be easily assessed recalling that $Pe_c = Pe_p \times L_{col}/d_p = 2J$.

Experimentally, hydrodynamic dispersion can be estimated by performing pulse injection experiments under dilute conditions at various fluid velocities. The y-intercept in the van Deemter plot (b_{VD}) is in fact inversely proportional to Pe_c (see Section 2.4.5). Such measurements can be useful to verify if a column is well packed. Indeed, it is expected that b_{VD} is in the order of two to four times the particle diameter, which corresponds to $Pe_p \approx 0.5$, as discussed in Section 2.4.5. A significant deviation from this result can be regarded as an indication that the chromatographic bed is not packed properly, suggesting, for example, the existence of channelling or dead volumes at the inlet or outlet of the column. For illustrative purposes, Figure 2.19 shows experimental van Deemter curves obtained with a monoclonal antibody on two SEC media characterized by two different particle diameters, namely 5 and 10 μm [210]. First, it is seen that the curves of the HETP as a function of the fluid velocity are relatively linear, which is expected from the analysis carried out in Section 2.4.5 for

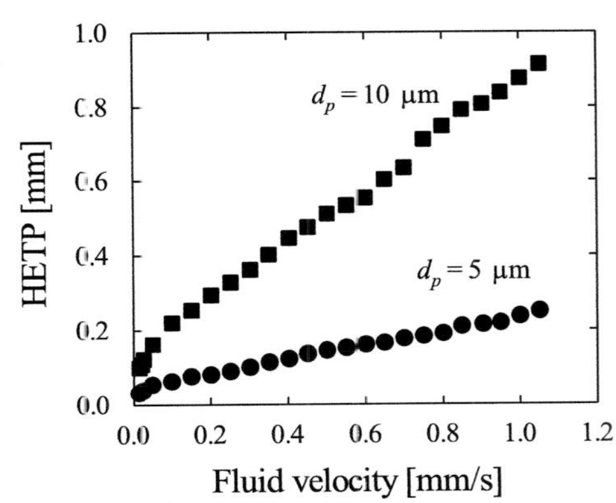

Figure 2.19 Experimental van Deemter curves obtained with a monoclonal antibody in two different SEC media characterized by different particle sizes. Adapted from [210] with permission.

proteins. Then, we observe that the y-intercept is roughly 0.04 mm and 0.1 mm for the column packed with 5 and 10 μm particles, respectively. These values are in the order of 10 times the particle diameter, which is a bit larger than what is commonly observed. This is most likely due to an imperfection in the packing, which is not unexpected for self-packed columns, as it was the case in this study. It is also interesting to note that an increase in the particle size by a factor of 2 leads to an increase in the y-intercept by a factor of about 2, as expected from Equation 2.78.

2.5.4 Kinetics of Mass Transfer

In this section, we discuss the determination of the model parameters describing the kinetics of mass transfer. Under linear conditions, the overall characteristic time for mass transfer can be estimated from the slope of the van Deemter plot, as described in Section 2.4.5. Considering the data shown in Figure 2.19, we find $c_{VD} \approx 0.2\ \mathrm{s}^{-1}$ for the column packed with particles of 5 μm diameter and $c_{VD} \approx 0.8\ \mathrm{s}^{-1}$ for the column packed with particles of 10 μm diameter. It is interesting to note that an increase in the particle size by a factor of 2 leads to an increase in c_{VD} (and thus of $t_{m,A}$) by a factor of 4. This is consistent with the expression of the characteristic time for mass transfer given in Section 2.4.4, where we have seen that $t_{m,A}$ approximately increases with d_p^2.

Obtaining an estimate of the overall mass transfer time is valuable information because it allows quantifying the importance of dispersive effects due to slow diffusion rates. However, because the characteristic times for external and internal mass transfer are impacted differently by the process variables (e.g., d_p, u, ε_i), it is useful to determine the individual contribution of t_i and t_e to the overall mass transfer process. It is relatively common that one step is substantially slower than the other, thus controlling the overall kinetics of mass transfer. This step is referred to as the *rate-controlling* or the *rate-limiting step*, and it should be the main focus of efforts aiming at reducing dispersive effects. Indeed, improving the other steps will have only a limited impact on the overall kinetics of mass transfer. For example, if the kinetics of mass transfer is limited by external diffusion, increasing the fluid velocity may be fruitful, but increasing the pore size is pointless. On the other hand, increasing the pore size may be beneficial if the kinetics of mass transfer is limited by internal diffusion.

Estimating $t_{i,A}$ and $t_{e,A}$ independently is a rather delicate task, especially under linear conditions where the two parameters contribute to the kinetics of mass transfer through $t_{m,A}$. Most often, the characteristic times for external and internal mass transfer are determined from experiments performed at various flow rates by the inverse method, i.e., by performing model simulations and fitting $t_{i,A}$ and $t_{e,A}$ to describe experimental chromatograms. It is also possible to use the empirical correlations given in Section 2.4.4 to obtain a rough approximation of $t_{e,A}$ and then estimate $t_{i,A}$ by subtracting $t_{e,A}\overline{K}_A$ to $t_{m,A}$.

2.5.5 Case Study

In this case study, the methods discussed in the previous sections are applied to an illustrative example. The system under consideration involves a protein (pI $\approx$ 9) injected in a cation exchange chromatographic column ($L_{col} = 2.5$ cm, $V_{col} = 1$ mL, $d_p = 90$ μm, $\varepsilon_e = 0.3$, $\varepsilon_i = 0.81$, $\overline{\overline{N}} = 2.7$ eq/L). As an eluent, an aqueous buffer at pH 7.0 with various concentrations of sodium chloride was used.

The total and external porosities of the chromatographic column were estimated from pulse injection experiments of different tracers, namely acetone and blue dextran, respectively. Then, the value of the intraparticle porosity was computed from $\varepsilon_i = (\varepsilon_t - \varepsilon_e)/(1 - \varepsilon_e)$. The ionic capacity $\overline{\overline{N}}$ was estimated from an independent titration experiment performed by the chromatographic media supplier.

We choose to model the chromatographic column as a plug flow with axial dispersion, to model the kinetics of external and internal mass transfer in the frame of the LDF approximation and to describe the fluid–solid equilibrium with the SMA model. This implies that the parameters $\overline{\overline{K}}_{0,A}$, z_A, ς_A, $t_{i,A}$, $t_{e,A}$, and Pe_p must be estimated. For the sake of simplicity, we focus here on the determination of model parameters of one solute only. However, in practice, it would be necessary to determine these parameters for all the solutes composing the feed mixture in order to properly simulate the separation process. Once all the model parameters are determined, the model can be used to predict the effect of various operating conditions, such as the fluid flow rate, the injection volume, the modifier concentration, etc., with a view to optimizing the process performances.

To assess the parameters necessary to describe the thermodynamic equilibrium ($\overline{\overline{K}}_{0,A}$, z_A, and ς_A), the hydrodynamics (Pe_p), and the kinetics of mass transfer ($t_{i,A}$ and $t_{e,A}$), experiments under different salt concentrations, fluid velocities, and solute concentrations must be performed. In this example, we choose to determine all the model parameters from pulse injection experiments under isocratic elution conditions. The selected set of experiments is reported in Table 2.4.

In order to better illustrate the parameter estimation procedure, the "experimental" chromatograms were generated with the selected model and the following model parameter values: $\overline{\overline{K}}_{0,A} = 1.1 \times 10^{-5} M^{z_A}$, $z_A = 5.7$, $\varsigma_A = 58$, $t_{i,A} = 5.5$ s, $t_{e,A} = 0.8$ s, and $Pe_p = 0.4$. By doing so, it is possible to assess the quality of the parameter determination method by comparing the estimated model parameters with the selected values reported earlier. In addition, this allows explaining the approach without discussing experimental errors (for example, due to random noise in the UV signal, nonlinearity in the detector response, dead volumes, and inaccuracy in the modifier concentration ...).

Table 2.4. Summary of the experimental conditions. For each of the six experiments, the injection volume, V^F was set to 0.5 mL.

	c_A^F [mmol/L]	u [cm/h]	C_M [mmol/L]
1	0.05	150	70
2	1.00	150	70
3	0.05	150	100
4	0.05	150	85
5	0.05	72	85
6	0.05	108	85

The model parameters are estimated from various combinations of the “experiments” given in Table 2.4 as follows:

1. The parameters $\overline{\overline{K}}_{0,A}$ and z_A are estimated from pulse injection experiments performed under dilute conditions at various modifier concentrations, i.e., experiments 1, 3, and 4. The corresponding chromatograms are shown in Figure 2.20(a).

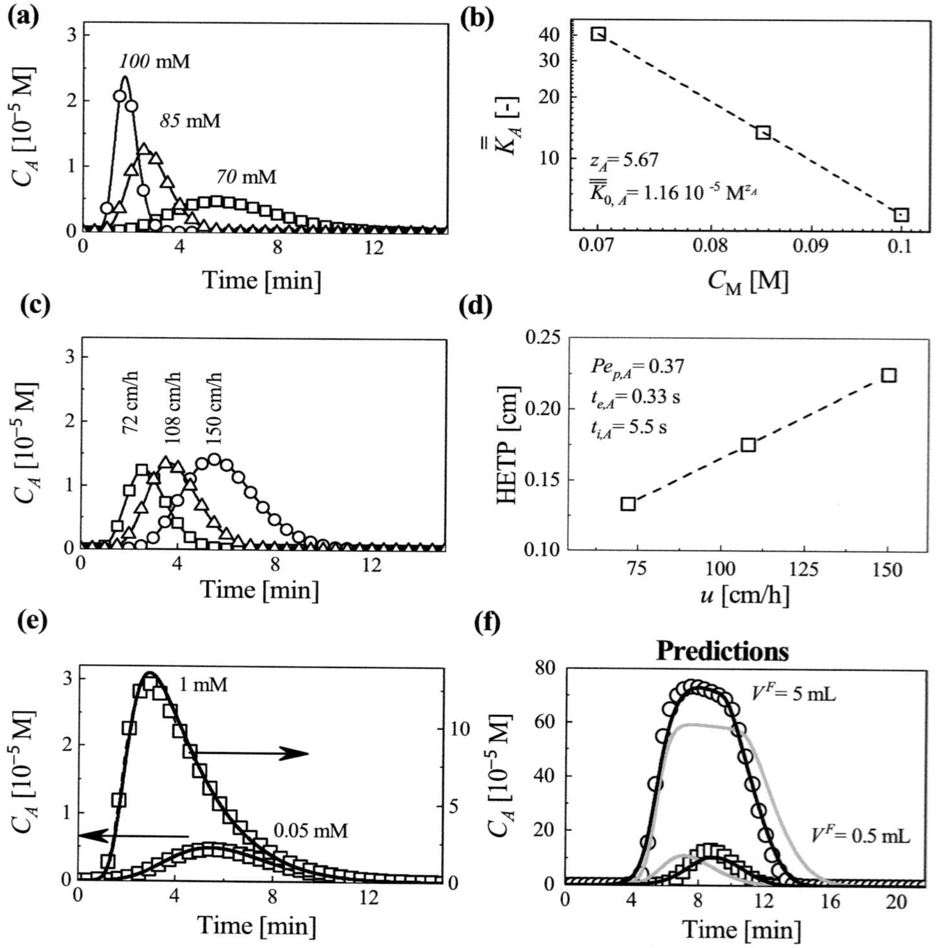

Figure 2.20 (a) Chromatograms from experiments 1, 3 and 4 to estimate the parameters of the Henry coefficient under linear equilibrium conditions. (b) Regression of the Henry coefficient at various modifier concentrations. (c) Chromatograms from experiments 4, 5, and 6 to estimate the hydrodynamic and mass transfer parameters. (d) Van Deemter plot. (e) Chromatograms from experiments 1 and 2 to estimate the steric parameter ς_A. (f) Comparison between experimental (open symbols) and predicted chromatograms with the step-by-step approach (thick black lines) and the brute force approach (thick gray lines). The following parameter values were obtained by the brute-force approach: $\overline{\overline{K}}_{0,A} = 6.8 \times 10^{-1} M^{z_A}$, $z_A = 1.40$, $\varsigma_A = 95.2$, $t_{i,A} = 6.84$ s, $t_{e,A} = 0.30$ s, and $Pe_{p,A} = 0.32$.

By estimating the average retention times and using Equations 2.6 and 2.7, the parameter $\overline{\overline{K}}_A$ can be computed for each modifier concentration. Then, the parameters $\overline{\overline{K}}_{0,A}$ and z_A are obtained from the regression with Equation 2.32, as shown in Figure 2.20(b). We obtain $\overline{\overline{K}}_{0,A} = 1.16 \times 10^{-5} M^{z_A}$ and $z_A = 5.67$.

2. The mass transfer and hydrodynamic parameters are estimated from pulse injection experiments performed under dilute conditions using different fluid velocities, namely experiments 4, 5, and 6. The corresponding chromatograms are shown in Figure 2.20(c). The first- and second-order moments were computed, and the results are presented in a van Deemter plot in Figure 2.20(d). The overall time for mass transfer and the particle Péclet number are estimated from the slope and the y-intercept of the van Deemter curve to $t_{m,A} = 6.6$ s and $Pe_p = 0.37$, respectively. The characteristic time for external mass transfer has been roughly estimated from Equation 2.62 and using the Carberry correlation (Equation 2.66). It is found that $t_{e,A}$ varies from 0.29 s at $u = 150$ cm/h to 0.37 s at $u = 72$ cm/h. At the considered modifier concentration, it was found from the previous set of experiments that $\overline{K}_A = 3.4$. Because $t_{m,A} = t_{i,A} + \overline{K}_A t_{e,A}$, the impact of $t_{e,A}$ on $t_{m,A}$ is expected to be relatively small with respect to that of $t_{i,A}$. It is thus reasonable to assume that $t_{m,A}$ is independent of the fluid velocity. For the sake of simplicity, we consider only the average value of the characteristic time of external mass transfer, namely $t_{e,A} = 0.33$ s. It follows that $t_{i,A} = 5.5$ s.
3. Finally, the steric parameter ς_A of the SMA model, whose contribution is only appreciable under nonlinear conditions, is estimated by the inverse method (all the other parameters being known) using experiments 1 and 2, which are performed at two different protein concentrations. Figure 2.20(e) compares experimental chromatograms (symbols) with model simulations obtained by using the model parameters previously estimated and fitting ς_A (lines). It is seen that a satisfactory agreement is obtained between experiments and model simulations for a value of the steric parameter $\varsigma_A = 56.1$.

Overall, the model parameters estimated with the proposed procedure are very close to the ones used to generate the "experimental" chromatograms. In order to further show the accuracy of the estimated model parameters, they were used to predict two chromatograms generated under different operating conditions (LGE instead of isocratic conditions, different feed concentrations and injection volumes). For these new "experiments," the feed concentration C_A^F was set to 0.5 mM, the fluid velocity to 90 cm/h, the modifier concentration was increased linearly from 50 to 100 mM in 12 min, and two injection volumes were chosen, namely 0.5 and 5 mL. The "experimental" chromatograms and the model predictions are compared in Figure 2.20(f) by open symbols and solid black lines, respectively. It is seen that the model predictions are in very good agreement with the "experimental" data, which demonstrates the reliability of the model parameter estimation procedure presented earlier.

Alternatively, there is often the tendency to think that it would be faster to use a "brute-force" approach, using all the chromatograms at once to regress the different model parameters simultaneously. For comparison purposes, we tried to fit the six chromatograms simultaneously with the six adjustable parameters, whose so-obtained values are reported in the caption of Figure 2.20. The model predictions using this brute-force approach are shown as solid gray lines in Figure 2.20(f). It can be seen that, in the present example, the model parameters obtained by the brute-force method fail at properly representing the new set of experimental data. This is most likely because the regression algorithm fell in a local minimum when fitting the chromatograms. Restarting the fitting procedure with different initial conditions or using another algorithm may improve the quality of the model predictions. In conclusion, we can give the following recommendations:

- Experiments at various modifier concentrations, various solute concentrations, and various fluid velocities must be performed in order to obtain reliable estimates of the different model parameters (thermodynamic equilibrium, hydrodynamics, and kinetics of mass transfer).
- It is advisable to estimate the various model parameters as independently as possible.

2.6 Process Design

2.6.1 Process Performance Criteria

In order to select the most appropriate process for a given separation task, one must define some parameters quantifying the process performance. As described in Chapter 1, biopharmaceutical products are subject to strict regulations in terms of *purity*, which is defined as the ratio of the quantity of the species of interest to the total quantity of solutes (i.e., the species of interest and the impurities) in the final product. The quantity of a species i at the outlet of a chromatographic system is computed as:

$$q_i = Q \int_0^{\Delta t_p} C_i \left(L_{col}, t\right) \mathrm{d}t \tag{2.87}$$

where Q denotes the flow rate at which the product is collected. The integration is performed over a given production duration Δt_p, e.g., one day or one cycle. The purity of the species A is then defined as $\chi_A = q_A / \sum_i q_i$, where the sum is computed over all the species present in solution, in general excluding the solvent and the buffer components.

Besides purity, another important process performance parameter is the *recovery* (also referred to as *yield*), which quantifies the fraction of the species of interest that is collected at the outlet of the process with respect to the amount introduced at the inlet. The quantity of species A introduced in the chromatographic system in a given time period is denoted q_A^F. The recovery of species A is then defined as $\vartheta_A = q_A / q_A^F$, and is expressed in percentage.

The *productivity*, which is key in the determination of the production cost, is typically defined as the amount of purified material per unit time and per unit column volume: $\varphi_A = q_A / \Delta t_p / V_{col}$. Alternative expressions of the productivity are sometimes adopted,

Table 2.5. Process performance parameters for a given species *A*.

	Common units	Expression
Purity, χ_A	$g_{product}/(g_{product}+g_{impurities})$	$q_A/\sum_i q_i$
Recovery, ϑ_A	$g_{product}/g_{product}^{feed}$	q_A/q_A^F
Productivity, φ_A	$g_{product}/L_{column}/day$	$q_A/\left(V_{col}\Delta t_p\right)$
Eluent consumption, γ_A	$L_{eluent}/g_{product}$	V_{eluent}/q_A

for example, by considering the volume of the chromatographic medium, or the diameter of the column instead of the column volume. In any case, it is essential to normalize the quantity of purified material by a parameter that reflects the size or cost of the installation.

Finally, *eluent consumption* is an essential parameter to consider in the cost estimation of large-scale chromatographic operations. Whether the eluent is regenerated or simply wasted, there is an obvious cost associated with its consumption. This is defined in volume of eluent per mass of purified material: $\gamma_A = V_{eluent}/q_A$, where V_{eluent} is the volume of eluent required to produce the quantity q_A of product A.

The definitions of the most relevant process performance parameters are summarized in Table 2.5.

2.6.2 Process Optimization

In this section, we briefly present basic concepts about process optimization, and the reader is referred to the following reference for more details about the optimization of chromatographic processes [158].

The first step in the optimization of a chromatographic process is the identification of a proper *objective function* to be maximized (e.g., productivity) or minimized (e.g., cost). A simple example of cost function (e.g., €/$g_{product}$) can be expressed as:

$$cost = \frac{a}{\varphi_A} + \frac{b}{\vartheta_A} + c \times \gamma_A \tag{2.88}$$

where a, b, c are three positive parameters. The first one refers to the operating cost of the process (e.g., €/h/L_{system}), the second is associated with the feed cost (e.g., €/g_{feed}) and the last one to the cost of eluent purchasing and processing (e.g., €/L_{eluent}). Equation 2.88 reflects that the cost associated with the purification of a certain quantity of product increases with the eluent consumption and decreases with the product recovery and productivity. A more complete estimation of the cost function would also account for the cost associated with the preprocessing and postprocessing steps.

After defining a proper objective function, two types of variables must be identified. On the one hand, the *manipulated variables* that can be tuned in order to improve the process performances, as, for example, the switch times, flow rates, column lengths, and particle diameter. On the other hand, the *constrained variables*, such as the purity and pressure drop, should meet certain specifications. Optimizing a chromatographic process then consists in finding the set of manipulated variables that provides the best value of the objective function while satisfying the requirements on the constrained variables.

Optimizing a process is a rather delicate task necessitating to find the best compromise. Indeed, one has to realize that it is impossible to maximize simultaneously productivity and recovery while minimizing the solvent consumption. As the popular saying goes: "You can't have your cake and eat it too".

2.6.3 Shortcut Design

The design of a chromatographic process is rather complex and requires taking a number of decisions about the equipment (i.e., which chromatographic medium, particle size, column dimensions) and the operating conditions (i.e., at which flow rate, eluent composition). This usually necessitates adopting an iterative approach, alternating between experimental characterization, model simulations, and optimization of the selected objective function.

In this section, we report a shortcut design strategy that allows obtaining a first reasonable operating point of a chromatographic process for which the chromatographic medium and eluent composition have been already selected. For the sake of clarity, we present here the method for a single-column process only. Nevertheless, this method is sufficiently general to be adapted to multicolumn processes, as discussed in Chapter 3.

The first step of this shortcut design strategy consists in estimating the normalized volume of feed mixture to be loaded in the column. To do so, it is assumed that the chromatographic column develops an infinite number of stages, i.e., only the fluid–solid thermodynamic equilibria are considered, while hydrodynamic and mass transfer limitations are neglected. Figure 2.21(a) shows a typical chromatogram obtained in the frame of the equilibrium theory for a binary mixture under linear equilibrium conditions. The chromatogram represented in this figure corresponds to the so-called touching-band

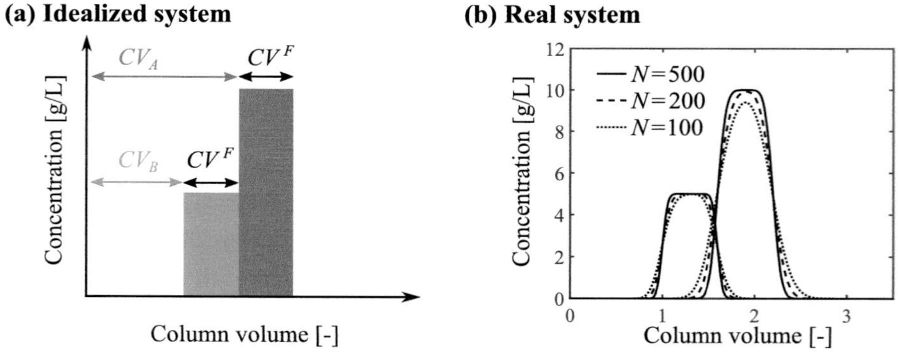

Figure 2.21 Illustration of the shortcut design strategy for a binary mixture separated in a single-column process under linear equilibrium conditions. (a) Estimation of the volume of feed in the frame of the equilibrium theory. (b) Estimation of the required number of plates using the MC model with $\overline{K}_A = 2$, $\overline{K}_B = 1$ and $\varepsilon_e = 0.4$.

situation, i.e., to the maximal volume of feed compatible with the complete separation of A and B. The optimal injection is quite often close to this maximal theoretical value and in practice, values ranging between 50 and 150% of this maximal injection volume are commonly used. Under linear conditions, the feed volume can be computed in terms of column volumes as follows so as to obtain an expression independent from the column dimensions:

$$CV^F = CV_A - CV_B = (1 - \varepsilon_e)(\overline{K}_A - \overline{K}_B) \tag{2.89}$$

The application of the equilibrium theory to nonlinear systems is discussed in the literature [158].

In a second step, we move from an "idealized" system (i.e., where dispersive effects are neglected) to a more realistic system. The goal now is to estimate the number of theoretical plates required to satisfy the separation objectives in terms of purity and recovery. To do so, it is convenient to use the MC model with a number of mixing cells equal to the number of theoretical plates assuming thermodynamic equilibrium. With this simple approach, dispersive effects due to hydrodynamics and mass transfer resistances are simultaneously accounted for in the value of the plate number N. Therefore, implementing this model requires only the knowledge of the thermodynamic equilibrium parameters and of the column external porosity. In addition, the results are independent of the column dimensions if the simulations are analyzed in terms of column volumes. As an illustration, Figure 2.21(b) shows model simulations performed with the MC model using three different plate numbers. It is seen that when N decreases, the peaks get broader and the quality of the separation decreases. Besides using the MC model, it is also possible to use simplified analytical expressions (valid only under linear equilibrium conditions) to predict chromatograms at various values of N [21]. Both approaches allow estimating the minimum number of plates required to obtain the desired process specifications, which we denote N^{min}. If it is crucial to achieve a minimum number of plates to collect the solutes at the desired purity and recovery, it is not recommended to design a chromatographic process with a number of plates much higher than N^{min} for economical reasons.

The constraint to set the number of plates equal to N^{min} imposes a first relation between the column length L_{col} and the fluid velocity u. According to the van Deemter equation (see Section 2.4.5), the following relation holds under linear conditions if dispersive effects are mainly associated with mass transfer limitations:

$$L_{col} \approx c_{VD} u N \tag{2.90}$$

Then, a physically meaningful constraint consists in setting the pressure at which the chromatographic system is operated. In an attempt to maximize productivity and make maximum use of the chromatographic unit, it is generally advisable to operate the unit close to the maximum pressure drop recommended by the column manufacturer (ΔP^{max}). We have seen in Section 2.4.3 that for noncompressible particles, the pressure drop increases linearly with the fluid velocity and the column length. Recalling the definition of the parameter δ_K, the Kozeny–Kármán equation writes as follows:

$$\Delta P = \delta_K L_{col} u \tag{2.91}$$

In the case of compressible particles, an appropriate expression relating the pressure drop with the fluid velocity should be selected, as discussed in Section 2.4.3.

By setting $N = N^{min}$ and $\Delta P = \Delta P^{max}$, the column length and fluid velocity can be computed as follows [158]:

$$L_{col}^{SC} = \sqrt{\frac{c_{VD} N^{min} \Delta P^{max}}{\delta_K}} \tag{2.92}$$

$$u^{SC} = \sqrt{\frac{\Delta P^{max}}{\delta_K c_{VD} N^{min}}} \tag{2.93}$$

where the superscripts SC designate "shortcut." These expressions are valid for non-compressible particles, under linear equilibrium conditions and in the case where the term c_{VD} dominates in the van Deemter equation. Equations 2.92 and 2.93 depend on thermodynamic and kinetic parameters through c_{VD}. If the solutes to be separated are characterized by similar numbers of plates, the expressions given above can be applied to any of these solutes, otherwise the solute leading to the lowest number of plate is usually selected.

It is also interesting to note that Equations 2.92 and 2.93 depend on the chromatographic medium (through δ_K) and on the elution conditions (through c_{VD}) but not on the column diameter. Therefore, a simple way to scale up a preparative chromatographic process is to increase the column diameter linearly with the volumetric flow rate, keeping the fluid velocity and the column length constant. As discussed in Section 2.4.3, this linear scaling with the column diameter is only valid for rigid particles and one should be extremely careful when using compressible gel particles.

The shortcut design strategy proposed in this section is summarized in Box 2.1. This methodology should provide a reasonable starting point for the design of chromatographic processes. Fine-tuning around this first guess with more detailed models is then always advisable.

Box 2.1 Shortcut Design of a Single-Column Chromatographic Unit

1. Estimate the normalized feed volume for an idealized system (i.e., for an infinite plate number).
2. Estimate the minimum number of plates required to perform the desired separation for a more realistic system (e.g., using the MC model).
3. Determine the column length and fluid velocity for the maximum pressure and minimum number of plates with Equations 2.92 and 2.93.
4. Determine the column diameter by setting the fluid flow rate, using the fluid velocity and the column length determined in point 3 so as to achieve the production requirement (e.g., in kg/day).

2.7 Conclusion

In this chapter, we presented the fundamentals of chromatography, with a particular focus on protein chromatography. For the purification of biomacromolecules, a large variety of chromatographic media is available, each one exploiting different physicochemical properties of the solutes to be separated, including size, hydrophobicity, charge, and biospecificity. These media can then be used in different chromatographic modes, which are classified as single-column processes (operated either in elution mode or in frontal mode) or as multicolumn processes (for continuous operations).

The key physical phenomena involved in any chromatographic process are: the *thermodynamics* of solute partitioning between the fluid and solid phases, the *hydrodynamics* in the chromatographic bed, and the *kinetics* of solute transfer from the liquid phase to the solid phase. While the thermodynamics of solute partitioning is the engine driving the separation of the different species, the hydrodynamic and kinetic limitations reduce the chromatographic column efficiency, which is reflected by a decrease in the effective number of theoretical plates. Chromatographic models are very helpful to elucidate the role of these phenomena on the separation process, and thus to design effective chromatographic units (i.e., select the appropriate column length, particle size, fluid flow rate). The implementation of chromatographic models requires the estimation of reliable model parameters, which necessitates to perform dedicated experiments at various feed concentrations and flow rates (and eluent compositions if one is interested in the impact of modifier concentration).

The performances of chromatographic processes can then be quantified by four main criteria: product purity, product recovery, productivity, and eluent consumption. It is impossible to optimize all these performances simultaneously, and it is therefore necessary to define an appropriate objective function to be maximized on a system specific basis. We also proposed a shortcut method that allows rapidly designing a chromatographic process based on the minimum number of required plates and the maximum recommended pressure drop. In the next chapters, we will use this shortcut method to compare the process performance criteria of single-column and multicolumn systems.

3 Countercurrrent Separation Processes

3.1 Introduction

In Chapter 1, we already highlighted the potential benefit of countercurrent processes. In particular, we discussed the countercurrent oxygen exchange in fish gills, but similar examples are numerous [211]. For instance, mammalian kidneys act as countercurrent exchangers to remove water from urea [212], duck legs' circulatory systems work as countercurrent heat exchangers to minimize the heat loss when standing in ice-cold water [213], and gazelles cool the bloodstream going to their brain through a countercurrent contact with the cold blood coming from their nasal cavity [214]. Just like these biological systems, industrial processes as heat exchangers, distillation and chromatographic columns, liquid–liquid extractors, etc. can benefit from the countercurrent technology.

In this chapter, we provide the key notions necessary to understand and design countercurrent separation processes. We start by considering simplistic systems characterized by linear fluid–solid thermodynamic equilibrium relations, negligible hydrodynamic dispersion, and infinitely fast mass transfer kinetics. Such simplifications allow introducing the basic concepts of countercurrent chromatography in a simple manner and provide suitable references for the design of multicolumn chromatographic units. Then, we increase progressively system complexity by introducing nonlinear thermodynamic equilibrium relations and dispersive effects to better describe real systems.

3.2 Countercurrent Separation for Idealized Systems

3.2.1 The Equilibrium Stage

Before introducing the concept of countercurrent separation, it is necessary to recall a notion extensively used by chemical engineers: the *equilibrium stage* or *equilibrium plate*. This notion is not restricted to chromatography, but applies to a wide range of separation techniques including liquid–liquid extraction, stripping, scrubbing, distillation, and others.

An equilibrium stage is an elementary volume inside which temperature, pressure, and chemical composition are at thermodynamic equilibrium. The schematic

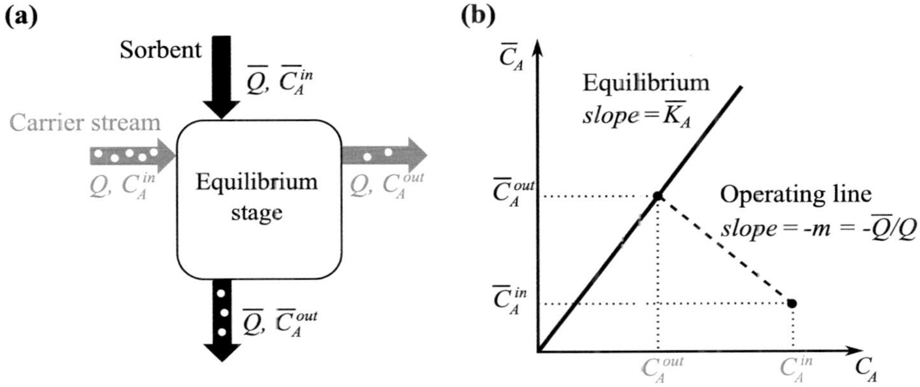

Figure 3.1 (a) Schematic drawing of an equilibrium stage, where a chemical or biological species (represented by white dots) is transferred from the carrier stream (in gray) to the sorbent stream (in black). (b) Graphical relation between the inlet and outlet concentrations after contact in the equilibrium stage.

representation of an equilibrium stage is shown in Figure 3.1(a), where a carrier stream containing certain chemical or biological species is brought in contact with a sorbent stream that can take up this species by sorption. These two streams are nonmiscible and may be of different kinds: two liquids for liquid–liquid extraction, one gas and one liquid for stripping or scrubbing, and one fluid and one solid for chromatography. The sorbent is more precisely referred to as an absorbent or an adsorbent when it is a fluid or a solid, respectively. In any case, once the thermodynamic equilibrium is reached between the carrier stream and the sorbent, the two phases are separated by some kind of settling and collected separately. An equilibrium stage can then be simply seen as a "mixer–settler" at thermodynamic equilibrium.

Let us focus first on the extraction of a single component, i.e., the transfer of a given species A from one phase to another. To do so, we denote C_A and $\overline{C}_A$ the concentration of this species in the carrier stream and in the sorbent stream, respectively, while the superscripts *in* and *out* refer to the inlets and outlets of the equilibrium stage. Because the unit is perfectly mixed, the outlet concentrations correspond to the concentrations inside the equilibrium stage. The volumetric flow rates of the carrier and sorbent streams are denoted Q and $\overline{Q}$, respectively, and are assumed to be constant due to the nonmiscibility of the two phases.

In order to characterize the extraction process, one typically quantifies the mass flux of the species of interest taken up by the sorbent ($\overline{Q}\overline{C}_A^{out}$) with respect to that remaining in the carrier stream (QC_A^{out}). Accordingly, we define the separation factor S_A as follows:

$$S_A = \frac{\overline{Q}\overline{C}_A^{out}}{QC_A^{out}} \tag{3.1}$$

The separation factor measures the amount of a given species taken up by the sorbent stream compared to that left in the carrier stream. For $S_A > 1$, most of A is extracted by the sorbent, while for $S_A < 1$, most of it remains in the carrier stream. Note that for binary separations, the key parameter is the ratio of the separation factors of the two solutes, also known as selectivity.

The concentrations in the two outlet streams leaving the equilibrium stage are linked through a mathematical relation representing the thermodynamic equilibrium condition. Assuming a linear equilibrium relation, we can define the partition coefficient $\overline{K}_A$ so that:

$$\overline{C}_A^{out} = \overline{K}_A C_A^{out} \tag{3.2}$$

Moreover, the mass balance on the equilibrium stage, which ensures the conservation of mass between the inlet and outlet streams, leads to:

$$QC_A^{in} + \overline{Q}\overline{C}_A^{in} = QC_A^{out} + \overline{Q}\overline{C}_A^{out} \tag{3.3}$$

It is seen that the outlet concentrations C_A^{out} and $\overline{C}_A^{out}$ can be computed for given values of C_A^{in}, $\overline{C}_A^{in}$, Q, $\overline{Q}$ by using Equations 3.2 and 3.3. A graphical resolution is also possible and is illustrated in Figure 3.1(b). The linear equilibrium condition given by Equation 3.2 is represented by the black solid line, of slope $\overline{K}_A$. The dashed line is instead called the operating line, and corresponds to a rearrangement of Equation 3.3 to obtain $\overline{C}_A^{out}$ as a function of C_A^{out}. The slope of the operating line is given by the opposite of the ratio between the carrier and sorbent flow rates, commonly referred to as the m factor:

$$\frac{\overline{C}_A^{in} - \overline{C}_A^{out}}{C_A^{in} - C_A^{out}} = -\frac{Q}{\overline{Q}} = -m \tag{3.4}$$

The outlet concentrations in the sorbent and carrier streams are given by the intersection of the operating line describing the mass balance on the equilibrium stage (see Equation 3.4) with the equilibrium line describing the thermodynamics of the system (see Equation 3.2).

From the definitions of $\overline{K}_A$ and m given by Equations 3.2 and 3.4, respectively, it follows that the separation factor introduced in Equation 3.1 can be computed according to:

$$S_A = \frac{\overline{K}_A}{m} \tag{3.5}$$

It appears that the separation factor of a system characterized by a given equilibrium constant $\overline{K}_A$ can only be increased by decreasing the m factor. In other words, the transfer from the carrier to the sorbent stream can be favored by increasing the sorbent flow rate with respect to that of the carrier stream. Once equilibrium is reached, mass transfer between the two streams stops. The only possibility to further enrich the sorbent stream is to bring it in contact with a carrier stream containing the species to be extracted at a concentration higher than the one reached at equilibrium. Similarly, the only possibility to further deplete the carrier stream is to bring it in contact with a sorbent stream containing the species to be removed at a concentration lower than the one reached at equilibrium.

3.2.2 Cascade of Equilibrium Stages

To achieve high levels of transfer of a species from the carrier to the sorbent stream, several equilibrium stages can be arranged in a cascade. Figure 3.2 shows the three configurations that are most commonly considered: (a) *cocurrent*, where the two streams flow in the same direction through the cascade, (b) *cross-current*, where the carrier stream is brought in contact with fresh sorbent at each equilibrium stage, and (c) *countercurrent*, where the two streams flow in opposite directions through the cascade.

By deriving the appropriate mass balance equations, it is possible to compute the outlet concentrations C_A^{out} and $\overline{C}_A^{out}$ in each cascade configuration as a function of the number of equilibrium stages and of the inlet concentrations C_A^{in} and $\overline{C}_A^{in}$ [215]. Numerical results are shown in Figure 3.3 for the illustrative example of $Q = \overline{Q}$ and $\overline{K}_A = 0.5$. It is seen that the three configurations are equivalent when considering only one equilibrium stage. However, as the number of equilibrium stages increases, the way to contact the two streams is shown to play a significant role.

In the cocurrent configuration, it is seen that increasing the number of stages does not allow reducing the concentration in the carrier stream. The two phases are in fact at equilibrium already at the outlet of the first stage, and therefore no further mass transfer from one stream to the other takes place in the subsequent stages. As mentioned earlier, the sole possibility to further deplete the carrier stream at the outlet of an equilibrium stage is to bring it in contact with a sorbent that contains the species of interest at a concentration lower than the one reached at equilibrium. This can be achieved by using the cross-current and countercurrent configurations shown in Figure 3.2(b) and (c),

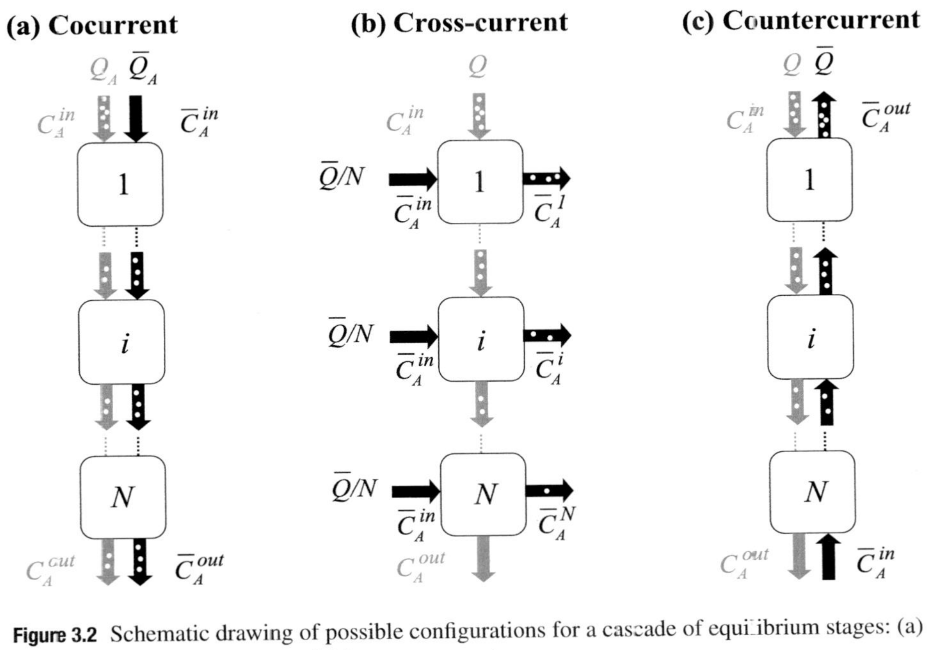

Figure 3.2 Schematic drawing of possible configurations for a cascade of equilibrium stages: (a) cocurrent, (b) cross-current, and (c) countercurrent.

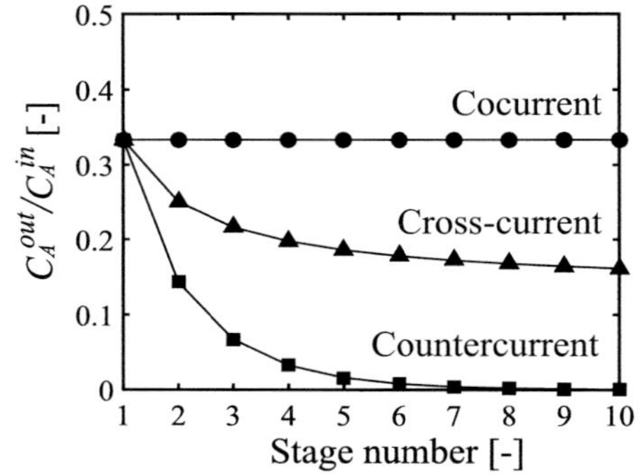

Figure 3.3 Comparison between cocurrent, cross-current, and countercurrent cascades in terms of concentration depletion in the carrier stream as a function of the number of equilibrium stages. Numerical results obtained for $Q = \overline{Q}$ and $\overline{K} = 0.5$.

respectively. It is worth noting that in the case of the cross-current cascade, the sorbent flowrate $\overline{Q}$ is split in N streams of flow rates $\overline{Q}/N$ in order to perform comparisons between the different configurations at the same total flow rate. It is seen in Figure 3.3 that as the number of equilibrium stages increases, the residual outlet concentration in the carrier stream decreases both in the cross-current and countercurrent configurations. However, for a given number of stages, the countercurrent configuration outperforms the cross-current one. This result holds true for any operating condition and equilibrium characteristic as it is only due to the relative orientation of the streams involved.

In addition, the notion of equilibrium stage is independent from the type of the separation unit, i.e., whether it is organized in actual stages, like a plate distillation or absorption column, or as a continuum, like a packed distillation or adsorption column. Chromatographic columns, characterized by a uniformly packed solid phase, typically fall in the latter category. Nevertheless, we have seen in Section 2.4.3 that, from a modeling point of view, a chromatographic column can be described as a series of stages or "cells" in series.

3.2.3 Two-Zone Countercurrent Separation Unit

Principle

In the previous section, we considered the case where a single species is transferred from one stream to another. This is for instance the case when one wants to remove a certain pollutant from some carrier stream by transferring it into a suitable sorbent stream. In the following, we analyze the separation of a binary mixture composed of two species, A and B, that we aim at recovering pure in two distinct streams. Compared to the case presented in Figure 3.1, three different streams are involved here: the feed stream (of flow rate Q^F), which contains the binary mixture to be separated ($A+B$), and two other streams in which A and B are eventually recovered separately. The separation is possible only if the species A and B have different affinities for the two streams, which

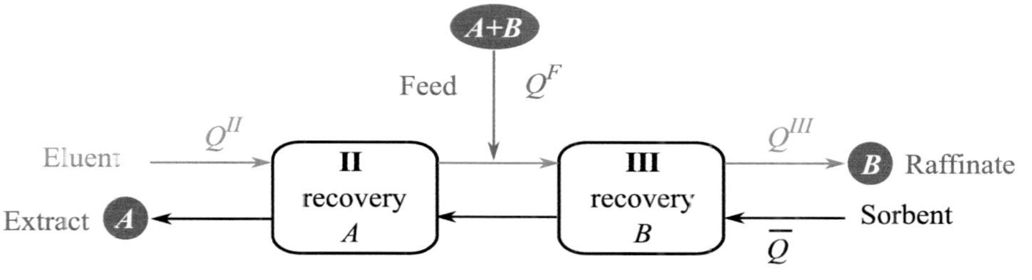

Figure 3.4 Scheme of the two-zone countercurrent separation unit.

are referred to as sorbent and desorbent (or eluent). Arbitrarily, we choose that A has a greater affinity for the sorbent (of flow rate $\overline{Q}$), and B for the eluent (of flow rate Q).

In order to perform this binary separation, we need the two-zone separation unit represented in Figure 3.4. Each of these two zones does not necessarily correspond to an equilibrium stage, but rather to a countercurrent cascade of equilibrium stages. The two zones of the separation unit are denoted II and III in order to anticipate the notation introduced in the next section. The sorbent stream enriched in A at the outlet of zone II is referred to as the *extract*, whereas the eluent stream enriched in B at the outlet of zone III is referred to as the *raffinate*.

In order to take up A preferentially in the sorbent stream and B preferentially in the eluent stream, the following constraints on the separation factors (defined by Equation 3.1) need to be satisfied [216]:

$$\begin{cases} S_A^{II} \text{ and } S_A^{III} \geq 1 \\ S_B^{II} \text{ and } S_B^{III} \leq 1 \end{cases} \tag{3.6}$$

Using Equation 3.5 and noticing that Q^{III} is necessarily larger than Q^{II} due to the addition of the feed stream leads to the following conditions on the m parameters in the two stages:

$$\overline{K}_B \leq m^{II} \leq m^{III} \leq \overline{K}_A \tag{3.7}$$

where $m^i = Q^i/\overline{Q}$.

Satisfying the constraint $m^{III} \leq \overline{K}_A$ ensures that A is dragged by the sorbent, while satisfying the constraint $\overline{K}_B \leq m^{II}$ ensures that B is dragged by the eluent. This concept of two-zone separation units is at the root of continuous countercurrent chromatography.

Complete Separation Region

Fulfilling the constraint given by Equation 3.7 ensures that A and B are separated and mainly recovered in the extract and raffinate, respectively. In order to reach a certain degree of separation, a sufficient number of stages is necessary (as seen in Figure 3.3 in the case of extraction). In the hypothetical limit of an infinite number of equilibrium stages, a complete separation between the two solutes can be achieved, i.e., no A leaves the unit in the raffinate stream and no B in the extract stream, so that each stream contains only the other component pure. This provides the basis of the equilibrium theory, which

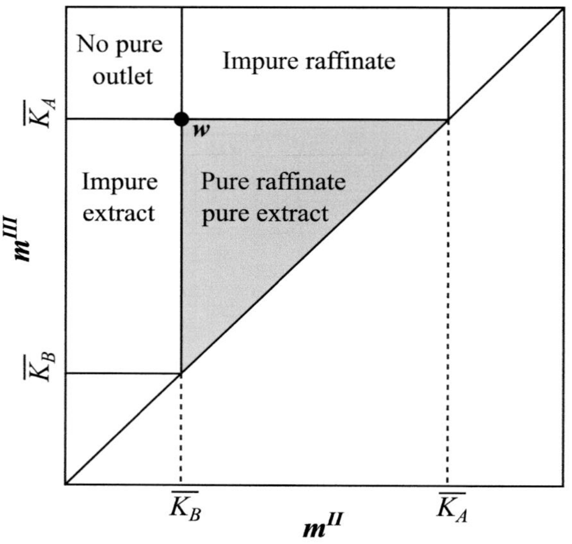

Figure 3.5 The triangle representation defines an operating space in the m^{II}–m^{III} plane providing a complete separation in a countercurrent process under linear thermodynamic conditions in the frame of the equilibrium theory.

allows a simple and didactically helpful description of the two-zone countercurrent separation unit.

The constraints of Equation 3.7 can be graphically represented in the m^{II}–m^{III} plane shown in Figure 3.5 [217, 218]. The pairs of m parameters allowing to recover both A and B pure in the outlet streams under the assumption of an infinite number of equilibrium stages are located inside the gray triangle. On the other hand, the operating points falling out of the triangle do not lead to a complete separation. In particular, if m^{III} is larger than $\overline{K}_A$ (i.e., the eluent flow rate is too high), A is dragged with the eluent flow together with B, and the raffinate is contaminated by the presence of A. Similarly, if m^{II} is smaller than $\overline{K}_B$ (i.e., the eluent flow rate is too low), B is dragged with the sorbent flow together with A, and the extract is contaminated by the presence of B. In the case where m^{II} is too small and m^{III} is too large, there is no pure outlet stream. The graphical representation shown in Figure 3.5 is often referred to as "the triangle theory." It was introduced in this chapter for linear thermodynamic equilibrium relations by a simple approach, while a more rigorous demonstration also valid for nonlinear equilibrium relations can be found elsewhere [217, 218].

A mass balance at the feed node in Figure 3.4 written in terms of the m ratios reads as $m^F = m^{III} - m^{II}$, where $m^F = Q^F/\overline{Q}$ represents the normalized feed flow rate. This shows that, even though all the points located in the triangular region in Figure 3.5 lead to a complete separation of the solutes A and B in the frame of the equilibrium theory, they are characterized by different m^F values. In particular, the point that maximizes the difference $m^{III} - m^{II}$ is the one treating the largest normalized feed flow rate. This point is found at the vertex of the triangle, i.e., $m^{II} = \overline{K}_B$ and $m^{III} = \overline{K}_A$, and denoted by w in Figure 3.5.

However, w does not represent a robust operating point, and it is relatively risky to operate a separation unit at the conditions given by the vertex. A slight deviation in the process conditions or a small error in the estimation of equilibrium parameters could in fact result in a deviation of the operating point outside of the triangle, thus compromising the separation. Depending on the direction of such deviations, one would observe completely different situations, i.e., only the extract is polluted, or only the raffinate, or both. Therefore, a more robust operating point located somewhere inside the triangle is usually selected at the expense of a slightly lower normalized feed flow rate [218].

3.2.4 Four-Zone Countercurrent Separation Unit

In order to complete the two-zone separation unit shown in Figure 3.4, the eluent and sorbent streams can be regenerated and recycled into the process by adding two supplementary zones. This leads to the design of the classical four-zone separation unit shown in Figure 3.6, where the role of each zone is as follows:

Zone I	regeneration of the sorbent, i.e., removal of the species A
Zone II	recovery of A pure in the sorbent
Zone III	recovery of B pure in the eluent
Zone IV	regeneration of the eluent, i.e., removal of the species B

The resulting unit allows separating a feed mixture in a continuous manner. It has two inlet streams (the feed and desorbent streams) and two outlet streams (the extract and the raffinate streams, which contain the species more retained by the sorbent and by the desorbent, respectively). Note that zones II and III are at the core of the separation, while zones I and IV allow obtaining a sorbent stream and an eluent stream free of A and free of B, respectively. This regeneration is necessary to recycle the sorbent and eluent

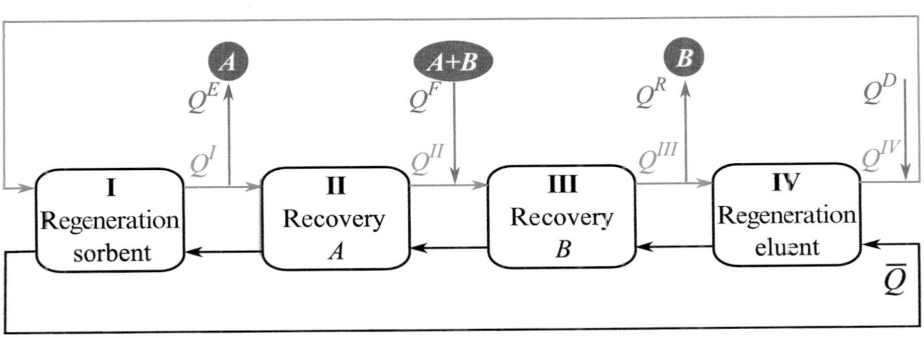

Figure 3.6 Scheme of the four-zone countercurrent separation unit, referred to as the true moving bed (TMB) process.

streams within the unit. The flow rates of feed mixture, extract, raffinate, and desorbent are denoted Q^F, Q^E, Q^R, and Q^D, respectively.

The constraints in the separation zones are the same than those given by Equation 3.6, while the following constraints apply to the regeneration zones:

$$\begin{cases} S_A^I \leq 1 \\ S_B^{IV} \geq 1 \end{cases} \tag{3.8}$$

Therefore, for the proper operation of the separation process, it is necessary to ensure that [216]:

$$\begin{cases} \overline{K}_A \leq m^I \\ \overline{K}_B \leq m^{II} \leq m^{III} \leq \overline{K}_A \\ m^{IV} \leq \overline{K}_B \end{cases} \tag{3.9}$$

As seen previously for the two-zone unit, the region of complete separation obtained in the frame of the equilibrium theory, i.e., assuming that each zone is constituted of an infinite number of equilibrium stages, is defined in the m^{II}–m^{III} plane by the triangle shown in Figure 3.5. In addition, the conditions of complete regeneration for the eluent and sorbent are met when $\overline{K}_A \leq m^I$ and $m^{IV} \leq \overline{K}_B$.

The mass balances at the four nodes between the zones in Figure 3.6 allow relating the different m parameters as follows:

$$m^F = m^{III} - m^{II} \tag{3.10}$$

$$m^E = m^I - m^{II} \tag{3.11}$$

$$m^R = m^{III} - m^{IV} \tag{3.12}$$

$$m^D = m^E + m^R - m^D = m^I - m^{IV} \tag{3.13}$$

with $m^i = Q^i/\overline{Q}$, for $i = I, II, III, IV, F, D, E$, and R.

Based on the preceding equations, the internal and external flow rates can be selected as described in the following. The feed flow rate Q^F is usually imposed by the upstream part of the production process. Then, the parameters m^{II} and m^{III} are selected in order to ensure complete separation (i.e., inside the triangular design space shown in Figure 3.5), and m^F is computed with Equation 3.10. We can then choose suitable values of the parameters m^I and m^{IV} so as to guarantee proper eluent and sorbent regeneration (i.e., $\overline{K}_A \leq m^I$ and $m^{IV} \leq \overline{K}_B$). It can be seen from Equation 3.13 that the lowest eluent consumption is obtained for the lowest value of $m^D = m^I - m^{IV}$, which is obtained for $m^I = \overline{K}_A$ and $m^{IV} = \overline{K}_B$. However, it must be realized that this operating point is not very robust because a small perturbation in the process or an imprecision in the estimated equilibrium parameters could compromise the purity of the recycled streams. Once suitable values of m^I and m^{IV} have been selected, m^E and m^R can be computed by Equations 3.11 and 3.12, respectively, while m^D can be computed with Equation 3.13. Finally, $\overline{Q}$ is easily estimated from the values of Q^F and m^F, which then allows evaluating all the Q values from the definition of the m parameters.

It is worth noting that in the case of chromatography, the sorbent is a solid, while the eluent is a fluid. Accordingly, the four-zone separation unit shown in Figure 3.6 involves the troublesome motion of the solid phase in the direction opposite to the one of the fluid phase. This process is commonly referred to as *true moving bed* (TMB) and constitutes a reference process in the field of countercurrent chromatography. However, its application is rather limited due to the technical difficulty to move the solid phase without generating too much channeling, multiple flow paths, and particle attrition. In the rare applications of the TMB that have been proposed, the chromatographic medium was moved under the effect of gravity, so that the four zones represented in Figure 3.6 were placed vertically with zone IV on top [219].

3.3 The Simulated Moving Bed Process

3.3.1 Principle

The technical difficulties inherent to the TMB process motivated Broughton and Gerhold to develop the *simulated moving bed* (SMB) process in the 1960s [123]. The idea is as bright as simple: instead of moving the chromatographic medium with respect to fixed inlets and outlets, the inlets and outlets are moved with respect to the chromatographic medium [158, 220]. In other words, in the reference frame of the laboratory or the plant, the chromatographic medium is fixed. But in the reference frame of the inlet and outlet streams, the chromatographic medium is in movement.

The impact of the reference frame on the perception of movement can be appreciated by considering the common example of a train about to leave the station. A passenger sitting in an immobile train next to the departing train may have the sensation of movement when looking out of the window (i.e., taking as a reference frame the departing train rather than the ground). Coming back to the SMB, the movement of the injection and extraction points opposite to the intended movement of the chromatographic medium (which in turn corresponds to the same direction than the eluent) allows simulating the countercurrent movement of the solid phase with respect to the eluent. Because the separation is not affected by a change in the reference frame, the process performances of TMB and SMB units are expected to be similar.

For technical reasons, the injection and extraction points cannot be moved continuously along the chromatographic column. Therefore, several columns are used in series, and the positions of the inlet and outlet streams are moved intermittently between the columns at regular time intervals. Figure 3.7 shows how the movement of the inlet and outlet streams is discretized from a true moving bed to a simulated moving bed process. The TMB process is theoretically reproduced at the limit of an infinite number of columns and of a switch time equal to zero. For each configuration, the SMB process comprises four zones delimited by four inlet/outlet streams: the feed, the eluent, the extract (containing the species with high affinity for the solid phase), and the raffinate (containing the species with high affinity for the eluent).

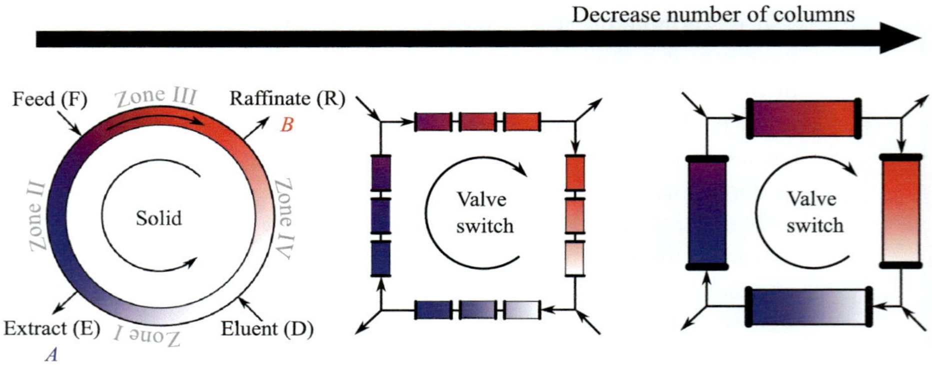

Figure 3.7 Schematic representation of countercurrent chromatographic processes, from the true moving bed (TMB) process on the left, to the simulating moving bed (SMB) process, here represented with three columns per zone and one column per zone.

From the first para-xylene SMB process that was realized with 24 columns in the original design of Broughton, the current trend is to decrease the number of columns. Because a SMB process requires the use of four distinct zones (see Figure 3.6), the theoretical minimum number of columns is four. Such a decrease in the number of columns makes the SMB to deviate more and more from the TMB process because the apparent movement of the solid is jerkier.

Because SMB and TMB are analogous processes seen by two different observers, the design procedure described in Section 3.2.4 can be used for both processes. The flow rates in the SMB process can simply be computed from the flow rates in the TMB process using the following relation, which reflects the change in the reference frame discussed earlier:

$$Q^i_{SMB} = Q^i_{TMB} + \frac{\varepsilon_e}{1-\varepsilon_e}\overline{Q} \tag{3.14}$$

where i is the zone number. The simulated flow rate of solid in the SMB process equivalent to that in the TMB is defined by:

$$\overline{Q} = \frac{\overline{V}}{\Delta t} = \frac{(1-\varepsilon_e)V_{col}}{\Delta t} \tag{3.15}$$

where $\overline{V}$ is the volume of the lumped solid, ε_e is the external porosity of the chromatographic bed, and V_{col} is the volume of one column. In agreement with the notation introduced in Chapter 2, $\overline{Q}$ and $\overline{V}$ are lumped quantities accounting for the intraparticle volume of liquid. Importantly, Δt is the time interval at which the positions of the inlet and outlet streams are switched. The SMB process is thus a periodic process that repeats itself cycle after cycle, where the cycle time is defined as the switch time Δt multiplied by the number of columns.

3.3.2 Dynamic Behavior

After having introduced the fundamentals of the SMB process, we investigate in this section its transient behavior using the mechanistic models described in Chapter 2.

For illustrative purposes, we consider here a four-column SMB unit with one column per zone. The hydrodynamics is described by a PD model (see Equations 2.43), whereas the kinetics of mass transfer is assumed to be limited by the internal diffusion and is modeled with the LDF approximation (see Equation 2.58). Moreover, the thermodynamic equilibrium is described by a Langmuir adsorption isotherm (Equation 2.21). These models are coupled with mass balance equations at the nodes (Equations 3.10–3.13), while the boundary conditions (Equations 2.44 and 2.45) are readjusted after each switch time, Δt.

The concentration profiles in the liquid phase along the column computed at various time points are shown in Figure 3.8(a–d). In Figure 3.8(a), the SMB unit is started: an equimassic feed mixture ($C_A^F = 10$ g/L, $C_B^F = 10$ g/L) is injected between columns 2 and 3, while column 3 is progressively filled. Following the notation previously introduced, A has more affinity for the solid phase than B ($\tilde{K}_A = 0.02$, $\tilde{K}_B = 0.01$), and thus moves slower through the chromatographic column. As the peak fronts approach the end of column 3, the inlets and outlets are moved by one column to the right: the solutes previously injected continue to elute through column 3, while the feed mixture is now injected in column 4 (Figure 3.8(b)). The column switching proceeds as shown in Figure 3.8(c) and (d), and the SMB unit is progressively filled with A and B. The concentration profiles at the end of a cycle (i.e., four switches in this case) are shown in Figure 3.8(e) for the first five cycles. After that, steady state conditions are reached, i.e., virtually no differences between the concentration profiles are observed.

The concentrations of A in the extract and of B in the raffinate are shown in Figure 3.8(f) as a function of time. It is seen that the outlet concentrations increase from zero to a periodic steady state, where they oscillate around an average value with a period equal to the switch time (5 min). This observation illustrates the general definitions introduced in Chapter 1: A chromatographic system is termed continuous when the mixture to be purified is fed at a constant flow rate in the separation unit. Even though the outlet flow rates are also constant, the outlet solute concentrations usually oscillate periodically due to the finite number of columns. Such oscillations would in fact become less significant for a larger number of columns, and the concentrations would approach constant values at the limit of an infinite number of columns.

3.4 Separation Region for More Realistic Systems

In Section 3.2.3, we determined the separation region of countercurrent chromatographic processes for idealized systems, i.e., with linear fluid–solid equilibria and where dispersive effects due to axial dispersion and mass transfer resistances are neglected. In this case, we obtained simple analytical expressions, and a graphical representations of the design space was presented. In particular, it was shown that the complete separation

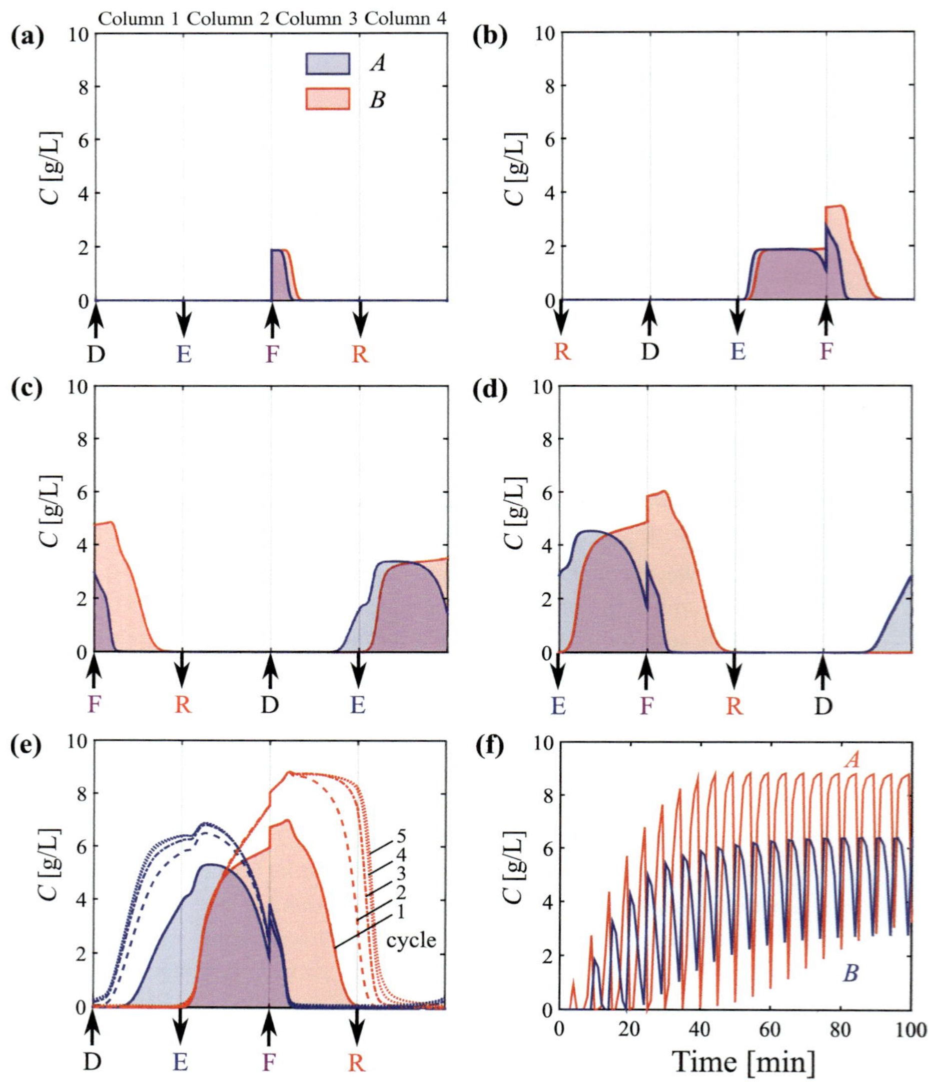

Figure 3.8 Model simulations of a four-column SMB unit. (a-d) Concentration profiles in the liquid phase along the four columns at various time points. The inlet and outlet positions are indicated in the abscissa. (e) Internal concentration profiles at various cycle numbers. (f) Concentrations of A in the extract and of B in the raffinate as a function of time. Simulation parameters: $\tilde{K}_A = 0.02$, $\tilde{K}_B = 0.01$, $C_A^F = C_B^F = 10$ g/L, $\overline{N} = 100$, $t_{i,A} = t_{i,B} = 0.3$ s, $Pe_c = 400$, $\varepsilon_e = 0.4$, $\varepsilon_i = 0.5$, $Q^E = 14.7$ mL/min, $Q^R = 8.5$ mL/min, $Q^F = 7.4$ mL/min, $Q^D = 15.9$ mL/min, $\Delta t = 5$ min, $L_{col} = 25$ cm, $S_{col} = 4.9$ cm^2.

region of a SMB process is defined by a triangle in the m^{II}–m^{III} plane (Figure 3.5). In the following, we investigate the impact of nonlinear equilibrium relations and dispersive effects on the design space of the SMB process.

3.4.1 Impact of Nonlinearities in the Fluid–Solid Equilibria

Let us start by analyzing the impact of nonlinear equilibrium relations in the case where dispersive effects are neglected. We focus on the common situation where the two solutes follow a binary Langmuir adsorption isotherm, but the analysis can be extended to other types of nonlinear adsorption isotherms [218, 221, 222].

The zone of complete separation can be described by five points denoted a, b, f, r, and w, which are linked by straight segments, except the portion from a to r, which is curved. Figure 3.9(a) shows how these five points are placed on a representative example of a zone of complete separation. The coordinates of these points are summarized in Table 3.1, where ω_1 and ω_2 denote the roots of the following equation:

$$\left(1+\tilde{K}_A C_A^F+\tilde{K}_B C_B^F\right)\omega^2-\left[\tilde{K}_B\overline{N}\left(1+\tilde{K}_A C_A^F\right)+\tilde{K}_A\overline{N}\left(1+\tilde{K}_B C_B^F\right)\right]\omega+\tilde{K}_A\tilde{K}_B\overline{N}^2=0 \tag{3.16}$$

and are defined such that:

$$0<\omega_1^F\le\tilde{K}_B\overline{N}\le\omega_2^F\le\tilde{K}_A\overline{N} \tag{3.17}$$

The equation of the curve going from point a to point r is given by:

$$m^{III}=m^{II}+\frac{\left(\sqrt{\tilde{K}_A\overline{N}}-\sqrt{m^{II}}\right)^2}{\tilde{K}_A C_A^F} \tag{3.18}$$

Using the preceding expressions, the region of complete separation is shown in Figure 3.9 for various sets of parameter values. It is seen in Figure 3.9(a–b) that the triangle distorts and shrinks with increasing feed concentration. It is also interesting to note that, as the feed concentration of the most retained solute C_A^F (Figure 3.9(a)) or of the least retained solute C_B^F (Figure 3.9(b)) increases, the region of complete separation shrinks in a different manner. The impact of the equilibrium constants can be appreciated in Figure 3.9(c–d). In Figure 3.9(c), $\tilde{K}_B = 0.01$ L/g and $\tilde{K}_A$ is varied, while in Figure 3.9(d),

Table 3.1. Points defining the zone of complete separation in the frame of the equilibrium theory and in the case of Langmuir adsorption isotherms [158, 218, 221, 222]. ω_1 and ω_2 denote the roots of Equation 3.16. The expressions were obtained assuming that there is no solute exclusion.

Point	m^{II}	m^{III}
a	$\overline{K}_A$	$\overline{K}_A$
b	$\overline{K}_B$	$\overline{K}_B$
f	$\omega_2+\varepsilon_i$	$\omega_2+\varepsilon_i$
r	$\dfrac{(\omega_2)^2}{\tilde{K}_A\overline{N}}+\varepsilon_i$	$\dfrac{\omega_2\left(\omega_2\left(\tilde{K}_B\overline{N}-\omega_1\right)+\omega_1\left(\tilde{K}_A\overline{N}-\tilde{K}_B\overline{N}\right)\right)}{\tilde{K}_B\overline{N}\left(\tilde{K}_A\overline{N}-\omega_1\right)}+\varepsilon_i$
w	$\dfrac{\omega_2\tilde{K}_B}{\tilde{K}_A}+\varepsilon_i$	$\dfrac{\omega_2\left(\tilde{K}_B\left(\tilde{K}_B\overline{N}-\omega_1\right)+\omega_1\left(\tilde{K}_A-\tilde{K}_B\right)\right)}{\tilde{K}_B\left(\tilde{K}_A\overline{N}-\omega_1\right)}+\varepsilon_i$

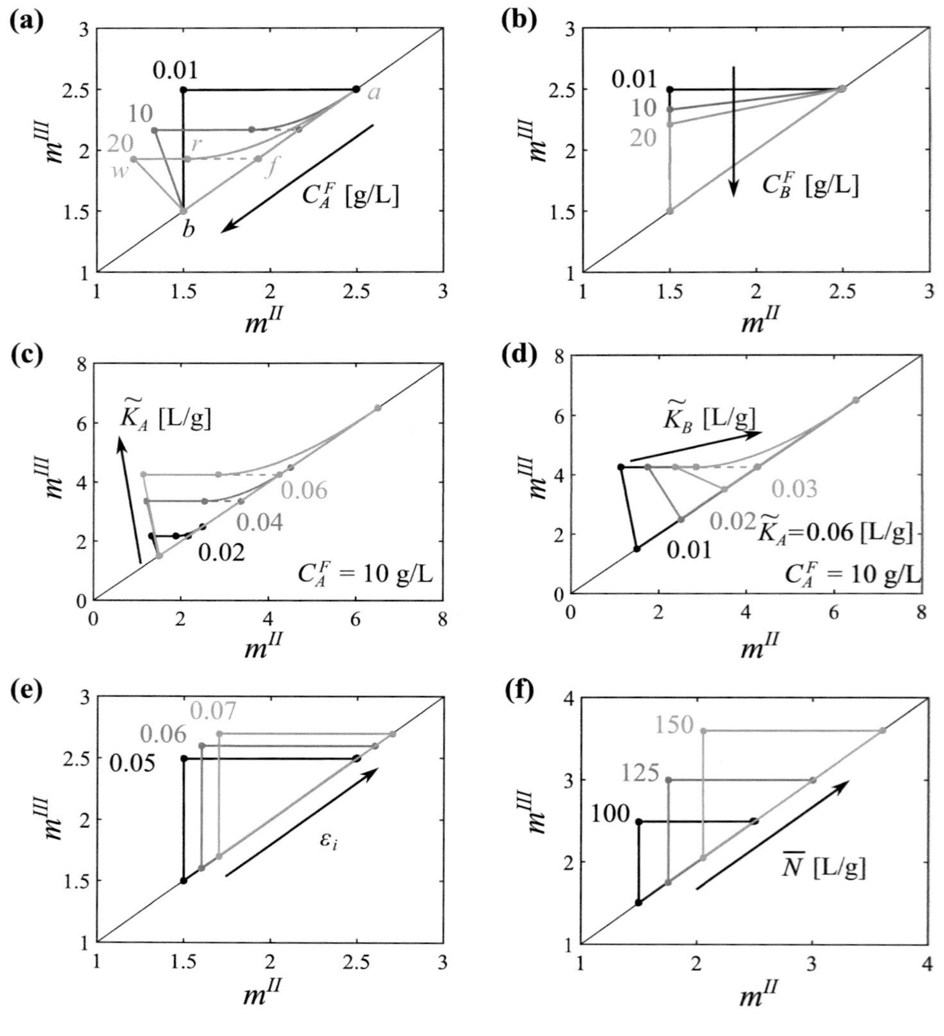

Figure 3.9 Impact of several parameters on the region of complete separation in the case of binary Langmuir adsorption isotherms with negligible dispersive phenomena: (a) concentration of the most retained solute, C_A^F, (b) concentration of the least retained solute, C_B^F, (c) equilibrium constant of the most retained solute, $\tilde{K}_A$, (d) equilibrium constant of the least retained solute, $\tilde{K}_B$ (e) internal porosity, ε_i, (f) lumped saturation capacity, $\overline{N}$. The following parameter values are used unless otherwise indicated in the figure: $\tilde{K}_A = 0.02$ L/g, $\tilde{K}_B = 0.01$ L/g, $\overline{N} = 100$ g/L, $C_A^F = C_B^F = 0.01$ g/L, $\varepsilon_i = 0.5$.

$\tilde{K}_A = 0.06$ L/g and $\tilde{K}_B$ is varied. As expected, the region of complete separation shrinks when the difference between $\tilde{K}_A$ and $\tilde{K}_B$ decreases because this corresponds to a decrease of the selectivity. Moreover, the complete separation region deviates more and more from a right triangle as $\tilde{K}_A$ and $\tilde{K}_B$ increase, i.e., when the nonlinearity of the adsorption isotherm is more pronounced. The impact of internal porosity can be easily understood because it corresponds to a simple shift of the zone of complete separation proportional to the change in ε_i, as shown in Figure 3.9(e) and Table 3.1. Finally, it is observed in Figure 3.9(f) that an increase in the saturation capacity enlarges the zone of complete separation and shifts it toward higher m values. This is consistent with the fact that an increase in $\overline{N}$ leads to an increase of both $\overline{K}_A$ and $\overline{K}_B$. The expressions given in this section help defining the region of complete separation in the m^{II}–m^{III} plane for Langmuir adsorption isotherms. Similar expressions have been derived to determine the region of complete regeneration in the m^{I}–m^{IV} plane [221, 222].

3.4.2 Impact of Dispersive Phenomena

In most applications, chromatographic systems exhibit dispersive effects due to axial dispersion and mass transfer limitations. It results that strong discrepancies are often observed between the equilibrium theory predictions and experimental results. Dispersive effects can be accounted for by considering that each of the zone of the countercurrent unit shown in Figure 3.6 is constituted by a finite number of stages. Note that columns located in different zones of a multicolumn system are characterized by a different number of plates because they are operated at different flow rates (see van Deemter equation in Chapter 2). Notably, columns operated at a high flow rate develop a lower number of stages, which is typically the case in zone I for an SMB process.

In the presence of dispersive phenomena, selecting an operating point in the complete separation region defined in the frame of the equilibrium theory does not guarantee to obtain a satisfactory purity. It is thus necessary to use mechanistic models to compute the purity in the extract and in the raffinate, which are necessarily lower than 100%. For illustrative purposes, the results of such simulations are shown in Figure 3.10 using the same model and parameter values as those reported in the caption of Figure 3.8. In the panels (a) and (b), the purity of the extract and of the raffinate, respectively, are presented for various combinations of the m^{II} and m^{III} parameters. Points with darker colors indicate higher purity values. To these simulation results are superimposed the zones where the equilibrium theory predicts pure extract and pure raffinate streams. It is seen that the simulated purity values tend to increase when moving inside the regions where the extract and raffinate are predicted to be pure in the frame of the equilibrium theory. From these results one can define some separation region where certain purity specifications are satisfied both in the extract and raffinate. Such regions are shown in Figure 3.10(c), where the purity specifications are set to 80%, 90%, and 95%. It appears that the separation region shrinks with more constraining purity specifications.

The analysis presented in Figure 3.10 was performed for various values of m^{II} and m^{III} using a constant column length. Figure 3.11 instead shows the purity in the extract and in the raffinate as a function of column length for the maximal feed flow rate leading to complete separation in the frame of the equilibrium theory (i.e., the vertex of the "triangle"). It is seen that the purity first dramatically increases with the column length and then flattens out.

Besides using model simulations, an alternative strategy to design a SMB unit in the presence of mass transfer limitations consists in applying the so-called standing

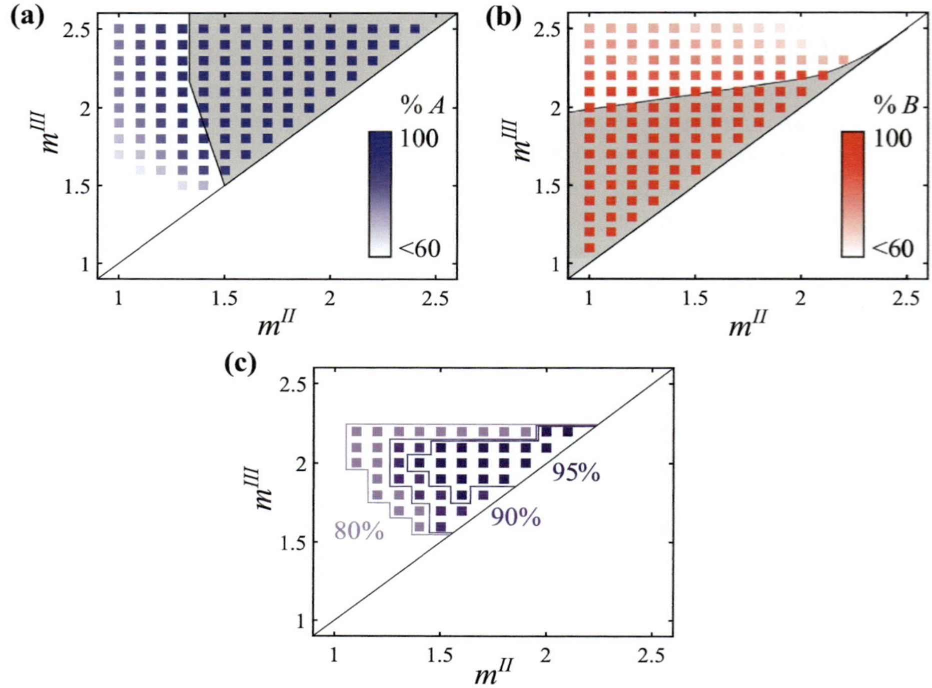

Figure 3.10 Simulated purity region (a) in the extract and (b) in the raffinate. Zones shaded in gray correspond to pure extract and raffinate, respectively, as computed with the equilibrium theory. (c) Zones allowing 80, 90, and 95% purity both in the extract and in the raffinate. The same model and parameter values as those reported in Figure 3.8 were used.

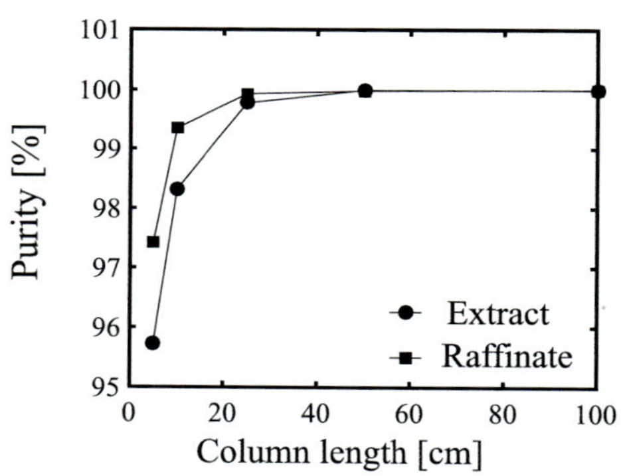

Figure 3.11 Purity in the extract and in the raffinate as a function of the column length. The model parameters are the same as in Figure 3.8, and the following internal flow rates were used: $m^I = 2.9$, $m^{II} = 1.8$, $m^{III} = 2$, $m^{IV} = 1.4$. The switch time was set to 5 min for a column of 25 cm and scaled linearly with the column length for the other conditions.

wave design. This approach allows estimating the m parameters for target purity values in the outlet streams at the maximum feed flow rate and minimum eluent flow rate. The standing wave design has been shown to provide an acceptable first guess of operating flow rates for systems characterized by adsorption isotherms that are linear [223, 224] or moderately nonlinear [225].

3.5 Design of Countercurrent Chromatographic Processes

3.5.1 Unified Design Approach

We have seen in the previous sections that the determination of the m parameters is crucial to design TMB and SMB processes. Here, we generalize the definition of these m parameters to properly extend them to single-column processes. This approach allows rationalizing the design of chromatographic processes within a unique framework and is thus referred to as the "unified design approach" [226]. The term *design* refers here to the determination of normalized operating flow rates (or volumes) for a given separation unit, i.e., for which the number of columns, column dimensions, packing material, elution conditions, etc., have already been selected.

Let us consider the simple case of a binary separation performed under linear equilibrium conditions in the absence of dispersive effects. A representative chromatogram of such situation is shown in Figure 3.12(a). Recalling the nomenclature introduced in Chapter 2, the solutes A and B start to elute at the times $t_{r,A}$ and $t_{r,B}$, respectively, while

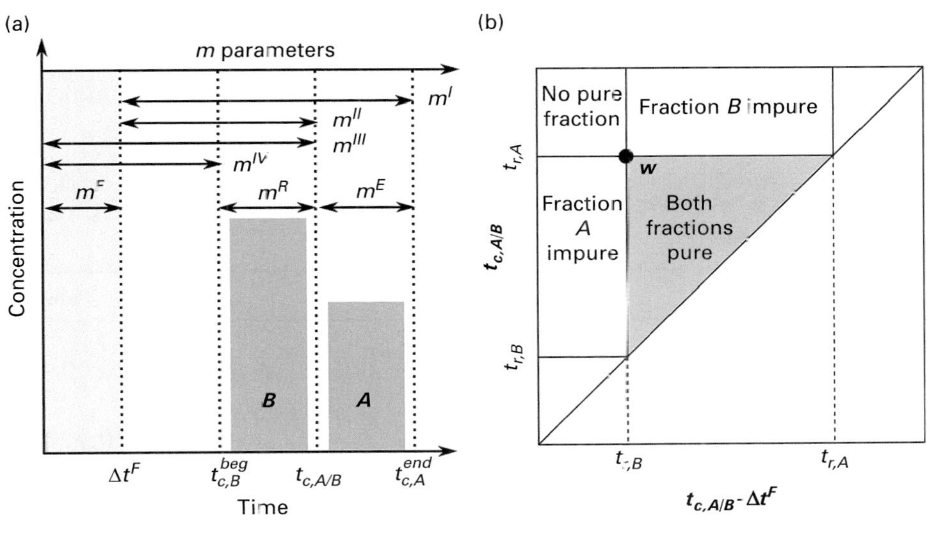

Figure 3.12 (a) Representative chromatogram obtained in a single-column process for a binary separation in the frame of the equilibrium theory under linear conditions. The arrows illustrate how to estimate the m parameters. (b) Zone of complete separation in a single-column process, analogous to the triangle shown in Figure 3.5 for TMB and SMB processes.

the width of each peak is equal to the injection time Δt^F. In addition, we define here $t_{c,A}^{beg}$ and $t_{c,A}^{end}$ as the beginning and the end of the collection window of A, respectively. Similarly, $t_{c,B}^{beg}$ and $t_{c,B}^{end}$ denote the beginning and the end of the collection window of B. We consider the case where the solute A is collected right after the solute B, and denote $t_{c,A/B}$ the cut time, so that $t_{c,B}^{end} = t_{c,A}^{beg} = t_{c,A/B}$. In the following, we identify the pairs of $t_{c,A/B}$ and Δt^F values that allow collecting the two solutes with 100% purity and 100% recovery.

Looking at the chromatogram shown in Figure 3.12(a), it is clear that if one wants to collect the two solutes pure, the injection time Δt^F should be selected such that A starts eluting only after B is completely eluted from the column. Moreover, the cut time $t_{c,A/B}$ should be comprised between the end of the elution of B and the beginning of the elution of A to ensure the complete recovery of both solutes. This means that one should have $t_{c,A/B} \geq t_{r,B} + \Delta t^F$ and $t_{c,A/B} \leq t_{r,A}$. Because $\Delta t^F \geq 0$, these two conditions lead to:

$$t_{r,B} \leq t_{c,A/B} - \Delta t^F \leq t_{c,A/B} \leq t_{r,A} \tag{3.19}$$

It is remarkable that Equation 3.19 can be written as $\overline{K}_B \leq m^{II} \leq m^{III} \leq \overline{K}_A$ provided that the parameters m^{II} and m^{III} of the single-column process are defined by the expressions given in Table 3.2. In other words, the complete separation of the single-column process is identical to the one of the TMB and SMB processes when using appropriate definitions of the m parameters.

The inequalities of Equation 3.19 can be graphically represented as shown in Figure 3.12(b), where any pair of $t_{c,A/B}$ and Δt^F located in the shaded triangle leads to the complete separation of A and B. The vertex of the triangle corresponds to the case where $t_{c,A/B} = t_{r,B} + \Delta t^F = t_{r,A}$, i.e., $\Delta t^F = t_{r,A} - t_{r,B}$. This means the injection time is set to the maximal value allowing complete separation between the two solutes. This is often referred to as the "touching bands" situation and is equivalent to the case where the

Table 3.2. Definitions of the m parameters for the TMB, SMB, and single-column processes [226] for an idealized system (i.e., without hydrodynamic and kinetic limitations and following a linear adsorption isotherm).

Parameter	TMB	SMB	Single-column
m^I	$\dfrac{Q^I}{\overline{Q}}$	$\dfrac{Q^I\Delta t - \varepsilon_e V_{col}}{(1-\varepsilon_e)V_{col}}$	$\dfrac{Q\left(t_{c,A}^{end} - \Delta t^F\right) - \varepsilon_e V_{col}}{(1-\varepsilon_e)V_{col}}$
m^{II}	$\dfrac{Q^{II}}{\overline{Q}}$	$\dfrac{Q^{II}\Delta t - \varepsilon_e V_{col}}{(1-\varepsilon_e)V_{col}}$	$\dfrac{Q\left(t_{c,A/B} - \Delta t^F\right) - \varepsilon_e V_{col}}{(1-\varepsilon_e)V_{col}}$
m^{III}	$\dfrac{Q^{III}}{\overline{Q}}$	$\dfrac{Q^{III}\Delta t - \varepsilon_e V_{col}}{(1-\varepsilon_e)V_{col}}$	$\dfrac{Qt_{c,A/B} - \varepsilon_e V_{col}}{(1-\varepsilon_e)V_{col}}$
m^{IV}	$\dfrac{Q^{IV}}{\overline{Q}}$	$\dfrac{Q^{IV}\Delta t - \varepsilon_e V_{col}}{(1-\varepsilon_e)V_{col}}$	$\dfrac{Qt_{c,B}^{beg} - \varepsilon_e V_{col}}{(1-\varepsilon_e)V_{col}}$

normalized feed flow rate of TMB/SMB processes is set to the maximal value allowing complete separation, which is also occurring at the vertex of the triangle shown in Figure 3.5.

Let us continue the analogy between single-column and multicolumn processes in terms of design space. In addition to the constraints given by Equation 3.19, the complete recovery of B necessitates that its collection begins before the solute starts eluting, and the complete recovery of A necessitates that its collection does not end before it completely elutes. It follows that the collection windows should be selected such that $t_{c,B}^{beg} \leq t_{r,B}$ and $t_{c,A}^{end} \geq t_{r,A} + \Delta t^F$. These two constraints can be rewritten as $\overline{K}_A \leq m^I$ and $m^{IV} \leq \overline{K}_B$ provided that the parameters m^I and m^{IV} are defined by the expressions given in Table 3.2.

Overall, the analysis carried out shows that the design of a single-column system is analogous to the design of TMB and SMB processes for linear systems in the frame of the equilibrium theory. In all cases, it is necessary to satisfy the constraints given by Equation 3.9, where the parameters m^I, m^{II}, m^{III}, and m^{IV} are defined by the expressions given in Table 3.2 [226]. The parameters m^F, m^E, m^R, and m^D can then be computed according to the mass balances shown in Equations 3.10–3.13.

Although the m parameters may look rather abstract at first sight, they simply represent flow rate ratios and volume ratios in the case of multicolumn and single-column processes, respectively. These ratios are defined as certain flow rates (or volumes) of fluid divided by the flow rate (or volume) of solid. These quantities can also be written in terms of column volumes, defined as a given volume V divided by the column volume, i.e., $CV = V/V_{col}$. Reasoning in terms of m parameters or in terms of column volumes is in fact equivalent because the m^i and CV^i are related through linear expressions. For the internal m parameters one has $CV^i = \varepsilon_e + (1 - \varepsilon_e)m^i$ for the zones I to IV. On the other hand, the inlet/outlet m parameters are simply proportional to the column volumes: $CV^i = (1 - \varepsilon_e)m^i$ for the feed, extract, raffinate, and desorbent. Because for single-column processes any volume is easily related to a time interval t^i through the volumetric flow rate Q, one can also write:

$$m^i = \frac{Qt^i - \varepsilon_e V_{col}}{(1 - \varepsilon_e)V_{col}} \quad \text{for the internal } m \text{ parameters} \tag{3.20}$$

$$m^i = \frac{Qt^i}{(1 - \varepsilon_e)V_{col}} \quad \text{for the inlet/outlet } m \text{ parameters} \tag{3.21}$$

Therefore, the m parameters of a single-column process can simply be computed from an experimental chromatogram. A graphical representation is proposed in Figure 3.12(a).

To conclude, the unified design approach allows rationalizing the design of various chromatographic processes within a generalized framework under the assumption of thermodynamic equilibrium. The design space identified in the frame of the equilibrium theory for linear thermodynamic equilibria appears to be a characteristic feature of the fluid–solid equilibrium and does not depend on the specific process adopted. The appropriate definitions of the parameters m^I to m^{IV} are summarized in Table 3.2 for the TMB, SMB and single-column processes. It has been shown that the equivalence of the parameters m^I, m^{II}, m^{III} between single-column and multicolumn processes can be extended to the case of Langmuir adsorption isotherms. However, no equivalence between the processes could be found for the parameter m^{IV} in the case of Langmuir fluid–solid thermodynamic equilibrium relations [158, 226].

The unified design approach has also been applied to other processes including cross-current processes and single-column processes with recycling [226]. The latter are particularly relevant in the case of partial peak overlapping: the unresolved fraction containing both solutes is collected and mixed with the fresh feed before being reinjected into the chromatographic column. Such situations will be examined in Chapter 5, where multicolumn processes with recycling of unresolved fractions will be introduced.

3.5.2 Empirical Design of a Multicolumn Process from a Single-Column Chromatogram

We have seen in Section 3.5.1 that there are some strong analogies in the design of single-column and multicolumn chromatographic processes. The unified design approach revealed that similar performances between multicolumn and single-column processes are expected when dispersive effects are negligible. However, in general, multicolumn processes accommodate better dispersive effects than their single-column equivalent. Therefore, the same set of m parameters applied to different chromatographic processes does not necessarily lead to the same performances, as further discussed in Section 3.5.4.

Although the unified design approach was strictly demonstrated only in the frame of the equilibrium theory, it can still help to rapidly set up a real multicolumn countercurrent process based on a chromatogram obtained from a single-column experiment. Indeed, an experimental chromatogram intrinsically contains relevant information about the thermodynamics, hydrodynamics, and kinetics of the process, which is then implicitly used to obtain an operating point leading to an acceptable separation. The methodology is given in Box 3.1, and it will be applied to practical cases in the next chapters.

Box 3.1 Empirical Design of a Multicolumn Process from a Single-Column Chromatogram.

1. Identify a set of conditions using a single-column process leading to a separation, which is considered to be satisfactory (e.g., the touching bands situation). This includes a proper selection of the chromatographic support, eluent, etc.
2. Compute the m parameters for the single-column process by reading the elution times directly on the selected chromatogram as shown in Figure 3.12.
3. From the obtained m values, compute the flow rates for the desired process (e.g., a multicolumn countercurrent process as the SMB) by using the conversion relations in Table 3.2 considering the same total column length and average fluid velocity as in the single-column process.
4. Run the desired process with the obtained flow rates.

This empirical design approach provides a reasonable starting point, but it is by no means expected to deliver the best process design. Optimization of process variables (e.g., flow rate, column length, column diameter) around this first guess by using mechanistic models is always advisable, especially in the case of strong mass transfer limitations typical for bioseparations.

3.5.3 Shortcut Design of Countercurrent Chromatographic Processes

In the previous section, we have presented an empirical design approach to transpose a single-column process into a countercurrent chromatographic process for a given column geometry, i.e., fixed column diameter and total column length. Here, we provide a simple model-based approach to design a countercurrent chromatographic process not only in terms of flow rates but also in terms of column diameter and column length. This method was developed throughout several studies [158, 227, 228] and a step-by-step procedure is summarized in Box 3.2. The procedure is similar to the shortcut design

Box 3.2 Shortcut Design of a Countercurrent Chromatographic Unit

1. Determine the region of complete separation in the m^{II}–m^{III} plane and the region of complete regeneration in the m^{I}–m^{IV} plane for an idealized system (i.e., for an infinite plate number). Analytical expressions exist for linear and generalized Langmuir adsorption isotherms, while numerical computations are required for other types of adsorption isotherms. The operating point is then selected to maximize the feed flow rate (i.e., the vertex of the triangle in the m^{II}–m^{III} plane) and minimize the eluent consumption (i.e., providing the lowest value of $m^{I} - m^{IV}$). A safety margin can be selected for more robust operations.
2. Estimate the minimum total number of plates required to perform the desired separation for a more realistic system (e.g., using the MC model) considering a given process configuration (e.g., two columns per zone) and assuming that all the columns have the same number of plates.
3. Determine the column length and fluid velocity in the zone where the m parameter is the largest (usually zone I for SMB) for the maximum recommended pressure and minimum number of plates per column with Equations 2.92 and 2.93. Alternatively, the total column length and average fluid velocity may be estimated considering the total pressure drop and total number of plates.
4. Compute Δt from the definition of the m parameters given in Table 3.2 considering the column length and fluid velocity estimated in point 3. Recalling that $Q^{i} = u^{i} S_{col}$ and $V_{col} = L_{col} S_{col}$ allows writing m^{i} in terms of u^{i} and L_{col}. Once one m parameter (or the average) has been determined, all the other internal velocities can then be computed using the expressions given in Table 3.2.
5. Determine the column diameter from Equation 3.21 by imposing the feed flow rate, Q^{F}, and using the values of m^{F}, Δt and L_{col} previously determined.

strategy presented in Box 2.1 and is given here for a SMB process, but can be extended to other countercurrent chromatographic processes. It will be used in Section 3.5.4 to compare the efficiency of single-column and countercurrent chromatographic processes. Furthermore, it will be applied to an illustrative example in Section 5.3.1 to design a SMB unit for the separation of two antibody variants.

3.5.4 Benefits from Countercurrent Chromatographic Processes

We have seen in Section 3.5.1 that the single-column and SMB processes are equivalent in the frame of the equilibrium theory, i.e., taking into account thermodynamic aspects only. This means that potential differences between single-column and countercurrent chromatographic processes must be related to column efficiency, which is impacted by hydrodynamic and kinetic limitations. Comparing different chromatographic processes is a quite delicate task due to the number of variables to be selected (e.g., column length, column diameter, flow rates, switch times) and the number of performance parameters that can be considered (e.g., purity, recovery, productivity, eluent consumption, overall cost). Here, we do not aim at performing a detailed analysis between single-column and countercurrent processes, but we adopt a rather simple approach to identify in which situations countercurrent processes are the most attractive. We illustrate our methodology with a four-column SMB process, but it could be extended to other multicolumn processes, as shown in Chapters 4 and 5 in the context of capture and polishing applications, respectively.

For the sake of simplicity, we consider a binary mixture whose solutes are characterized by linear thermodynamic equilibrium relations. Following the shortcut design strategy presented in Box 2.1 (single-column) and Box 3.2 (SMB), we start by setting the normalized feed volume/flow rate to the maximum value allowing complete separation considering an infinite number of plates. This corresponds to the vertex of the triangle defining the zone of complete separation in the m^{II}–m^{III} plane, which corresponds to the touching-bands situation for the single-column process. In addition, we select the conditions that minimize eluent consumption while ensuring complete regeneration of the solid and liquid phases, i.e., $m^{I} = \overline{K}_A$ and $m^{IV} = \overline{K}_B$. To allow a fair comparison between the two processes, the collection windows in the single-column process are determined in the frame of the equilibrium theory, i.e., the collection volume is equal to the feed volume.

Then, we perform model simulations by using the chromatographic models introduced in Chapter 2. Briefly, the hydrodynamic dispersion is described by a finite number of mixing cells connected in series, and it is assumed that thermodynamic equilibrium between the fluid and solid phases is reached instantaneously. Note that in this frame, all dispersive effects (hydrodynamic dispersion and mass transfer limitations) are attributed to hydrodynamic dispersion through an apparent number of mixing cells. This apparent number of mixing cells corresponds to the number of plates N, which quantifies the column efficiency. The higher the number of plates, the higher the column efficiency and the sharper the fronts. Our aim here is to determine the number of plates needed to obtain given separation performances, both for the single-column and SMB processes. We have seen in Chapter 2 that the purity and recovery are interconnected, a high purity being often associated with a low recovery and vice versa. To allow a simple comparison of the single-column and SMB processes, we define a "purification factor" as the average between the purity and recovery of the solute A. In particular, a purification factor of 100% can be obtained only if both the purity and recovery of A are equal to 100%. A purification factor below 100% indicates that either the purity, or the recovery, or both are below 100%. Although such definition of the purification factor only gives a rough estimation of the process performance, it is sufficient for the purpose of this section.

For numerical applications, let us consider two systems: one system for which $\overline{K}_A = 1.5$ and $\overline{K}_B = 1$ and one system for which $\overline{K}_A = 1.2$ and $\overline{K}_B = 1$ The separation performed in the second case is more difficult since the selectivity, defined as $\overline{K}_A/\overline{K}_B$, is lower. Figure 3.13(a) and (b) show the purification factor as a function of the number of plates for a selectivity of 1.5 and 1.2, respectively. Note that in the case of the SMB, N corresponds to the total plate number, i.e., it is the sum of the plate numbers of the four columns. It is seen in Figure 3.13 that the purification factor increases with the number of plates and tends toward a value of 100% for sufficiently large plate numbers. An important observation is that at a given plate number, the purification factor is larger for the SMB than for the single-column process in all conditions. In other words, the number of plates required to reach a target purification factor is lower in the SMB than in the single-column process. It is also striking that the difference between the single-column and SMB processes is more pronounced for the most difficult separation. For example, considering the example with $\overline{K}_A/\overline{K}_B = 1.2$, more than 1400 plates are required to reach a purification factor of 85% in the single-column process, while only 440 plates are required for the SMB process. The same methodology was applied to compare various multicolumn processes in the context of sugar separation [229]. It was

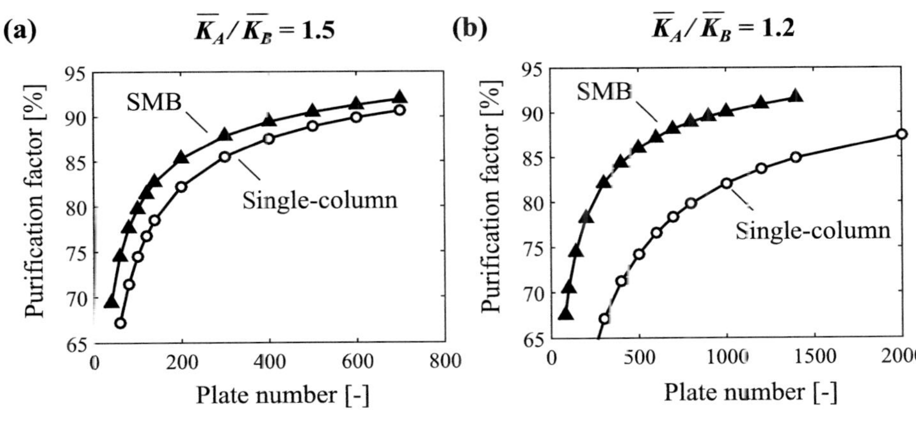

Figure 3.13 Comparison between single-column and SMB processes considering a binary mixture whose solutes follow linear thermodynamic equilibrium relations. Two cases are considered: (a) $\overline{K}_A = 1.5$ and $\overline{K}_B = 1$ and (b) $\overline{K}_A = 1.2$ and $\overline{K}_B = 1$. The purification factor is defined as the average between the purity and recovery of A. Simulations were performed by using the MC model and assuming that thermodynamic equilibrium is reached instantaneously.

Table 3.3. Comparison between single-column and SMB processes to obtain a purification factor of 85% for two different selectivity values. The following values were considered: $\Delta P = 5$ bars, $d_p = 50$ μm, $\eta = 10^{-3}$ Pa.s, $t_m = 1$ s, $\varepsilon_e = 0.4$, $h_k = 150$. This leads to $c_{VD} = 1.15$ s and $\delta_K = 3375$ bar.s.m^{-2}.

	$\overline{K}_A/\overline{K}_B = 1.5$		$\overline{K}_A/\overline{K}_B = 1.2$	
	Single-column	SMB	Single-column	SMB
Total plate number [-]	285	195	1400	440
Total column length [m]	0.67	0.55	1.54	0.87
Fluid velocity [m/s]	8.0	9.6	3.5	6.2

found that modifications of the SMB, such as the Sequential SMB or Improved SMB, allow further decreasing the required number of plates.

Let us now examine the impact of a decrease in the number of plates on the selection of the column length L_{col} and fluid velocity u. To do so, we apply the shortcut design strategy presented in Section 2.6.3. Briefly, we have seen that the van Deemter equation and the Kozeny–Kármán equation provide two useful relations to rapidly design chromatographic processes, namely $L_{col} \approx c_{VD}uN$ and $\Delta P = \delta_K L_{col}u$. The expressions of c_{VD} and δ_K as a function of physical parameters can be found in Section 2.4. Setting N to the total plate number required to obtain a given purification factor and setting ΔP to the total maximum recommended pressure drop allows obtaining an estimate of L_{col} and u. It is worth noting that in the case of the SMB process, the fluid velocity varies between the different zones, and the so-obtained value of u therefore corresponds to a fluid velocity averaged between the different zones. Numerical results are presented in Table 3.3 for a target purification factor of 85% considering $\Delta P = 5$ bars. The values of c_{VD} and δ_K are given in the caption of Table 3.3. Importantly, it is seen that shorter column lengths and larger fluid velocities can be used with the SMB process than with the single-column process. This gain is particularly advantageous for the example with a low selectivity because the SMB process allows reducing the total column length and increasing the fluid velocity almost by a factor of 2 to obtain the same purification factor. Therefore, the SMB process is expected to lead to a significant increase in productivity as compared to the single-column process. Note that the actual improvement is connected to the value of c_{VD}.

In the light of the discussion presented in this section, it can be concluded that countercurrent chromatography is very attractive for separations that require a large number of plates (e.g., characterized by a low selectivity or a high target purity/recovery) and for separations where it is difficult to obtain plates (e.g., characterized by important mass transfer limitations).

3.6 Conclusion

In this chapter, we introduced key notions about countercurrent separation. We presented the TMB and SMB processes, which are reference cases in the field of multi-column chromatography. The design of these processes relies on the proper selection of the so-called m parameters, which are dimensionless flow ratios between the fluid and solid phases. In the frame of the equilibrium theory, the region of complete separation in the m^{II}–m^{III} plane can be computed analytically for linear and generalized Langmuir adsorption isotherms. On the other hand, in order to take into account dispersive effects, it is necessary to perform detailed simulations. In this case, the design space of a chromatographic process is defined as the region where target purity values are met in the raffinate and in the extract. In particular, the design space shrinks with increasing purity constraints and increasing dispersive effects.

The unified design approach rationalizes the design of single- and multicolumn processes within a unique framework by generalizing the definition of the m parameters in the frame of the equilibrium theory. This provides the basis of the empirical design of multicolumn processes based on a single-column chromatogram. For real systems that are subject to hydrodynamic and kinetic limitations, countercurrent chromatographic processes allow reducing the number of plates required to obtain a target separation performance. Importantly, a decrease in the required number of plates is associated with a decrease in column length and an increase in the fluid velocity, which are related to an increase in productivity. The comparison between single-column and multicolumn processes was illustrated considering linear equilibrium relations. Nevertheless, the conclusions remain valid also under nonlinear conditions. Overall, countercurrent chromatography is especially attractive for difficult separations, that is, for high purity/recovery targets, low selectivity, or very strong dispersive effects.

We will see in the next chapters how to modify the classical SMB process so as to cope with some difficulties encountered during the downstream processing of therapeutic proteins, such as the separation of ternary mixtures, the removal of strongly adsorbed solutes by chemical modulation, or the recycling of partly purified fractions. We will also see that it is possible to design countercurrent chromatographic processes with only three, or even two columns.

4 Countercurrent Chromatography for the Capture Step

4.1 Introduction

Recovering proteins from biological feedstocks, be they cell supernatant, cell lysate, or a natural source, is a complex task that requires multiple steps to remove impurities of different nature, varying in size and chemical properties. As discussed in Chapter 1, the role of the capture step is to extract the protein of interest from a clarified fluid with a high product recovery. Most of the impurities (such as culture media components, host cell proteins, DNA, and endotoxin in the case of cell culture) should be considerably depleted during this step, while achieving a several-fold reduction in volume of the protein solution so as to reduce the quantities to be treated in the subsequent units.

When dealing with therapeutic proteins, the capture step is most often performed with chromatography operated in frontal mode. This means that operating conditions are selected so that the target protein strongly binds onto the chromatographic medium, whereas impurities simply flow through the column. Then, a change in the elution conditions allows recovering the protein of interest. A high selectivity of the chromatographic medium for the target protein is of primary importance, which explains the large success of affinity chromatography. Nevertheless, protein capture is sometimes performed with less-specific chromatographic media, for example using ion exchange chromatography.

In this chapter, we start by analyzing the two main process operations involved in the capture step, namely loading and protein recovery, in terms of the key physical phenomena identified in Chapter 2: thermodynamics of the fluid–solid equilibrium, hydrodynamics, and kinetics of mass transfer. We treat protein capture as a particular case of chromatography where: *(i)* the feed mixture can be regarded as a binary solution (target protein/impurities), *(ii)* the target protein strongly binds to the chromatographic medium, and *(iii)* the impurities do not compete with the target protein at the surface of the chromatographic medium. In the second part of this chapter, we demonstrate how the capture step can benefit from multicolumn countercurrent chromatographic technologies. Based on the notions developed in Chapter 3, the concepts necessary to design such processes are presented and applied to practical examples.

4.2 Process Operations and Associated Physical Phenomena

4.2.1 Typical Sequence of Process Operations

A representative chromatogram obtained during the protein capture step is presented in Figure 4.1. A typical sequence of operations starts with a rather long load phase in frontal mode, during which the elution of impurities leads to an increase of the signal at the column outlet until a plateau is reached. In order to limit the loss of product, the load phase ends when the target protein starts breaking through the column, which is indicated by a slight increase in the UV signal (barely appreciable in Figure 4.1).

Then, the chromatographic medium is washed to remove nonadsorbed solutes present in the liquid phase, resulting in the decrease of the UV signal. A subsequent wash is sometimes applied to remove moderately retained impurities, which leads here to a small peak at around 125 min. These impurities may either interact with the chromatographic medium or with the adsorbed proteins, as it is sometimes the case for certain HCPs binding to antibodies. Particular attention is devoted to the development of efficient wash protocols to disrupt target protein/HCPs interactions and thus reach a high purity [230, 231]. Once the column is properly washed, the recovery of the protein can be triggered by applying a step-gradient elution. This corresponds to the sharp and high peak observed at around 165 min in Figure 4.1. Then the column regeneration phase starts, which involves various wash and cleaning protocols to remove impurities that strongly interact with the backbone of the chromatographic medium. Because these strongly adsorbed impurities may lead to media fouling and potential contamination for the subsequent injections, harsh cleaning procedures may be required to maintain safe process operations. For instance, high salt concentrations or dilute sodium hydroxide solutions are commonly used at this stage [21]. Special care should be given to the choice of the contact time because too harsh cleaning conditions may cause the irreversible modification of the chromatographic medium or induce leaching of the affinity ligands whose covalent binding with the solid phase could be hydrolyzed. Finally, a

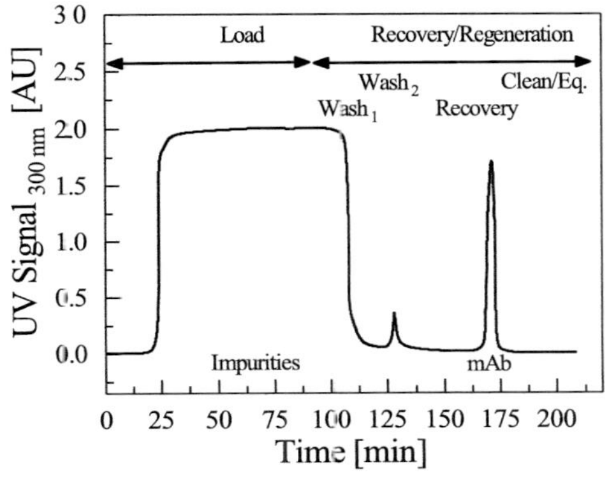

Figure 4.1 Typical chromatogram observed in protein capture applications. Two main phases are identified: the load phase and the recovery/regeneration phase.

re-equilibration phase is performed to prepare the chromatographic medium for a new run. Note that Figure 4.1 is only an illustrative example of a protein capture process, but the actual implementation of the different phases, notably the number and type of wash/cleaning procedures, is protein and medium specific.

Overall, the process operations described above can be classified in two main phases: *(i)* a *load* phase during which the feed solution (e.g., harvested cell culture fluid) is loaded into the column and *(ii)* a *recovery/regeneration* phase to remove unbound solutes, recover the loaded product, and regenerate the chromatographic medium. In the following, the load and recovery phases are thoroughly analyzed with a view to identifying the underlying physical mechanisms.

4.2.2 Thermodynamics of Fluid–Solid Equilibrium

Load Phase

Figure 4.2 shows typical adsorption isotherms obtained with an antibody on Protein A affinity chromatographic media under conditions relevant for the load phase [202]. These adsorption isotherms were measured from batch uptake experiments (filled symbols) and breakthrough experiments (open symbols), as described in Section 2.5. Very similar adsorption isotherms are observed with the different chromatographic media, all showing a sharp increase at low protein concentration followed by a plateau reached for a protein concentration in the liquid phase above about 1 g/L. This means that at the low protein concentrations commonly observed in the harvested cell culture fluid (usually in the range of few g/L), the protein concentration inside the particles is already close to the lumped saturation capacity.

The slope at zero concentration and the value of the plateau at large concentrations are related to the lumped Henry coefficient and the lumped saturation capacity, respectively (see Section 2.4.2). Lumped Henry coefficients in the range 10^2–10^3 are obtained for

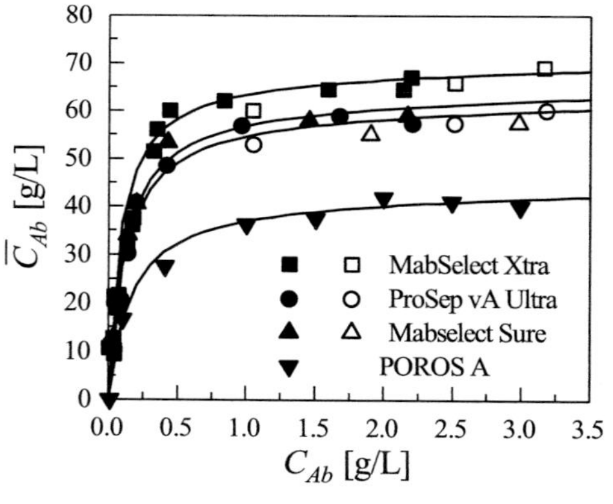

Figure 4.2 Adsorption isotherms of an antibody on various Protein A chromatographic media. The bound protein concentrations were estimated from breakthrough experiments (open symbols) and batch uptake experiments (filled symbols). Adapted from [232], copyright 2003, and from [202], copyright 2005, with permission from Elsevier.

the data shown in Figure 4.2, which is common for antibodies on Protein A chromatographic media. As a comparison, the Henry coefficients of impurities, as, for example, the host cell proteins, are often very low, in the range of unity because they exhibit very little interactions with affinity ligands. In addition, it is seen in Figure 4.2 that lumped saturation capacities are in the order of $\overline{N} = 50-70$ g/L for all the considered chromatographic media. The POROS A medium deserves a particular comment because it is a perfusive material, i.e., it contains large pores allowing a certain extent of intraparticle convection (see Section 2.4.5). This leads to a reduction of the surface area and thus to a decrease in the saturation capacity with respect to the other chromatographic media. More details about the use of perfusive materials for the capture step are given later in Section 4.2.3.

Assuming that the protein of interest is the only solute strongly interacting with the affinity ligands, the fluid–solid equilibrium relation is usually modeled by a non-competitive Langmuir adsorption isotherm (see Equation 2.18):

$$\overline{C}_A = \varepsilon_i C_A + \frac{\overline{N}\tilde{K}_A C_A}{1 + \tilde{K}_A C_A} \tag{4.1}$$

from which one derives the expression of the lumped Henry coefficient $\overline{K}_A = \varepsilon_i + \overline{N}\tilde{K}_A$. The experimental data of Figure 4.2 were fitted with Equation 4.1, and the results are shown with solid lines. It is observed that the Langmuir adsorption isotherm describes well the set of experimental data. Variations of the classical Langmuir adsorption isotherms have also been proposed to account for the presence of multiple binding domains on the Protein A ligands. In particular, a dual-site model is sometimes used [233].

In the case of a single-solute system, the adsorption equilibrium relation allows estimating the mean position of the breakthrough curve of a solute A as follows [158]:

$$t_{r,A} = t_0 \left(1 + \phi_e \frac{\overline{C}_A^*}{C_A^F}\right) \tag{4.2}$$

where $\phi_e = (1 - \varepsilon_e)/\varepsilon_e$ is the phase ratio and $\overline{C}_A^*$ is the lumped concentration in the particles at equilibrium with the feed solution of concentration C_A^F. For a linear thermodynamic equilibrium relation, it is given by $\overline{C}_A^* = \overline{K}_A C_A^F$. On the other hand, for a rectangular adsorption isotherm, it simplifies to $\overline{C}_A^* = \varepsilon_i C_A^F + \overline{N}_A$. It is seen from Equation 4.2 that the mean position of the breakthrough curve exclusively depends upon the adsorption isotherm and is therefore independent from the hydrodynamic and kinetic limitations.

Recovery Phase

After the load and wash steps, protein recovery can be obtained by two main different ways, which are sometimes termed *specific* and *nonspecific* elution [234]. Under specific elution conditions, protein desorption is caused by competitive binding (without affecting the equilibrium constant $\tilde{K}_A$), whereas under nonspecific elution conditions, the equilibrium constant $\tilde{K}_A$ is reduced (but the modifier does not adsorb on the chromatographic medium). This classification is certainly an idealization of what happens in reality, and it is introduced here only to illustrate possible mechanisms triggering the elution of the product.

The specific elution mode involves the use of a solute in the eluent competing with the target protein (also called displacer). In the presence of the competing modifier, the Langmuir adsorption isotherm given by Equation 4.1 is modified as follows:

$$\overline{C}_A = \varepsilon_i C_A + \frac{\overline{N}\tilde{K}_A C_A}{1 + \tilde{K}_A C_A + \tilde{K}_M C_M + \cdots} \tag{4.3}$$

where C_M is the concentration of the competing modifier (with $\tilde{K}_M \gg 0$), and the sum in the denominator should account for all the species possibly competing with the protein to be recovered. It can be seen from Equation 4.3 that the concentration of bound protein decreases with increasing the concentration of modifier. For instance, we have seen in Section 2.2 that with IMAC, the elution of proteins can be obtained by injecting complexing agents such as EDTA, which strongly complexes the metal ions of the chromatographic support. It is worth mentioning that in this case, it is necessary to separate the displacer from the protein in a subsequent step.

For the nonspecific elution mode, the eluent composition is varied in order to affect the equilibrium constant $\tilde{K}_A$. In this case, the physical interactions between the chromatographic medium and the solute (electrostatic, hydrophobic, H-bond...) are modified to weaken the strength of the binding. For instance, the solution pH is usually lowered from 7 to below 4 to promote the elution of IgG bound to Protein A ligands.

In order to get an idea about the variations of the binding constant with the solution pH in Protein A chromatography, data obtained from batch uptake experiments with various media and pH values ranging between 4 and 8 have been collected [232, 235, 236]. The binding constants $\tilde{K}_{Ab}$ of antibodies were fitted for the different pH values assuming a Langmuir adsorption isotherm. For the sake of simplicity, it was assumed that the value of $\overline{N}$ is independent of the solution pH and that the adsorbed concentration at equilibrium equals the saturation capacity at pH 7 (i.e., $\overline{C}_{Ab} \approx \overline{N}$). The binding constants $\tilde{K}_{Ab}$ were computed using Equation 4.1 and knowing C_{Ab}, $\overline{C}_{Ab}$ and $\overline{N}$. The obtained values are reported as a function of pH in Figure 4.3 for POROS A, rProtein A Sepharose and MabSelect media.

These experimental data were fitted with the equation presented in Section 2.4.2 and recalled here:

$$\tilde{K}_{Ab} = \tilde{K}_{0,Ab} \times \left(1 + \frac{\tilde{K}_{0,Ab}}{\tilde{K}_H} 10^{-\text{pH}}\right)^{-1} \tag{4.4}$$

It is seen that this correlation describes relatively well the experimental results shown in Figure 4.3. In the investigated range of pH, the binding constant $\tilde{K}_{Ab}$ is the highest at neutral pH values, which corresponds to the conditions observed at the outlet of bioreactors. When decreasing the pH of the solution, $\tilde{K}_{Ab}$ drops by two to three orders of magnitude. The drastic reduction of the Henry coefficient when lowering the solution pH enables to trigger the desorption of the target protein and to increase significantly its concentration in the eluate.

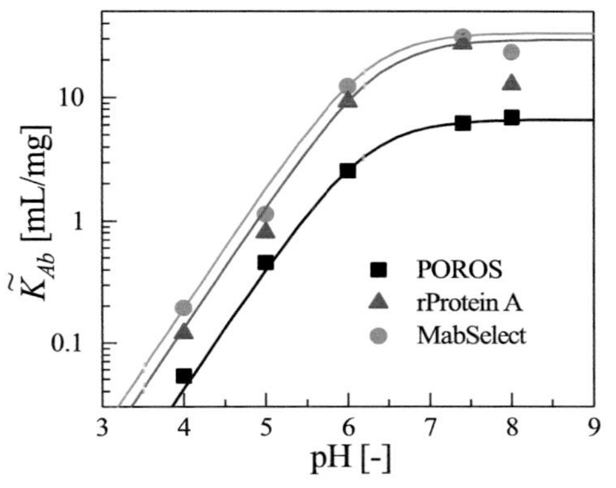

Figure 4.3 Binding constants of antibodies, $\tilde{K}_{Ab}$, as a function of the solution pH for different Protein A media. Experimental data (symbols) taken from [232, 235, 236]. The experimental results were fitted with Equation 4.4 (lines). The parameter $\tilde{K}_{0,Ab}$ was found equal to 6.6, 29.4 and 33.4, while $\log(\tilde{K}_H)$ was estimated to −5.38, −6.34 and −6.24 for POROS A, rProtein A Sepharose FF, and MabSelect, respectively.

To illustrate this point, let us consider the simple case of batch uptake chromatography. In such experiments, the general mass balance equation applied to a solute A writes as follows:

$$C_A^F V + \overline{C}_A^F \overline{V} = C_A V + \overline{C}_A \overline{V} \tag{4.5}$$

where C_A^F and $\overline{C}_A^F$ are the concentrations initially fed in the liquid and solid phases, respectively, while C_A and $\overline{C}_A$ are the corresponding solute concentrations at a given time. The volume of the liquid and solid phases are denoted V and $\overline{V}$, respectively.

In a first step, a solute-free Protein A chromatographic medium ($\overline{C}_A^{F,1} = 0$) is put in contact with a solution of an antibody A. We consider that the antibody concentration in this solution is $C_A^{F,1} = 2$ g/L and that the solution pH is 7.4, which are typical conditions observed in the harvested cell culture fluid. In addition, we consider that the volume of liquid is 25 times larger than the volume of solid. With these numerical values, Equation 4.5 is rearranged to give the *capture operating line*:

$$\overline{C}_A = -25 \times C_A + 25 \times C_A^{F,1} \tag{4.6}$$

The protein concentration in the liquid and solid phases at equilibrium is given by the intersection between the capture operating line and the adsorption isotherm at pH 7.4. This is graphically represented in Figure 4.4, where the starting and equilibrium compositions are indicated with the symbols ⓵ and (1*), respectively.

In a second step, the liquid phase is withdrawn, and the particles are immersed in a protein-free solution at pH 4.0, so that $C_A^{F,2} = 0$ and $\overline{C}_A^{F,2} = \overline{C}_A^{*,1}$. The protein recovery is performed with $V/\overline{V} = 5$. Similar to Equation 4.6, the *recovery operating line* is derived:

$$\overline{C}_A = C_A^{F,2} - 5 \times C_A \tag{4.7}$$

The concentration in the two phases at equilibrium is given by the intersection between the recovery operating line and the adsorption isotherm at pH 4.0. This is

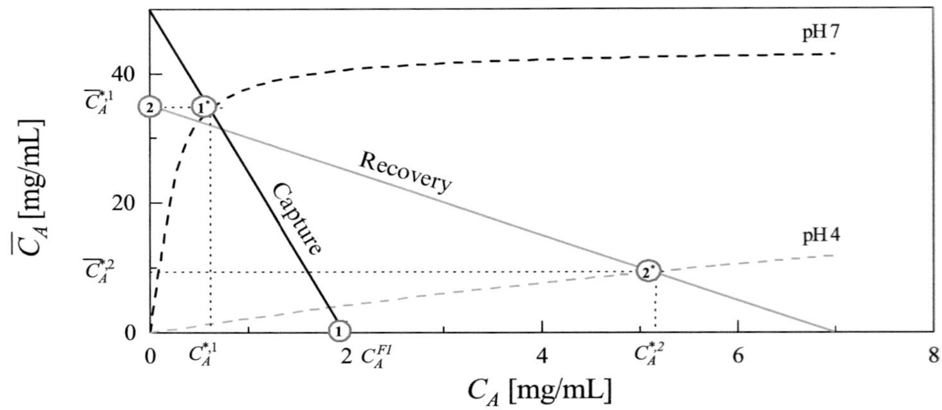

Figure 4.4 Capture/recovery process of a mAb on a Protein A medium in a batch uptake experiment. A Langmuir adsorption isotherm is considered with $\overline{N} = 43.8$ mg/mL, $\tilde{K}_A$(pH 7.4) = 6.25 mL/mg, and $\tilde{K}_A$(pH 4.0) = 0.05 mL/mg (values taken from [232]). The points (1) and (2) indicate the initial protein concentrations for the capture step at pH 7.4 and the recovery step at pH 4.0, respectively, while the points (1*) and (2*) indicate the corresponding equilibrium concentrations.

shown in Figure 4.4, where the new starting point and equilibrium composition are indicated with the symbols (2) and (2*), respectively. It is seen that the protein is recovered in the liquid phase at a concentration of $C_A^{*,2} = 5.2$ mg/mL, which is a bit more that twice the initial concentration of 2 mg/mL.

This simple analysis shows how a more concentrated solution can be obtained by capturing the protein at neutral pH and recovering it at low pH using a smaller volume of eluent. This is also observed in column experiments when sequentially performing long injections at neutral pH and a short elution step at low pH. The ratio between the load volume (usually corresponding to several tens of column volumes) and the elution volume (typically between 1 and 10 column volumes) is giving the expected concentration factor. This increase in concentration allows reducing the volume of solution to be further processed in the subsequent units.

However, a too high protein concentration induces an increase in the fluid viscosity, which in turn increases the pressure drop. Moreover, the difference in viscosity between the feed solution and the eluent used during protein recovery may cause viscous fingering, leading to significant asymmetric band broadening, with a sharp front and a tailing rear [237]. Therefore, the elution conditions should be carefully selected in order to obtain a reasonably high solute concentration in the eluate, while preventing these detrimental effects.

4.2.3 Hydrodynamics and Kinetics of Mass Transfer

Load Phase

Characteristic Time for Mass Transfer

The characteristic time for mass transfer can, for instance, be estimated from batch uptake experiments. An example of such experiment is reported in Figure 4.5(a), where

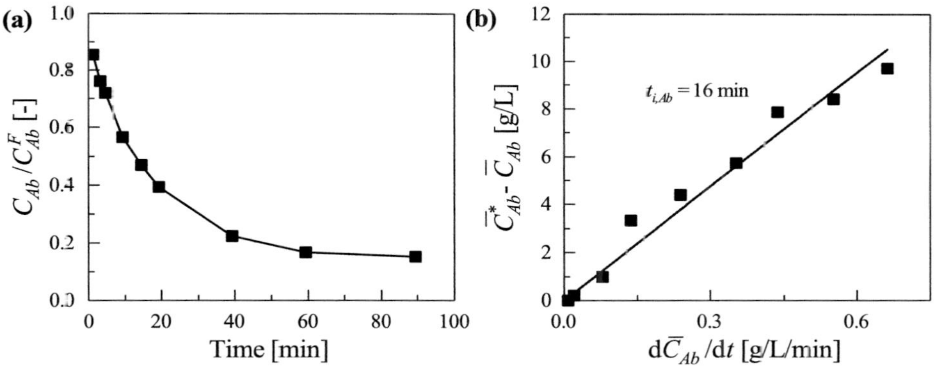

Figure 4.5 (a) Batch uptake kinetics of an IgG on a Protein A chromatographic medium obtained with $C^F_{Ab} = 0.1$ mg/mL, $d_p = 75$ μm and $V/\overline{V} = 140$. Adapted from [202], copyright 2005, with permission from Elsevier. (b) Linearization providing the characteristic time for internal mass transfer.

the concentration of an antibody (Ab) at neutral pH in contact with an affinity chromatographic medium is shown as a function of time. It can be seen that the concentration of antibody in the liquid phase slowly decreases until reaching the equilibrium concentration after slightly more than 1 h. Note that this type of experiment also allows obtaining the adsorption isotherm from the equilibrium values obtained with different initial protein concentrations.

The characteristic time for internal mass transfer was determined by fitting these experimental data with Equation 2.59 and approximating the derivative $d\overline{C}_{Ab}/dt$ by a finite difference. Figure 4.5(b) shows the linear regression, whose slope provides the value of $t_{i,Ab} \approx 16$ min. This value is in line with typical characteristic times for internal mass transfer observed during mAb capture with Protein A ligands, which often exceed tens of minutes till hours during the load phase. Therefore, the kinetics of internal mass transfer is usually the main contribution to band broadening.

This very slow kinetics is not only due to the large protein size, which inevitably leads to a small diffusion coefficient, but also to the very strong interactions between the solute and the affinity ligands. It is worth recalling that for a solute A, $t_{i,A}$ scales with the lumped Henry coefficient and the square of the particle diameter as follows (see Equation 2.62):

$$t_{i,A} = \frac{d_p^2}{60\overline{D}_A} \tag{4.8}$$

where the lumped diffusion coefficient is given by $\overline{D}_A = (\varepsilon_i \grave{D}_A + (\overline{K}_A - \varepsilon_i)\overline{\overline{D}}_A)/\overline{K}_A$ under linear adsorption conditions. It averages the contributions from the diffusion of the solutes in the liquid inside the pores (with the diffusion coefficient $\grave{D}_A$) and from the diffusion of the solutes bound to the chromatographic medium (with the diffusion coefficient $\overline{\overline{D}}_A$). In the case where $\varepsilon_i \neq 0$, $\overline{D}_A$ decreases when increasing the adsorption strength from $\grave{D}_A$ in the absence of interactions (i.e., $\overline{K}_A = \varepsilon_i$), to $\overline{\overline{D}}_A$ for sufficiently large $\overline{K}_A$ values. It results that strongly bound solutes have difficulties diffusing through the pores, as diffusion on the solid surface is expected to be much slower than diffusion in the liquid phase. For the sake of simplicity, the considerations above were drawn assuming linear adsorption conditions. Under nonlinear conditions, it can be shown that $\overline{D}_A$ depends on the derivative of the adsorption isotherm [158]. A simplified expression of the characteristic time for internal mass transfer under nonlinear conditions will be given in the next paragraph (see Equation 4.11).

Concentration Profiles inside the Particles

So far, we have considered the solute concentration within the particles of the chromatographic medium as uniform, both in the liquid and adsorbed phases. Although very simple, this approach is most often sufficient to describe properly the chromatograms observed experimentally. If one is interested in gaining a better understanding of the phenomena occurring within the particles, imaging techniques can be used to visualize the time evolution of the concentration profiles along the particle radius [238–240]. As an illustration, confocal laser scanning microscopy pictures of particles immersed in a mAb solution are shown in Figure 4.6(a) [240]. It is seen that the mAb progressively diffuses toward the center of a particle, starting from a thin ring at the outer surface, gradually filling the entire particle. The diffusing fronts are very sharp, indicating that as soon as a protein reaches the surface of the chromatographic medium, it is rapidly

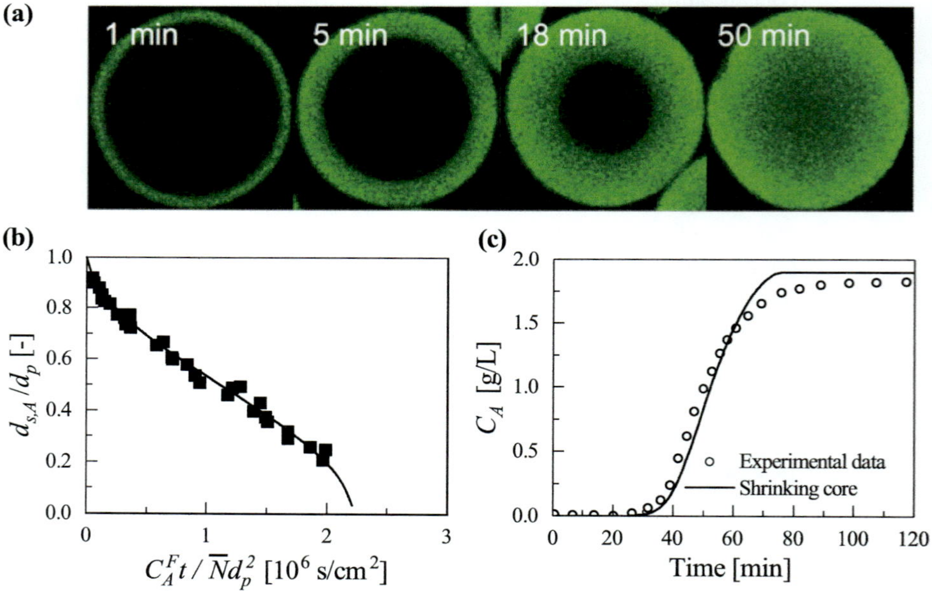

Figure 4.6 Adsorption of a mAb on a multimodal HIC/IEC chromatographic medium at pH 5. (a) Confocal laser scanning microscopy images at different times in a batch uptake experiment. (b) Progression of the diffusing front as a function of time. Symbols correspond to experimental data, while the solid line is the prediction of the shrinking core model with Equation 4.9. (c) Experimental breakthrough curve and predictions from the shrinking core model. Adapted from [240] with permission of Springer.

adsorbed. This is due to the strong adsorption on the chromatographic medium, as it is typically the case in protein capture applications. Figure 4.6(b) shows the position of the diffusing front measured from the center of the particle as a function of time, denoted $d_{s,A}$ and represented with symbols. In other words, $d_{s,A}$ is the diameter of the inner black core still free of solute on Figure 4.6(a).

Such experimental radial concentration profiles could be simulated using models describing internal diffusion based on Fick's law or Stefan-Maxwell equations (see Section 2.4.4). A simplified model, commonly known as the *shrinking core model*, has been derived from Fick's law under the following assumptions [161]: *(i)* only one solute is adsorbing on the chromatographic medium, *(ii)* the adsorption isotherm is rectangular, *(iii)* the solute concentration in the pores is negligible with respect to the concentration of solute adsorbed, and *(iv)* there is no surface diffusion. In the case where the solute concentration surrounding the particles is sufficiently high to be considered constant with time, the following equation can be used to describe the progression of the diffusive front observed with imaging techniques during batch uptake experiments:

$$2\left(\frac{d_{s,A}}{d_p}\right)^3 - 3\left(\frac{d_{s,A}}{d_p}\right)^2 + 1 = \frac{24\varepsilon_i \grave{D}_A C_A t}{\overline{N} d_p^2} \tag{4.9}$$

In the example under consideration, all the parameters were assessed independently except the diffusion coefficient, which has been estimated to $\grave{D}_A = 1.5 \times 10^{-6}$ cm^2/s by fitting the experimental data of Figure 4.6(b) with Equation 4.9. The simulated progression of the diffusion front is shown with a solid line in Figure 4.6(b), where it is seen that the shrinking core model represents well the experimental data.

In the frame of the lumped approach used throughout this book, the radial concentration profiles are averaged over the particle radius to compute the lumped solute concentration. In this case, the shrinking core model leads to the following equation describing the kinetics of mass transfer [241]:

$$\frac{d\overline{C}_A}{dt} = \frac{12\varepsilon_i \grave{D}_A}{d_p^2} \frac{\left(1 - \overline{C}_A/\overline{N}\right)^{1/3}}{1 - \left(1 - \overline{C}_A/\overline{N}\right)^{1/3}} \times C_A \tag{4.10}$$

The right-hand side of Equation 4.10 can be seen as the ratio of the driving force given by $(C_A - 0)$, where C_A is the solute concentration at the surface of the particle and 0 the concentration in the particle at the position $d_{s,A}$, over the characteristic time for internal mass transfer, which now depends upon the adsorbed solute concentration. In order to describe breakthrough experiments, Equation 4.10 can be coupled with a hydrodynamic model (such as the PD or MC models presented in Section 2.4.3). Note that Equation 4.10 is limited to loading conditions and thus cannot be used to model desorption during column regeneration.

Figure 4.6(c) shows an experimental breakthrough curve (symbols) together with the simulated curve obtained using Equation 4.10 to describe the kinetics of mass transfer (solid line). For this simulation, $\grave{D}_A$ was set to the value estimated from the imaging experiments reported in Figure 4.6. It is seen that the prediction from the shrinking core model is in good agreement with the experimental breakthrough curve.

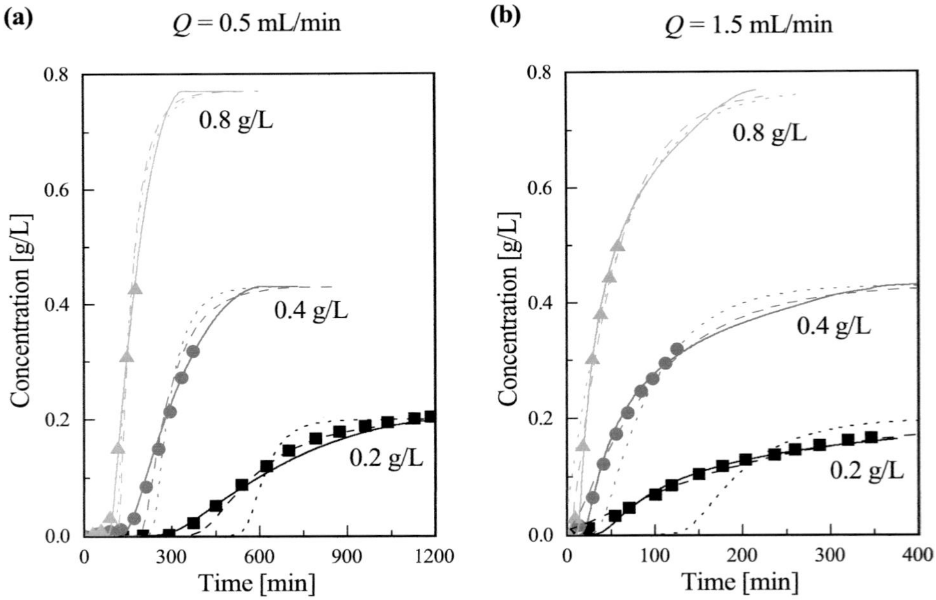

Figure 4.7 Comparison between experimental breakthrough curves (symbols) and simulations from three models: the shrinking core model (Equation 4.10, solid lines), the LDF approximation with $t_{i,A}$ varying with the protein concentration (Equations 2.58 and 4.11, dashed lines), the LDF approximation with constant $t_{i,A}$ (Equation 2.58, dotted lines). Two flow rates, namely (a) 0.5 mL/min and (b) 1.5 mL/min, and three feed concentrations, namely 0.2 g/L, 0.4 g/L and 0.8 g/L, were considered. The following parameters were used: $\varepsilon_i = 0.52$, $\varepsilon_e = 0.4$, $\tilde{K}_A = 64$ L/g, $\overline{N} = 108$ g/L, $L_{col} = 5$ cm, $D_{col} = 0.5$ cm. For the shrinking core model, the ratio $d_p^2/(12\varepsilon_i \dot{D}_A)$ was estimated to 2.25 min. For the LDF with variable $t_{i,A}$, the parameter $t_{i,A}(C_A = 0)$ was estimated to 42.6 h, which leads to $t_{i,A} = 185$ min at 0.2 g/L, 90 min at 0.4 g/L and 51 min at 0.8 g/L. For the LDF with constant characteristic time for mass transfer, $t_{i,A}$ was estimated to 59 min. Experimental data taken from [242], copyright 2016, with permission from Elsevier.

In the following, we compare the results of the shrinking core model with the LDF approximation considering a set of experimental breakthrough curves obtained at different flow rates and protein concentrations. We have seen in Section 2.4.4 that the LDF approximation can still be used under nonlinear conditions provided that a proper dependence of $t_{i,A}$ with the protein concentration is accounted for [158]. In the particular case of breakthrough curves obtained with a single solute, the characteristic time for internal mass transfer is given by [21]:

$$t_{i,A}(C_A) = t_{i,A}(C_A = 0) \times \frac{\overline{C}_A^*}{C_A \overline{K}_A} \tag{4.11}$$

where the concentration C_A is taken equal to the concentration in the feed C_A^F and $\overline{C}_A^*$ is the lumped concentration in the particles at equilibrium with the feed solution. Under sufficiently dilute conditions, $\overline{C}_A^* \approx \overline{K}_A C_A$, so that $t_{i,A} = t_{i,A}(C_A = 0)$ is independent of the protein concentration as given by Equation 2.62.

Figure 4.7 shows experimental breakthrough curves (symbols) obtained with a mAb on a Protein A chromatographic medium together with the results of three different models, namely the shrinking core (solid lines), the LDF approximation with $t_{i,A}$ varying with the protein concentration according to Equation 4.11 (dashed lines) and the LDF approximation with a constant $t_{i,A}$ (dotted lines). Panels (a) and (b) of Figure 4.7 correspond to two different flow rates, 0.5 mL/min and 1.5 mL/min, respectively. For each flow rate, three protein concentrations were investigated, namely 0.2 g/L, 0.4 g/L and 0.8 g/L. For each model, the parameters were determined by fitting the simulations to the experimental data at the lowest and highest protein concentration for the two flow rates (i.e., considering four experimental breakthrough curves). It is seen in Figure 4.7 that the LDF approximation with a constant $t_{i,A}$ value fails to describe the impact of protein concentration on the shape of the breakthrough curves. On the other hand, the shrinking core model and the LDF approximation with variable $t_{i,A}$ give similar results in good agreement with experimental data.

Impact of the Fluid Flow Rate on Breakthrough Curves

Let us now discuss in more detail the impact of the fluid flow rate on the shape of the breakthrough curves. As an illustration, experimental breakthrough curves of an antibody obtained with a Protein A chromatographic medium at different fluid flow rates are shown in Figure 4.8 [202]. First, it is seen that all the curves have the same mean position when plotted as a function of the elution volume. This mean position results only from the fluid–solid equilibrium relation and does not depend upon mass transfer limitations. It can be computed in the frame of the equilibrium theory by using Equation 4.2 and the parameter values of the fluid–solid equilibrium previously determined from the data shown in Figure 4.2. The breakthrough curve predicted by the equilibrium theory is represented by a solid line in Figure 4.8. Second, it can be observed that shallower and shallower breakthrough curves are obtained when increasing the flow

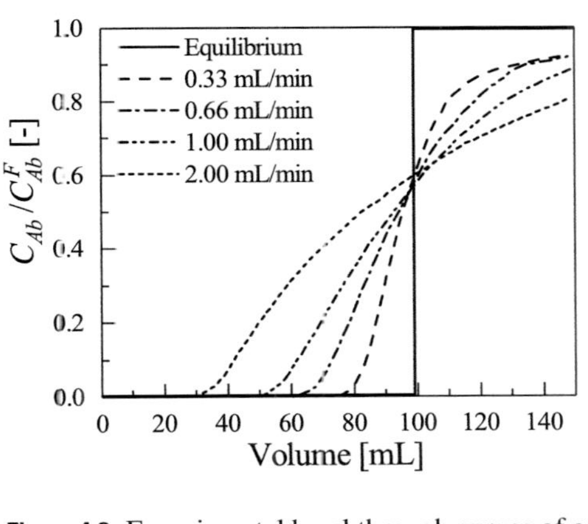

Figure 4.8 Experimental breakthrough curves of an antibody obtained on a Protein A chromatographic medium at various flow rates (keeping the column cross section constant). Adapted from [202], copyright 2005, with permission from Elsevier. The prediction of the equilibrium theory is also reported (solid line).

rate. The shape of the fronts depends mostly on the ratio between the zero-retention time (given by $t_0 = \varepsilon_e V_{col}/Q$) and the time it takes to reach equilibrium (approximately given by $t_{i,A}$ because dispersive effects are dominated by the kinetics of internal mass transfer). At high flow rate, the time spent by the solute in the column is too short as compared to the time it takes to diffuse in the particles, and shallow breakthrough curves are obtained. On the other hand, at low flow rate, the breakthrough curves get steeper and thus closer to the prediction from the equilibrium theory.

It is seen in Figure 4.8 that for an injection volume of 100 mL, which corresponds to the retention volume predicted in the frame of the equilibrium theory, high flow rates lead to important protein losses. To improve the protein recovery, the flow rate can be reduced, which could, however, be detrimental in terms of productivity. Therefore, it can already be seen from this example that there is a trade-off between recovery and productivity.

Perfusive Materials

In order to overcome the strong internal mass transfer limitations described earlier, perfusive materials can be envisaged. As described in Section 2.4.5, internal mass transfer in perfusive chromatographic media is facilitated by the connection of the particle pores to a network of wider pores, where solute transport takes place by convection rather than diffusion. The characteristic time for internal mass transfer in perfusive particles can be estimated as follows [158, 197]:

$$t_{i,A} \approx \frac{d_p^2}{60\bar{D}_A}\left(1 + \frac{Pe_i}{3}\right)^{-1} \tag{4.12}$$

where Pe_i is the intraparticular Péclet number defined as $Pe_i = u_i d_p/(6\grave{D}_A)$. The intraparticle velocity u_i is estimated as a fraction of the superficial velocity: $u_i = F_i u$, where F_i is usually much smaller than unity. At high velocity, it comes from Equation 4.12 that $t_{i,A} \propto d_p/u$, which implies that the characteristic time for internal mass transfer decreases when increasing the fluid velocity. Therefore, thanks to the internal convective flow, larger fluid flow rates can be used without reducing the column efficiency.

Figure 4.9 shows breakthrough curves obtained with an antibody on two affinity media (HyperD in panel (a) and POROS in panel (b)) at different flow rates and at low protein concentration (0.4 mg/mL, so that the adsorption isotherms can be assumed linear). Both media have the same average particle diameter (d_p = 50 μm), and the lumped Henry coefficients are very similar in both cases ($\bar{K}_{Ab} \approx 330$) [232]. In contrast to HyperD, which is composed of rigid particles whose pores are filled with hydrogels, the POROS chromatographic medium is composed of perfusive particles. It is seen in Figure 4.9(a–b) that the breakthrough curves measured with the POROS medium are sharper than the ones obtained with the HyperD medium. Moreover, the breakthrough curves obtained with the POROS medium are little affected by the increase of the fluid velocity as compared to those obtained with the HyperD medium.

The characteristic times for overall mass transfer were estimated by the inverse method, fitting the breakthrough curves with the linear driving force model for both

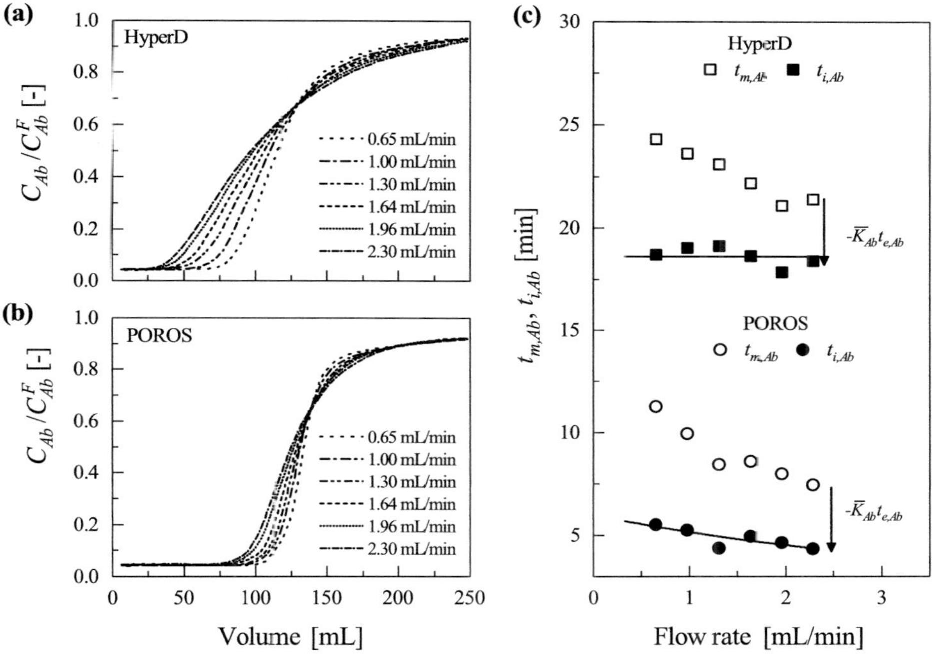

Figure 4.9 Breakthrough curves of an antibody obtained with (a) Protein A Hyper D medium and (b) with POROS A medium. Adapted from [232], copyright 2003, with permission from Elsevier. (c) Overall characteristic times of mass transfer, $t_{m,Ab}$, obtained by the inverse method from the breakthrough curves measured at different flow rates (open symbols). The characteristic times for internal mass transfer (filled symbols) were computed as $t_{i,Ab} = t_{m,Ab} - \overline{K}t_{e,Ab}$, with $t_{e,Ab}$ estimated from the Carberry's correlation (see Equation 2.66).

media. The resulting $t_{m,Ab}$ values are shown with open symbols in Figure 4.9(c) as a function of the fluid flow rate. First, it is seen that 3 to 4 times smaller values of $t_{m,Ab}$ were obtained with the perfusive material as compared to the nonperfusive one, which explains that sharper fronts are obtained. Moreover, a decrease of the characteristic time for mass transfer is observed in both cases when increasing the fluid velocity. This is mostly due to the contribution of the external mass transfer resistance to the overall mass transfer rate. Indeed, the characteristic time for external mass transfer $t_{e,Ab}$ diminishes when increasing the fluid velocity due to a decrease of the film layer thickness. The characteristic time for external mass transfer was evaluated at each flow rate by using the Carberry's correlation (see Equation 2.66) and the values of $\overline{K}_{Ab}t_{e,Ab}$ were then subtracted to $t_{m,Ab}$ in order to estimate $t_{i,Ab}$. The obtained characteristic times for internal mass transfer are presented in Figure 4.9(c) as filled symbols for both media. It can be seen that the characteristic time for internal mass transfer is almost constant for the HyperD medium, which is consistent with Equation 4.8, while it decreases for the POROS medium, which is consistent with Equation 4.12. The latter equation was further used to regress the experimental data obtained with the POROS medium and estimate the parameters $\overline{D}_{Ab}$ and F_i.

Values of 3.6×10^{-12} m^2/s and 0.024% were obtained, respectively, which is in line with those reported in the literature for the same material: 3.1×10^{-12} m^2/s and 0.050% [200].

Recovery Phase

Because of the coupling between internal mass transfer kinetics and adsorption strength, $t_{i,A}$ is strongly affected by the elution conditions (see Equation 4.8). For the sake of simplicity, let us assume that $\overline{\overline{D}}_A \approx 0$, so that Equation 4.8 leads to $t_{i,A} = d_p^2 \overline{K}_A/(60\varepsilon_i \grave{D}_A)$. Because $\grave{D}_A$ can be assumed invariant with $\overline{K}_A$, it follows that $t_{i,A}/\overline{K}_A$ becomes rather independent from the elution conditions. Therefore, one can write:

$$\left(t_{i,A}/\overline{K}_A\right)^F \approx \left(t_{i,A}/\overline{K}_A\right)^R \tag{4.13}$$

where the superscript F and R denote the load phase and the recovery phases, respectively. Considering that the protein recovery phase is performed under nonbinding conditions, the lumped Henry coefficient simplifies to its steric contribution, i.e., $\overline{K}_A^R = \varepsilon_i$, thus leading to:

$$t_{i,A}^F \approx t_{i,A}^R \frac{\overline{K}_A^F}{\varepsilon_i} \tag{4.14}$$

As an example, let us consider the following parameter values from the literature for an antibody Ab captured on a Protein A chromatographic medium: $\overline{K}_{Ab}^F = 780$, $t_{i,Ab}^F = 340$ s, and $\varepsilon_i = 0.88$ [205, 243]. With the preceding equation, $t_{i,Ab}^R$ is then estimated to 0.4 s, which is significantly lower than $t_{i,Ab}^F$. Figure 4.10 shows experimental data (open circles) of a single-column mAb capture process [243] together with the

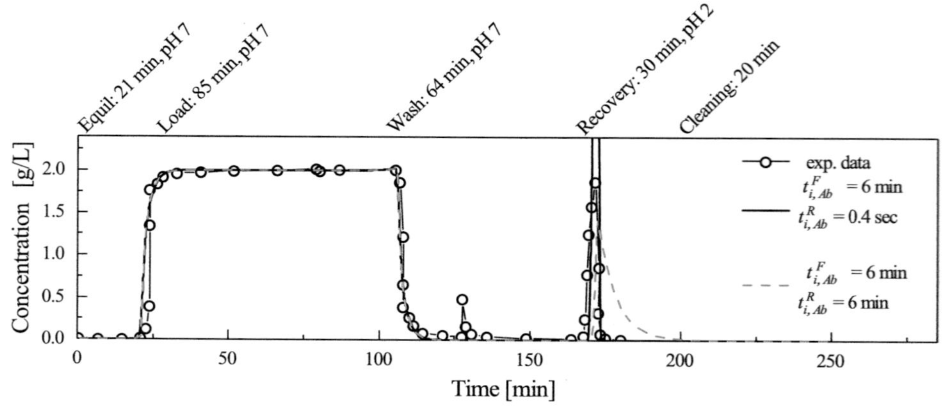

Figure 4.10 Experimental chromatogram obtained when capturing a mAb on a Protein A chromatographic medium (open symbols) and comparison with two types of simulations: using a single characteristic time for mass transfer (dotted gray line) and using different mass transfer kinetics for the load and elution phases (solid black line), as indicated in the legend. Adapted from [243], copyright 2014, with permission from Elsevier. Details concerning the model parameters can be found in [205].

simulation results using the LDF approximation (lines). Two sets of model parameters were considered, one for which the same characteristic time for internal mass transfer is used for both the load and the recovery phases (i.e., $t^F_{i,Ab} = t^R_{i,Ab} = 6$ min, dashed line), and a second one where the characteristic time for internal mass transfer during the recovery step was computed using Equation 4.14 (i.e., $t^F_{i,Ab} = 6$ min and $t^R_{i,Ab} = 0.4$ s, solid line). It can be seen in Figure 4.10 that the simulation performed with $t^F_{i,Ab} = t^R_{i,Ab}$ leads to an elution peak significantly broader than the one observed experimentally. In fact, $t^R_{i,Ab}$ has to be decreased by more than a factor 10 with respect to $t^F_{i,Ab}$ to describe the experimental data. This confirms that a significant reduction of the characteristic time for internal mass transfer is observed under recovery conditions compared to loading conditions due to the interplay between mass transfer kinetics and the adsorption thermodynamics.

It is noteworthy that, although the expression of $t^R_{i,Ab}$ was derived under linear conditions, the simple analysis presented above was sufficient to describe the experimental data presented in Figure 4.10. For a more detailed treatment, it would be necessary to account for the dependence of the characteristic time for internal mass transfer with protein concentration.

4.3 Design of Countercurrent Chromatographic Processes

4.3.1 Principle

In this section, the fundamental notions developed in Chapter 3 are applied to design countercurrent chromatographic units for the capture of therapeutic proteins. From a conceptual viewpoint, going from a single-column to a continuous chromatographic process necessitates to transpose our reasoning from a temporal to a spatial sequence of operations. This is no longer a question of doing one thing and *then* another one, but rather to do something here, and *there* something else. In order to illustrate this point, Figure 4.11(a) shows the sequence of operations typically performed in a single-column process during protein capture: load, wash, recovery, wash, cleaning, and equilibration. This example is given for illustrative purposes only and different processes may include various numbers of wash/cleaning operations, which does not affect the validity of the discussion.

An intuitive design of a countercurrent capture process would consist in dedicating one zone to each of these operations, so that they can all take place simultaneously. Such process is shown in Figure 4.11(b) in the TMB-like representation and has originally been implemented experimentally using an eight-column process consisting of four different zones (load, wash, recovery, and cleaning) [244]. Note that the order of the different steps is read from left to right for the single-column process, whereas the zones are read from right to left in the multicolumn process. This is due to the countercurrent movement of the solid phase with respect to the liquid phase. Because each operation requires different elution conditions, recycling the liquid phase is not always possible, and multicolumn processes are often operated in an open-loop configuration.

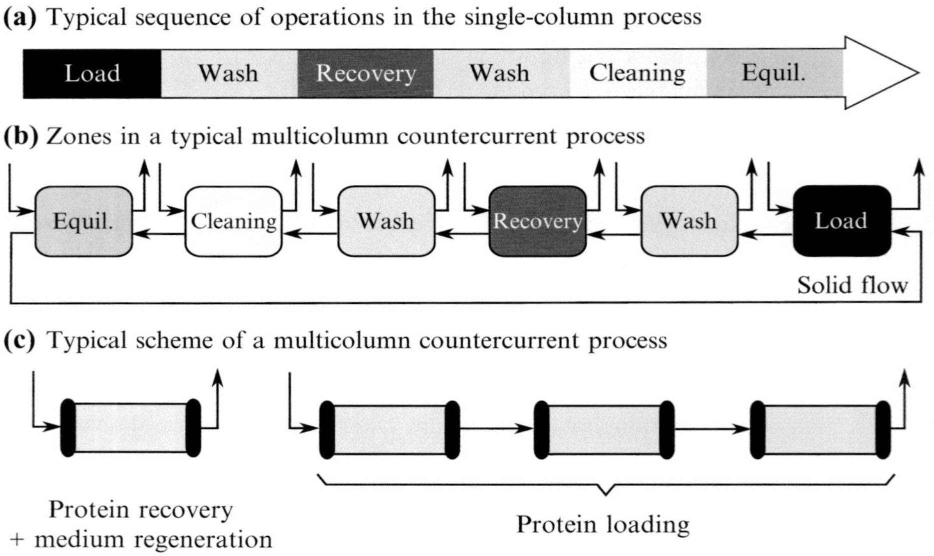

Figure 4.11 Typical chromatographic process for protein capture. (a) Sequence of operations in a single-column process. (b) Zones of the equivalent countercurrent process in the TMB-like representation. (c) Practical implementation of a countercurrent process for the protein capture step.

This means that there is no continuity in the liquid flow between the zones and that each zone is operated independent from the others.

In practice, several columns are often connected in series in the load zone, whereas all the other operations (wash, protein recovery, cleaning, and equilibration) are performed sequentially in one column or in several columns operated in parallel. A representative scheme of a countercurrent process for protein capture, with several columns in the load zone and one column for the protein recovery/medium regeneration zone, is shown in Figure 4.11(c).

With illustrative purposes, we detail the sequence of operations considering the three-column process shown in Figure 4.12. Initially, two columns connected in series are loaded with the feed mixture, while the third one successively goes through the recovery and regeneration phases. After some time, the positions of the inlet/outlet streams are switched, so that the first column of the load zone enters the recovery/regeneration zone. Meanwhile, the freshly re-equilibrated column is connected at the end of the load zone. This movement of the inlet/outlet streams is repeated once more to complete a cycle. The selection of the switch time is key for the proper operation of the system and depends on the target recovery/productivity, as discussed in the next sections.

Figure 4.12 presents the basic ideas behind multicolumn chromatographic processes for protein capture, but many variations have been introduced by different suppliers to improve protein recovery and productivity. For instance, various numbers of columns per zone, different regeneration procedures and alternative loading protocols allowing to feed several columns in parallel can be implemented [245–249]. Although the

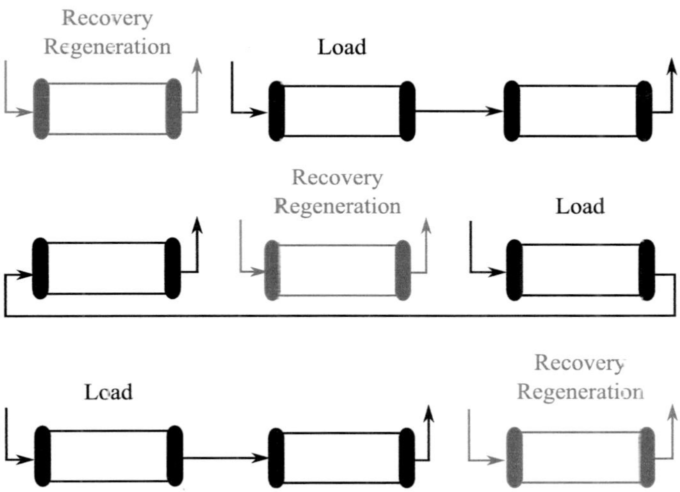

Figure 4.12 Schematic representation of the different phases of a three-column chromatographic capture system with two columns in the load zone and one in the recovery/regeneration zone.

commercially available chromatographic units present their own subtleties, they are all conceived according to similar ideas. Our intention is not to discuss the differences between all possible processes, but rather to draw general conclusions that will help the reader to understand the key concepts behind countercurrent chromatography for protein capture. For a review of the existing processes, the reader is referred to [83].

4.3.2 Definition of the Process Variables

With a view to designing multicolumn chromatographic processes for protein capture, the following process variables must be defined:

- Q^j the fluid flow rate during the operation j. In particular, Q^F is the feed flow rate.
- n^F the number of columns in the load zone, n^R the number of columns in the recovery/regeneration zone and $n^T = n^R + n^F$ the total number of columns.
- Δt^j the duration associated with the operation j. In particular, Δt^F is the feed time and $\Delta t^R = \Delta t^{wash} + \Delta t^{recovery} + \Delta t^{cleaning} + \Delta t^{equilibration} + \cdots$ is the recovery/regeneration time.

In a single-column process, the load and recovery/regeneration phases are performed sequentially. One advantage of multicolumn processes is the possibility to perform load and recovery/regeneration operations simultaneously. The switch time is then given by the slowest operation, i.e., $\Delta t = \max(\Delta t^F, \Delta t^R)$. There are therefore three possible cases that can be envisaged:

$\Delta t^R < \Delta t^F$ There is always a "fresh" column ready to be connected to the load zone, and some idle time is necessary after the recovery/regeneration phase. The switch time is given by $\Delta t = \Delta t^F$ and the process is said to be *loading-controlled*.

$\Delta t^R = \Delta t^F$ The load and recovery/regeneration phases are synchronized and $\Delta t = \Delta t^F = \Delta t^R$.

$\Delta t^R > \Delta t^F$ Some idle time is necessary after the load phase before a "fresh" column is ready to be connected. The switch time is given by $\Delta t = \Delta t^R$, and the process is said to be *regeneration-controlled*.

In all cases, the cycle time is equal to $n^T \Delta t$.

As introduced in Section 2.6.1, the recovery ϑ_A corresponds to the quantity of solute A recovered at the process outlet divided by the quantity of solute A injected at the process inlet over one cycle. This overall recovery can be divided into two contributions: one associated with the load phase ϑ_A^F and one associated with the recovery/regeneration phase ϑ_A^R, so that $\vartheta_A = \vartheta_A^F \times \vartheta_A^R$. The recovery ϑ_A^F corresponds to the amount of protein contained in the column at the end of the load phase over the amount that has been injected. A value of ϑ_A^F lower than 1 indicates that some protein was lost at the column outlet during the load phase. On the other hand, ϑ_A^R corresponds to the amount of protein recovered over the amount of protein present in the column at the end of the load phase. A value of ϑ_A^R lower than 1 indicates some product loss during the recovery phase, for example due to the collection time selection, irreversible denaturation, or aggregation. In the following, we assume that $\vartheta_A^R = 1$, so that $\vartheta_A = \vartheta_A^F$.

Besides the recovery, another key process performance parameter is the productivity, which is defined as the amount of solute A recovered at the process outlet per unit time and per unit column volume. The productivity of a multicolumn process is given here:

$$\varphi_A = \frac{\vartheta_A C_A^F Q^F \Delta t^F}{n^T V_{col} \Delta t} = \vartheta_A \frac{\varepsilon_e C_A^F}{t_0^T} \frac{\Delta t^F}{\Delta t} \tag{4.15}$$

where $t_0^T = n^T t_0 = n^T \varepsilon_e V_{col} / Q^F$.

For comparison purposes, the productivity of a single-column process whose column volume is identical to the one of the multicolumn process ($n^T V_{col}$) is given by:

$$\varphi_A = \frac{\vartheta_A C_A^F Q^F \Delta t^F}{n^T V_{col} (\Delta t^F + \Delta t^R)} = \vartheta_A \frac{\varepsilon_e C_A^F}{t_0^T} \frac{\Delta t^F}{\Delta t^F + \Delta t^R} \tag{4.16}$$

A third parameter known as the capacity utilization CU will be introduced later (see Equation 4.20). It quantifies the fraction of the column effectively used to capture the protein. In other words, it corresponds to the amount of protein stored in the column at the end of the load phase over the maximum amount that can theoretically be stored in the frame of the equilibrium theory. For a given process, working at values of CU close to 1 ensures minimizing the number of cycles necessary to treat a certain amount of feed solution, which may be relevant for expensive chromatographic media with a limited life time.

4.3.3 Processes with a Number of Columns per Zone Constant during One Cycle

Empirical Design

In this section, we show how to select certain process variables (namely Δt^F, n^F, and Q^F) for a multicolumn process based on experiments performed with a single-column process consisting of n^F columns connected in series. This empirical approach implies that the column dimensions, the feed composition, and the recovery/regeneration protocol are predetermined. First, we discuss the selection of the feed time Δt^F for given values of n^F and Q^F. Second, we show how to select experimentally n^F and Q^F.

Selection of the Feed Time

To illustrate the approach, let us consider a process with $n^F = 3$. Schematic protein concentration profiles inside the columns at a given time are shown in Figure 4.13(a). The quantity of protein in each column is proportional to the light gray, dark gray and black areas for Column 1, 2, and 3, respectively.

For expensive proteins, Δt^F is commonly selected to maximize the recovery ϑ_A, which implies maintaining the amount of protein breaking through the last column of the load zone below some specifications. Evaluating ϑ_A requires integrating the protein concentration profile recorded at the outlet of the load zone. For practical reasons, specifications on the maximum acceptable product loss are often given as a percentage of the feed concentration measured at the column outlet (x% with x typically between 1 and 10).

In the empirical approach described in this section, the feed time Δt^F of the multicolumn process is estimated from a breakthrough experiment performed with a single-column process having n^F columns connected in series. The feed time of the

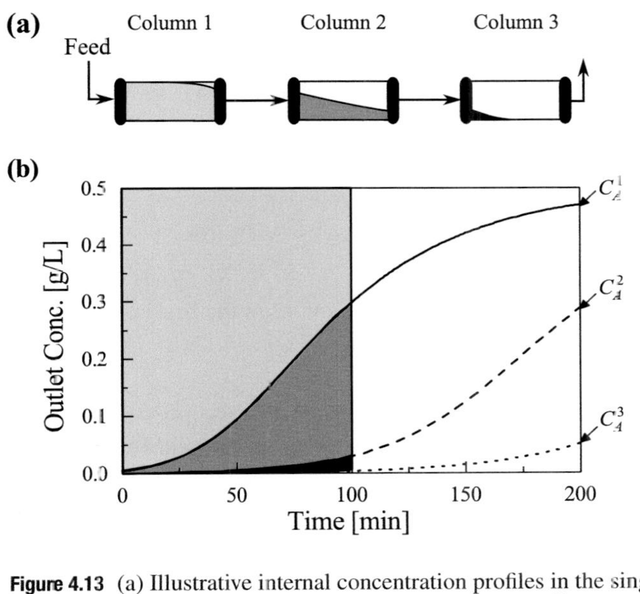

Figure 4.13 (a) Illustrative internal concentration profiles in the single-column experiment used for the empirical design of a multicolumn process with $n^F = 3$. (b) Concentration profiles measured at the outlet of the first (solid line), second (dashed line), and third (dotted line) column. The following values were considered: $V_{col} = 1$ mL, $Q^F = 1$ mL/min, $\varepsilon_e = 0.4$, $\varepsilon_i = 0.7$, $\tilde{K}_A = 6.4$ mL/mg, $\overline{N}_A = 85$ mg/mL, and $C_A^F = 0.5$ mg/mL. The numbering of the columns corresponds to the situation shown in panel (a). The amount of protein inside each column is proportional to the filled areas.

single-column process for which the selected threshold concentration value is reached at the outlet of the last column is called t_{BT}, where BT stands for breakthrough. If we denote C_A^{nF} the protein concentration at the outlet of the last column, we can relate ϑ_A and t_{BT} as follows:

$$\vartheta_A = 1 - \int_0^{t_{BT}} \frac{C_A^{nF}(t)}{C_A^F} \frac{dt}{t_{BT}} \tag{4.17}$$

The preceding equation shows that it is equivalent to set a minimum recovery ϑ_A, a maximum acceptable outlet protein concentration C_A^{nF}, or the breakthrough time t_{BT}. If the feed time of the single-column process having n^F columns connected in series is:

- larger than t_{BT}, some protein is lost from the last column of the load zone, thus leading to a recovery lower than the set value of ϑ_A.
- smaller than t_{BT}, the last column of the load zone is underused.

Following our example with $n^F = 3$, a representative concentration profile measured at the outlet of the last column is shown by the dotted line in Figure 4.13(b) for the parameter values given in the caption. As a comparison, the breakthrough curves at the outlet of the first and second columns are also shown with solid and dashed lines, respectively. In this example, the breakthrough time is selected to reach 99% recovery, leading to $t_{BT} = 100$ min according to Figure 4.13(b).

Once the breakthrough time is determined from a single-column experiment, the maximum amount of protein that can be injected during each switch time of the multicolumn process can be estimated. To do so, we start by computing the quantity of solute A present inside the first column of the n^F columns connected in series after a time t_{BT}. It is denoted q_A^1 and is estimated from the area comprised between the breakthrough curve exiting the first column (solid line in Figure 4.13) and the feed concentration (0.5 mg/mL in this example), that is:

$$q_A^1 = \int_0^{Q^F \times t_{BT}} (C_A^F - C_A^1)\mathrm{d}V \tag{4.18}$$

where C_A^1 is the protein concentration at the outlet of the first column. The result of this integral is proportional to the light gray area in Figure 4.13(b). In this example, the value of q_A^1 was estimated to 38.8 mg.

Similarly to q_A^1, the amount of solute inside the second column (q_A^2) is related to the area between the breakthrough curves of the first (solid line) and second (dashed line) columns in Figure 4.13(b). The amount of protein inside the last column (q_A^3) is estimated from the area comprised between the second (dashed line) and the third (dotted line) breakthrough curves. Finally, the amount of protein that has eluted from the last column is proportional to the area below the breakthrough curve of the third column.

Once q_A^1 has been computed, the feed time Δt^F of the multicolumn process can be estimated for a given flow rate and feed concentration as follows:

$$q_A^1 = \vartheta_A Q^F C_A^F \Delta t^F \tag{4.19}$$

Box 4.1 Empirical Selection of the Feed Time for a Multicolumn Capture Process for a Target Recovery

1. Perform one breakthrough experiment in a single-column process consisting of n^F columns connected in series at the selected flow rate Q^F, recording the protein concentration at the outlet of the first and last columns.
2. Select an appropriate breakthrough time, t_{BT}, based on the target protein recovery.
3. Compute q_A^1 for the selected t_{BT} from Equation 4.18.
4. Estimate the feed time Δt^F from Equation 4.19.

With $\vartheta_A = 99\%$ and the parameter values given in the caption of Figure 4.13, Δt^F is estimated to 78 min. Let us stress that the feed time Δt^F and the breakthrough time t_{BT} should not be confused. While t_{BT} is selected from a single-column experiment to obtain a given recovery (here equal to 100 min), Δt^F is a process variable of the multicolumn process (here equal to 78 min), which is necessarily smaller than t_{BT}.

Box 4.1 summarizes how to select the feed time Δt^F for a given multicolumn countercurrent process, i.e., given number of columns in the load zone, n^F, column volume, V_{col}, feed concentration, C_A^F, and feed flow rate, Q^F.

With the definition of q_A^1 in hand, we can also define the *capacity utilization*, CU, which corresponds to the amount of protein contained in the column over the maximum amount that can theoretically be stored:

$$CU = \frac{q_A^1}{\int_0^\infty \left(C_A^F - C_A^1\right)\mathrm{d}V} = \frac{\vartheta_A \Delta t^F}{t_0\left[1 + \phi_e \overline{C}_A^*/C_A^F\right]} \tag{4.20}$$

It is worth noting that the denominator in the right-hand side of Equation 4.20 is equal to the mean position of the breakthrough curve (see Equation 4.2). This represents the theoretical maximum feed time that could be used with an infinitely efficient column to obtain a recovery of 100%. Therefore, the capacity utilization can be seen as the ratio between the actual feed time and the maximal one defined as $t_0\left[1 + \phi_e \overline{C}_A^*/C_A^F\right]$.

The maximum amount of protein that can be loaded in the column can either be estimated from a complete breakthrough curve (see central term of Equation 4.20) or from the knowledge of the adsorption isotherm (see right-hand side of Equation 4.20). In the latter case, $\overline{C}_A^*$ is the lumped solid concentration in equilibrium with the feed solution of concentration C_A^F. In the considered example, the denominator of Equation 4.20 is estimated to 49.1 mg, which leads to a capacity utilization of the first column of about 80%. If one desires to improve the capacity utilization in the first column of the load zone while maintaining the same recovery, two options can be envisaged: increase the number of columns in the load zone (while keeping Q^F constant) or decrease the feed flow rate (while keeping n^F constant). The selection of n^F and Q^F does not only impact the capacity utilization but also the productivity, which is the topic of the next section.

Selection of the Number of Columns and Feed Flow Rate

We have seen in Section 4.2.3 that the fluid velocity strongly influences the shape of the breakthrough curves. Here, we propose to investigate experimentally the influence of the feed flow rate Q^F on the design of the load zone for a given column geometry (L_{col}, D_{col}). To do so, it is useful to realize that when dispersive effects are dominated by the kinetics of internal mass transfer, which is the case in most of the protein capture applications, the shape of the breakthrough curve depends almost exclusively on the zero-retention time $t_0 = \varepsilon_e V_{col}/Q^F$ and on the characteristic time for internal mass transfer $t_{i,A}$. This is illustrated with experimental results in Figure 4.14, where the breakthrough curves measured using two columns in series at 1 mL/min and only one column at 0.5 mL/min superimpose. As a comparison, the breakthrough curve measured with one column at a flow rate of 1 mL/min is also shown.

The observation that the same breakthrough curves are obtained as long as the ratio V_{col}/Q^F is kept constant has practical consequences. Indeed, it is equivalent to measure the breakthrough curve at the outlet of n^F columns connected in series or to measure the breakthrough curve at the outlet of only one column but using a reduced feed flow rate equal to Q^F/n^F. The productivity can then be computed from Equation 4.15 for each investigated condition, so as to select the feed flow rate delivering the highest productivity (or the best value of another objective function).

To sum up, the complete procedure to estimate Q^F and n^F with only one column goes as described in Box 4.2. The proposed methodology provides a comprehensive picture of the variations of the productivity as a function of the feed flow rate and of the number of columns in the load zone. It gives very satisfactory results as long as dispersive effects are mainly due to the kinetics of internal mass transfer [250].

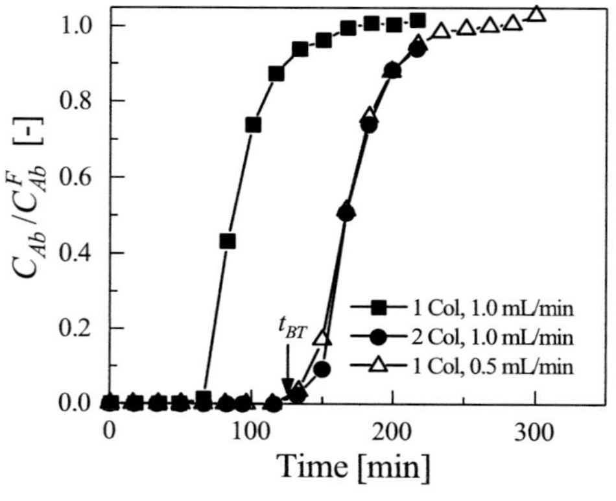

Figure 4.14 Breakthrough curves of a monoclonal antibody obtained on 5 × 100 mm columns packed with a Protein A chromatographic medium. The experiments were performed either with one column at the flow rates of 0.5 mL/min (open triangles) and 1 mL/min (filled squares) or with two columns connected in series at 1 mL/min (filled circles).

Box 4.2 Empirical Selection of the Feed Flow Rate and of the Number of Columns for a Multicolumn Capture Process for a Target Protein Recovery

1. Select a number of columns in the load zone n^F, which typically varies between two and five, and a feed flow rate Q^F.
2. Record experimentally the breakthrough curves with one column at the selected Q^F and at Q^F/n^F.
3. Estimate t_{BT} for a given protein loss from the breakthrough curve obtained at the flow rate Q^F/n^F, which emulates the breakthrough curve that would be obtained with n^F columns at the flow rate Q^F.
4. Compute the amount of solute q_A^1 stored in the first column after each feed time with Equation 4.18 from the breakthrough curve obtained at the flow rate Q^F.
5. Estimate the feed time Δt^F from Equation 4.19.
6. Repeat the procedure for different flow rates Q^F and for different numbers of columns n^F, and plot the productivity as a function of the total zero-retention time $n^T t_0$.
7. Select the flow rate and the number of columns that deliver the best performance.

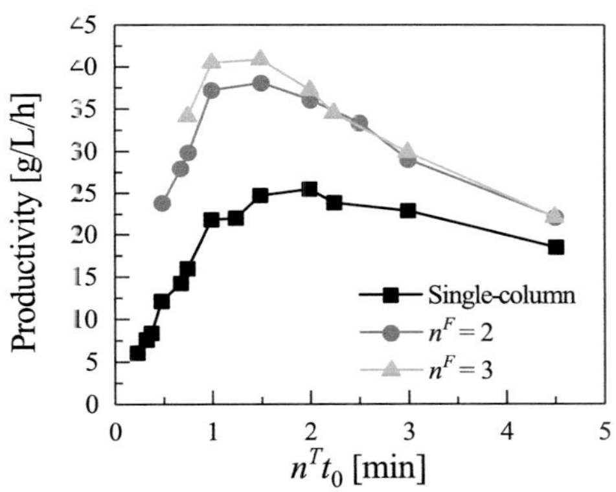

Figure 4.15 Productivity of a mAb capture step as a function of the total zero-retention time for a single-column process and multicolumn processes with two and three columns in the load zone. The columns were filled with a Protein A chromatographic medium and a pure mAb solution was used for the feed. These data were obtained following the procedure summarized in Box 4.2. Adapted from [250], copyright 2015, with permission from Elsevier.

Figure 4.15 shows productivity data obtained with a single-column process and using the procedure given in Box 4.2 to emulate multicolumn processes with two or three columns in the load zone [250]. In this example, the breakthrough time was selected as the largest value ensuring a recovery of 99%. The productivity curves show a clear maximum at a total zero-retention time around 1.5 min for the three considered configurations. It can be seen that at this optimal zero-retention time, the productivity can be greatly improved in the multicolumn processes as compared to the single-column process. However, the gain in productivity when increasing the number of columns in the load zone above two becomes less advantageous. Indeed, although the addition of a column in the load zone has a positive impact on the amount of protein that can be stored in the column, it also leads to an increase of the cycle time, which has a negative impact on the productivity.

Model-Based Design

The empirical design approach presented in the previous section allows selecting the number of columns and feed flow rate based on a few breakthrough experiments performed with a single-column process. However, with this method, the column dimensions of the multicolumn process are imposed equal to those of the single-column process, which are not necessarily the most suited ones. On the other hand, model-based approaches allow exploring a wide range of operating conditions to identify more appropriate values of the process variables. In this section, we discuss the use of model-based approaches to design chromatographic processes in terms of column dimensions and operating conditions (flow rates, injection time . . .).

We have seen in Chapter 2 that the mean position of the eluting fronts is dictated by the thermodynamics of the fluid–solid equilibrium, whereas their shape is controlled by the hydrodynamics and the kinetics of mass transfer. Besides the parameters describing the adsorption isotherm, three model parameters are usually sufficient to describe the shape of the fronts: the zero-retention time in one column t_0, the number of mixing cells J (or alternatively the column Péclet number Pe_c), and the characteristic time for internal mass transfer t_i. Most often in protein capture applications, the contribution of the hydrodynamics to the overall peak broadening is negligible as compared to that of the kinetics of internal mass transfer, so that the value of J has only little impact on the shape of the fronts. Consequently, the model predictions become independent from the column geometry (i.e., from D_{col}, L_{col}) for a given column volume and given fluid flow rate. The characteristic time for internal mass transfer is imposed by the choice of the chromatographic medium (e.g., ligand density, particle size, pore size) and of the loading conditions (e.g., solution pH, ionic strength). Once these have been selected, designing a chromatographic process for protein capture consists in properly determining the zero-retention time t_0 delivering the product at the target recovery and at the highest productivity.

Determination of the Number of Columns and Zero-Retention Time for $n^R = 1$

In the following, we investigate the influence of the zero-retention time t_0 on the protein recovery and productivity for multicolumn processes with different numbers of columns. To do so, we consider chromatographic processes with n^F columns connected in series in the load zone and one column in the recovery/regeneration zone. Such process is illustrated for $n^F = 2$ in Figure 4.12. We investigate values of n^F ranging from one to four and compare the results with those obtained with a single-column process. For the sake of simplicity, we consider fixed values of the process variables for the recovery/regeneration zone. In general, the recovery/regeneration protocols are developed for single-column processes and kept unchanged for multicolumn processes. Chromatographic media suppliers typically recommend using about 20 column volumes of different eluents to perform the recovery/regeneration operations (e.g., 5 CV for wash, 5 CV for recovery, 5 CV for cleaning, and 5 CV for equilibration). Considering a fluid flow rate of about 0.5 CV/min leads to $\Delta t^R \approx 40$ min for $n^R = 1$. The impact of the number of columns in the regeneration zone will be the topic of a separate discussion.

The first step of the proposed design approach consists in evaluating the feed time based on the equilibrium theory, which provides a reasonable estimate to start with. For favorable adsorption isotherms, the maximum feed time ensuring a recovery of 100% with an infinitely efficient column is given by $\Delta t^F = t_{r,A}$ estimated from Equation 4.2, which explicitly depends on t_0. The ratio between the actual feed time and the one predicted in the frame of the equilibrium theory for a solute A is denoted by:

$$\beta = \frac{\Delta t^F}{t_0\left(1 + \phi_e \bar{C}_A^* / C_A^F\right)} \tag{4.21}$$

The optimal value of β is expected to be close to 1: if β is significantly lower than 1, the amount of protein injected into the column is too low to ensure a decent productivity, whereas if β is significantly greater than 1, a low recovery value is obtained due to the important amount of product that breaks through the column during the load phase. Note that according to Equations 4.20 and 4.21, the parameters β, ϑ_A, and CU are related by $CU = \vartheta_A \times \beta$. In the following, we select $\beta = 1$ for illustrative purposes, which implies that the recovery and capacity utilization are equal. This choice of β allows a simple comparison of different processes, and choosing another value would not qualitatively change the conclusions drawn in this section. However, the proper estimation of this parameter for a specific process should be part of a rigorous optimization strategy.

In a second step, simulations are performed at various t_0 values using model parameters determined from prior experiments [251]. Here, the simulations were performed with a MC model using the LDF approximation, while the fluid–solid equilibrium was described by a Langmuir adsorption isotherm. The model parameters are reported in the caption of Figure 4.16. In particular, the recovery/regeneration time was kept constant equal to $\Delta t^R = 40$ min and the numerical evaluation of Equation 4.21 led to $\Delta t^F = 148 t_0$. The recovery as a function of the zero-retention time in the load zone ($n^F t_0$ for the multicolumn and t_0 for the single-column) is shown in Figure 4.16(a) for the different processes under investigation. It is seen that for each process, the recovery increases with the zero-retention time, which corresponds to longer columns and/or smaller fluid velocities. At long residence times, the contribution of the kinetics of mass transfer to the band broadening becomes less significant, thus resulting in sharper breakthrough curves. The observation that the recovery tends to 100% at large $n^F t_0$ values indicates that hydrodynamic dispersion plays a minor role in the present example. Importantly, it is seen in Figure 4.16(a) that for a given product recovery, a smaller total zero-retention time in the load zone is required with multicolumn processes (for $n^F > 1$) as compared to the single-column process. However, this reduction of the zero-retention time with the number of columns becomes less and less pronounced when $n^F \geq 2$.

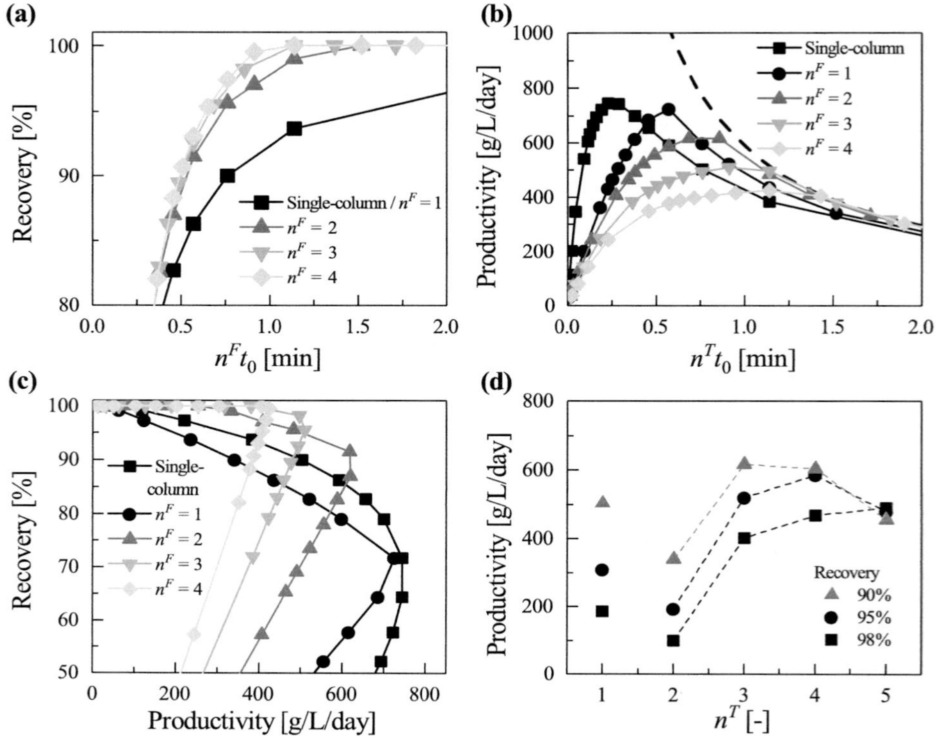

Figure 4.16 Performance parameters for protein capture processes with various column numbers: single-column process and multicolumn processes with n^F varying from 1 to 4 and with $n^R = 1$. (a) Protein recovery as a function of the zero-retention time in the load zone ($n^F t_0$ for the multicolumn process and t_0 for the single-column process). (b) Productivity as a function of the total zero-retention time. The dashed line corresponds to the maximum theoretical productivity. (c) Recovery as a function of the productivity. (d) Productivity as a function of the total number of columns for different target recovery values. The simulations were performed considering a MC model ($J = 50$, $\varepsilon_e = 0.38$), using the LDF approximation ($t_{i,A} = 31$ min) and assuming a Langmuir adsorption isotherm ($\overline{N} = 97$ g/L, $\tilde{K}_A = 11.3$ L/g, $\varepsilon_i = 0.52$). The following values of the process variables were used: $\Delta t^R = 40$ min, $\Delta t^F = 148 t_0$, $C_A^F = 1$ g/L. Adapted from [251], copyright 2017, with permission from Elsevier.

Figure 4.16(b) shows the productivity as a function of the total zero-retention time ($n^T t_0$) for the five processes under consideration. The productivity was computed from Equation 4.15 using the simulated values of ϑ_A. It can be seen that the productivity goes through a maximum and that two regions can therefore be observed:

- At low values of $n^T t_0$, the fluid velocity is too high and/or the columns are too short, so that shallow breakthrough curves are obtained. This leads to the loss of large quantities of product during the load phase and consequently to a low protein recovery, which in turn results in low productivity values (see Equation 4.15).

Moreover, at short zero-retention times, the productivity is likely to be limited by the regeneration of the column ($\Delta t = \Delta t^R$) so that some idle time is necessary at the end of the load phase, which further decreases the productivity of the multicolumn processes. In this case, it may be advantageous to have more than one column in the recovery/regeneration zone, as discussed in the next paragraph.

- At high values of $n^T t_0$, the fluid velocity is low and/or the columns are long, which results in low productivity values. In that region, sharp breakthrough curves are observed thus leading to a protein recovery close to 1. The productivity of the multicolumn systems is limited by the load phase ($\Delta t = \Delta t^F$) and becomes inversely proportional to the total zero-retention time. The maximum theoretical productivity (i.e., when $\vartheta_A = 1$) is given by $\varphi_A = \varepsilon_e C_A^F / t_0^T$ and is presented with the dashed line as a comparison. It can be noted that all the processes, irrespective of the number of columns, converge to this theoretical limit at high total zero-retention time.

At each t_0 value in Figure 4.16 corresponds a value of the feed time given by Equation 4.21. By analyzing the simulation results, it was found that, for a given process, the maximum productivity is obtained when the zero-retention time is selected such that the recovery/regeneration and the load phases are perfectly synchronized, i.e., $\Delta t^F = \Delta t^R$.

Interestingly, among all the processes under investigation, the highest productivity value is obtained for the single-column. However, at such high productivity, the recovery is quite low. This is shown in Figure 4.16(c), where the ϑ_A values presented in Figure 4.16(a) are plotted as a function of the φ_A values of Figure 4.16(b). It is seen that the high productivity of the single-column process is obtained at the expense of a 30% loss of product during the load phase. When the number of columns increases, the maximal achievable productivity decreases but the corresponding recovery increases. Similarly to results shown in Figure 4.16(b), two regions can be identified in Figure 4.16(c):

- At low recovery values, the protein recovery increases linearly with the productivity. In that region, the switch time is imposed by the recovery/regeneration time ($\Delta t = \Delta t^R$), so that $\varphi_A \propto \vartheta_A$ independent of t_0 according to Equation 4.15. Increasing the productivity requires increasing Δt^F toward Δt^R, which is achieved by decreasing the fluid velocity and/or increasing the column length, which in turn improves the protein recovery.
- At high recovery values, the recovery decreases when increasing the productivity. In that region, the switch time is imposed by the feed time ($\Delta t = \Delta t^F$). Increasing the productivity requires increasing the fluid velocity and/or decreasing the column length, which results in shallow breakthrough curves and thus in a lower recovery.

It appears in Figure 4.16(c) that, depending on the target protein recovery, different chromatographic processes should be selected in order to reach the maximum productivity. Figure 4.16(d) shows the productivity as a function of the number of columns for three target values of ϑ_A, which are in this example requested to be equal to *CU* because we selected $\beta = 1$. It is observed that, if a minimum of 98% recovery is targeted, the five-column system (i.e., $n^F = 4$) is the most productive. But if only 95% recovery is acceptable, the four-column system allows reaching the highest productivity. Finally, if a 90% recovery is sufficient, the most productive process only requires three columns. These results illustrate that the most productive chromatographic process is case-specific. Nevertheless, for each recovery value above 90%, it was possible to design a multicolumn chromatographic process that outperforms the single-column process in terms of productivity. Besides productivity, the final choice of the process also depends on other factors such as the eluent consumption, investment costs, process robustness and process complexity.

Impact of the Number of Columns in the Regeneration Zone

When a process is regeneration-controlled (i.e., $\Delta t = \Delta t^R$), it might be advantageous to increase the number of columns in the recovery/regeneration zone. Even though the total time necessary to perform all the steps of the recovery/regeneration protocol determined from single-column experiments is fixed, this time can be distributed over multiple columns in the multicolumn process. For instance, one may think of recovering the product in one column, while regenerating another column simultaneously. In a simplistic scenario, one can consider that the step time Δt^R can be reduced to $\Delta t^R/n^R$, where n^R is the number of columns in the recovery/regeneration zone. This might require to adapt the recovery/regeneration protocol so that the various operations performed in the different columns are well coordinated. For a given feed flow rate, it is expected that increasing n^R allows reducing the volume of one column V_{col}. Whether or not this will lead to an increase in productivity depends on the relative change in the total column volume given by $(n^R + n^F)V_{col}$.

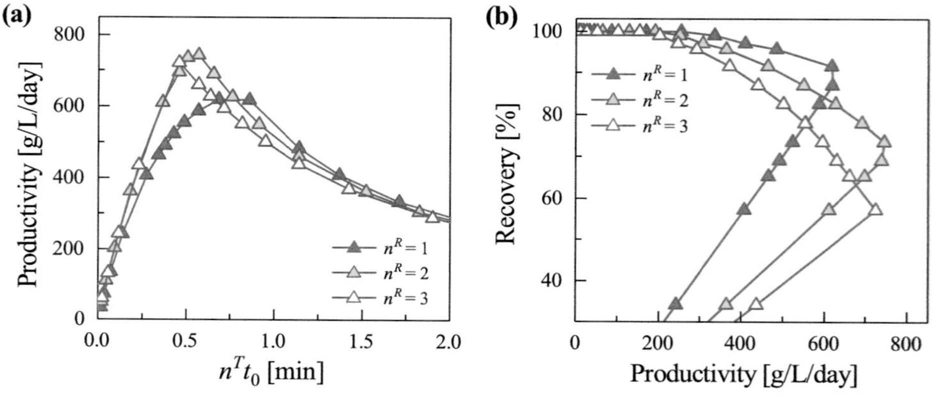

Figure 4.17 Impact of the number of columns in the recovery/regeneration zone for a multicolumn process with $n^F = 2$. (a) Productivity as a function of the total zero-retention time. (b) Recovery as a function of productivity. The model parameters are reported in the caption of Figure 4.16.

As an example, Figure 4.17(a) shows the productivity as a function of the total zero-retention time for a multicolumn process with $n^F = 2$, n^R ranging from 1 to 3 and Δt^R values of 40, 20, and 13 min, respectively. The data obtained with $n^R = 1$ (filled dark gray triangles) are identical to those shown in Figure 4.16(b). It is seen in Figure 4.17(a) that increasing the number of columns in the regeneration zone from 1 to 2 allows increasing the productivity from around 600 to 800 g/L/day by decreasing the zero-retention time t_0. However, increasing n^R above 2 does not lead to a further increase in productivity in the case under consideration.

The potential improvement in productivity with increasing n^R has to be balanced with the reduction in recovery due to shorter zero-retention times. This is shown in Figure 4.17(b), where the recovery is plotted as a function of productivity. It is seen that when increasing n^R from 1 to 2, the productivity increased from around 600 to 800 g/L/day but the recovery drops from approximately 90% to 70%. Note that these results were obtained from $\beta = 1$, and other values should be investigated to obtain the best compromise between productivity and recovery.

Once the process variables n^F, n^R, and t_0 have been selected, two types of design can be envisaged: either *(i)* the feed flow rate Q^F is imposed by the production requirement (e.g., in kg/day) and the column dimensions D_{col}, L_{col} are to be determined, or *(ii)* the column dimensions are given for an already existing unit and the feed flow rate is computed from $Q^F = \varepsilon_e V_{col}/t_0$. Case *(ii)* is straightforward and only requires to verify that the so-estimated value of Q^F leads to a pressure drop lower than the maximal acceptable value, otherwise Q^F has to be lowered to satisfy the pressure drop constraint. Case *(i)* requires some discussion, and more details are given in the following.

Determination of the Column Dimensions for a Given Feed Flow Rate

To illustrate the methodology, let us continue the example presented in Figure 4.16, where n^R was set to 1. We consider a target product recovery of 98% and compare the single-column and the five-column processes in more details for a daily production requirement of $\dot{q}_A = 1$ kg/day. According to Figure 4.16(c), a recovery of 98% corresponds to a productivity of 164 and 397 g/L/day in the two processes, respectively. The values of t_0 associated with the target recovery can be read on Figure 4.16(a), where a zero-retention time per column of 3 and 0.2 min are obtained for the single-column and the five-column processes, respectively. Recalling that in this example the feed time is equal to $148t_0$, the values of $\Delta t^F = 444$ min and $\Delta t^F = 29.6$ min are obtained, for the single-column and multicolumn process, respectively. Because the recovery/regeneration time was set to 40 min, it turns out that the five-column process is limited by the recovery/regeneration phase.

The feed flow rate can be computed from the production requirement $\dot{q}_A$ as $Q^F \Delta t^F C_A^F \vartheta_A = \dot{q}_A \Delta t$ for the multicolumn process and as $Q^F \Delta t^F C_A^F \vartheta_A = \dot{q}_A(\Delta t^F + \Delta t^R)$ for the single-column process. Finally, the volume of one column is estimated from $V_{col} = Q^F t_0/\varepsilon_e$. This leads to column volumes of 6.1 L and 0.5 L for the single-column and the five-column processes, respectively. Notably, the total volume of the

Table 4.1. Comparison between the single-column and five-column processes in terms of column dimensions for a target recovery of 98% and a daily production requirement of 1 kg/day. Adapted from [251], copyright 2017, with permission from Elsevier.

	Single-column	Five-column
φ_A [g/L/day]	164	397
t_0 [min]	3	0.2
Δt^F [min]	444	29.6
Q^F [L/min]	0.77	0.95
V_{col} [L]	6.1	0.5
L_{col} [cm]	21.7	5.6
D_{col} [cm]	18.9	10.7
L_{col}/D_{col} [-]	1.15	0.53

multicolumn process (2.5 L) is less than half the one of the single-column process. All the numerical values are summarized in Table 4.1.

Similar to the approach presented in Box 2.1, the column dimensions (L_{col}, D_{col}) are estimated from the maximal acceptable pressure drop once the column volume is determined. To do so, the Kozeny–Kármán expression (Equation 2.50) is rearranged as:

$$L_{col} = \sqrt{\frac{\Delta P^{max} t_0}{\delta_K \varepsilon_e}} \tag{4.22}$$

For our calculations, the maximal pressure drop is set to $\Delta P^{max} = 3$ bar and the parameter δ_K to 0.05 bar.min/cm^2. Recalling that $V_{col} = \pi D_{col}^2 L_{col}/4$, the column diameter is easily computed. The results of L_{col} and D_{col} for the single-column and five-column processes are summarized in Table 4.1. Note that the obtained value of L_{col} is independent of the production scale, whereas D_{col} is adapted to meet the production requirement. Consequently, the ratio L_{col}/D_{col} is expected to decrease when scaling up a chromatographic process. With rigid particles, this ratio has only little importance, but it becomes critical in the case of soft particles, as discussed in Section 2.4.3. The smaller the ratio L_{col}/D_{col}, the lower the fluid velocity that can be withstood by the chromatographic bed before it compresses. It is interesting to note in Table 4.1 that for a production requirement of 1 kg/day, the ratio L_{col}/D_{col} is equal to 1.15 for the single-column process and to 0.53 for the five-column process. Therefore, the column geometry is less favourable for multicolumn processes when using soft gels. In that case, one should be particularly careful when scaling up multicolumn processes (or consider rigid particles).

The complete shortcut design approach for single and multicolumn capture processes is summarized in Box 4.3.

Box 4.3 Shortcut Design of Single and Multicolumn Capture Processes

1. Select the value of β to be used in Equation 4.21. Values close to 1 usually provide a good starting point.
2. Perform simulations at different t_0 values until the target recovery is met for processes with various numbers of columns.
3. Select the process that provides the best value of the objective function, as, for example, the productivity computed with Equation 4.15 (Equation 4.16 for the single-column process) for the t_0 value providing the target recovery.
4. Determine the column length from the maximal pressure drop using Equation 4.22.
5. Determine the column diameter from the selected zero-retention time t_0 and production requirement.
6. This procedure can be repeated to optimize the value of β leading to the optimal value of the selected objective function.

4.3.4 Processes with a Number of Columns per Zone Changing during One Cycle

In the previous section, we discussed processes where the number of columns per zone, i.e., n^F and n^R, does not change during the cycle. Here, we introduce a new class of processes, where the cycle is divided in two subphases characterized by a different distribution of the total number of columns between the load and the recovery/regeneration phases. The CaptureSMB process [243] implements this concept in the case of a total number of two columns, and an example at the pilot scale has been reported for the capture of two mAbs [252]. Its schematic representation is shown in Figure 4.18, where it is seen that the load phase is divided into two subphases:

- A disconnected phase of duration $\Delta t^{F,DC}$, during which the process consists of two distinct zones with one column per zone: a load zone and a recovery/regeneration zone. The first column is loaded while the second one undergoes wash, protein recovery, cleaning, and regeneration during a time Δt^R.
- An interconnected phase of duration $\Delta t^{F,IC}$, during which the regeneration zone disappears and the two columns are connected in series in the load zone. By doing so, the freshly regenerated column captures the protein breaking through the first column.

At the end of this sequence, the position of the columns are switched and the disconnected/interconnected phases are repeated to complete a cycle. The average number of columns in the load zone during the total feed time $\Delta t^F = \Delta t^{F,DC} + \Delta t^{F,IC}$ is given by $n^F = (1 \times \Delta t^{F,DC} + 2 \times \Delta t^{F,IC})/(\Delta t^{F,DC} + \Delta t^{F,IC})$. Accordingly, depending upon the selected duration of the two subphases, this process can operate with an average number

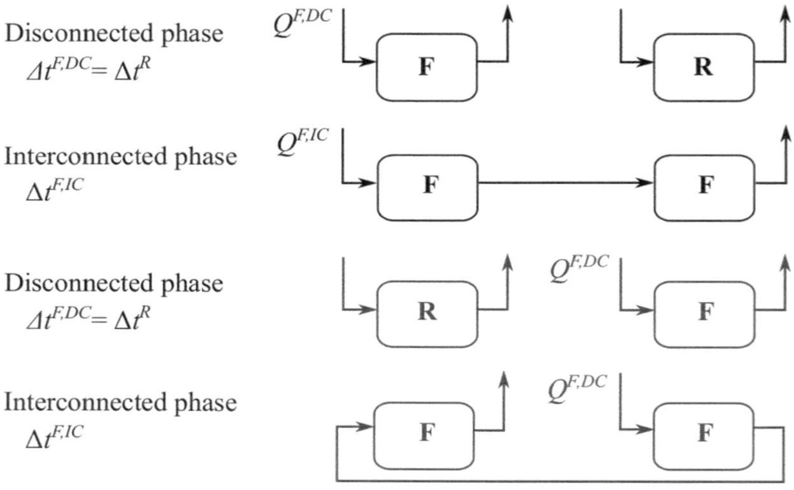

Figure 4.18 Schematic representation of the different phases of the CaptureSMB process. To ease the representation, it is considered that $\Delta t^{F,DC} = \Delta t^R$.

of columns in the load zone such that $1 \leq n^F < 2$, while in the regeneration zone the effective number of columns is such that $0 < n^R \leq 1$.

Empirical Design

In the following, we extend the approach described in Section 4.3.3 to the CaptureSMB process and illustrate it with a case study, where a mAb was captured from the clarified cell culture harvest with Protein A chromatography. Designing the load zone of a CaptureSMB process consists in selecting four process variables: two for the disconnected phase ($Q^{F,DC}$ and $\Delta t^{F,DC}$) and two for the interconnected phase ($Q^{F,IC}$ and $\Delta t^{F,IC}$). In this example, the feed flow rates were fixed so that only the feed times had to be determined. The feed flow rate during the interconnected phase $Q^{F,IC}$ was set to 1 mL/min, whereas the feed flow rate during the disconnected phase $Q^{F,DC}$ was set to 0.5 mL/min. Setting $Q^{F,DC} = Q^{F,IC}/2$ implies that the total zero-retention time in the load zone $n^F t_0$ is kept constant between the two subphases. By doing so, the CaptureSMB process behaves as if there were always two columns connected in series in the load zone operated at the feed flow rate $Q^{F,IC}$. This allows applying in a rather straightforward manner the empirical design approach described in Box 4.1.

The breakthrough curves measured at the outlet of the first and second column connected in series are shown in Figure 4.19. Here, a breakthrough time t_{BT} of 92 min was selected, which leads to a recovery close to 100%. The amount of protein inside the first column of the load zone at the end of the load phase (i.e., at the end of the interconnected and disconnected subphases) is computed from Equation 4.18. With $Q^{F,IC} = 1$ mL/min and $C_A^F = 1.2$ mg/mL, we obtain $q_A^1 = 81$ mg. This quantity is proportional to the light gray area in Figure 4.19. On the other hand, the dark gray area is proportional to the

Table 4.2. Design of the CaptureSMB process based on the breakthrough curves shown in Figure 4.19 [243].

	Position 1	Position 2
$\Delta t^{F,DC} = \Delta t^{R} = 61$ min	Recov/Regen at 1 mL/min	Load at 0.5 mL/min
$\Delta t^{F,IC} = 37$ min	Interconnected load at 1 mL/min	
$\Delta t^{F,DC} = \Delta t^{R} = 61$ min	Load at 0.5 mL/min	Recov/Regen at 1 mL/min
$\Delta t^{F,IC} = 37$ min	Interconnected load at 1 mL/min	

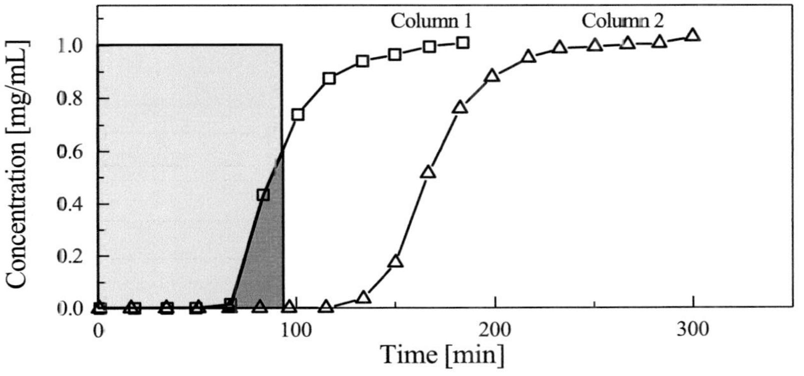

Figure 4.19 Breakthrough curves obtained at 1 mL/min measured at the outlet of the first and second column of the load zone during the interconnected phase. The experiment was performed with two identical columns (5 × 100 mm) packed with a Protein A chromatographic medium used to capture a mAb at 1.2 mg/mL from the clarified cell culture harvest. The light gray area is proportional to the amount of solute present in the first column at the end of the interconnected phase. The dark gray area is proportional to the amount captured in the second column of the load zone during the same time interval.

mass of protein in the second column, which was estimated to $q_A^2 = 28.9$ mg from the integral between the two breakthrough curves.

Then, one has to decide which fraction of the mass q_A^1 (or equivalently which fraction of the volume $q_A^1/C_A^F = 67.5$ mL) should be attributed to the two subphases. In this example, it has been decided to inject 45% of the feed volume during the disconnected phase and the rest during the interconnected phase. This leads to $\Delta t^{F,DC} = 61$ min and $\Delta t^{F,IC} = 37$ min. This fraction of q_A^1 attributed to each subphase is an optimization variable that should be carefully selected to get the best productivity out of this two-column process without losing product during the disconnected phase. This aspect is discussed in more details in Section 4.3.4. Finally, to avoid idle time during the disconnected phase, the flow rates for the recovery and regeneration operations are selected so that $\Delta t^{R} = \Delta t^{F,DC} = 61$ min. The recovery/regeneration protocol was determined based on single-column experiments. The obtained values of the process variables are summarized in Table 4.2.

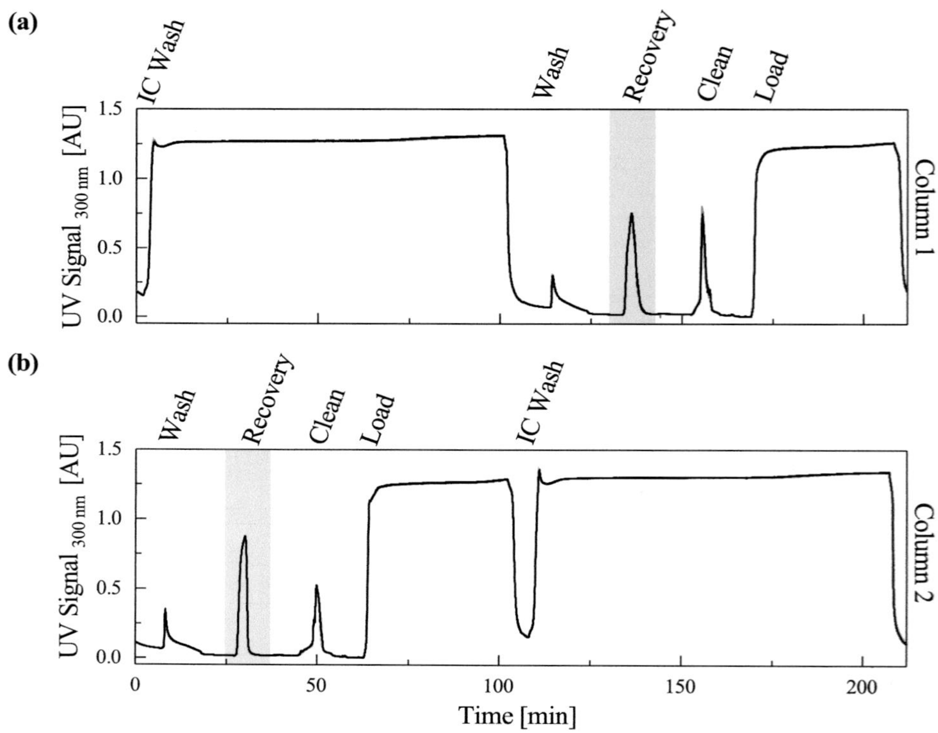

Figure 4.20 Experimental UV signal measured at the outlet of the two columns of the CaptureSMB process: (a) Column 1, (b) Column 2. The operating conditions are described in Table 4.2.

The chromatograms measured at the outlet of the two columns at steady state are shown in Figure 4.20, where the different operations are also indicated.

Overall, this approach provides a simple way to adapt a single-column process to the CaptureSMB process for given column dimensions. However, this process is most likely not optimal, meaning that a better design could lead to improved process performances.

Model-Based Design

In this section, we focus on the model-based design of the CaptureSMB process. We propose here a design approach in line with the methodology described in Section 4.3.3, but other approaches have been proposed in the literature [242, 253]. As compared to the processes described in Section 4.3.3, the total feed time $\Delta t^F = \Delta t^{F,IC} + \Delta t^{F,DC}$ is split into an interconnected and a disconnected phase. There is therefore an additional process variable, namely the fraction of the feed volume allocated to each subphase.

The model-based design approach described in Section 4.3.3 is applied to the CaptureSMB process for three different repartitions of the feed volume: 0%, 33% and 50% allocated to the interconnected phase. Just like in Section 4.3.4, the feed flow

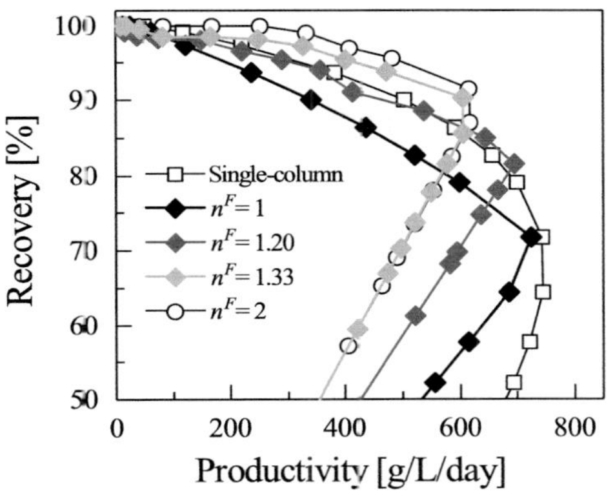

Figure 4.21 Protein recovery as a function of the productivity for two-column processes with non-integer numbers of columns in the load zone given by $n^F = (\Delta t^{F,DC} + 2 \times \Delta t^{F,IC})/\Delta t^F$. As a comparison, the results obtained with the single-column and the three-column ($n^F = 2$, $n^R = 1$) processes are also included. The model parameters are reported in the caption of Figure 4.16.

rates are selected such that $Q^{F,IC} = 2Q^{F,DC}$, which leads to 1, 1.2, and 1.33 columns in the load zone on average, respectively. The simulation results are shown in Figure 4.21 with filled symbols. As a comparison, the results obtained with the single-column and the three-column ($n^F = 2$, $n^R = 1$) processes are also reported with open symbols.

It is seen that using the two-column process without an interconnected phase (i.e., $n^F = 1$) presents no interest as it is always less performant than the single-column process in terms of recovery and productivity. Then, when the fraction of the feed volume allocated to the interconnected phase increases, it becomes possible to reach higher productivities than with the single-column process for comparable recovery values. Interestingly, when 50% of the feed volume is allocated to the interconnected phase (i.e., $n^F = 1.33$), the process performances are quite close to those of the three-column process (i.e., $n^F = 2$). Therefore, the CaptureSMB with its two columns outperforms the process with a constant number of column per zone (i.e., $n^F = n^R = 1$) for reasonable recovery values. It approaches very closely the three-column process with an additional column in the load zone (i.e., $n^F = 2$, $n^R = 1$). Nevertheless, in the upper region where the feed time is the limiting step ($\Delta t^F > \Delta t^R$), the recovery obtained for a given productivity is slightly below than that obtained with the three-column process. This is due to the absence of the second column in the load zone during the disconnected phase.

Overall these results show that processes with a number of columns per zone changing during a cycle allow a better use of the columns. Indeed, for a given total number of columns, a higher productivity can be reached for a target recovery. This was illustrated here for $n^T = 2$, but the method can be extended to similar processes with a larger total number of columns.

4.3.5 Process Optimization

The design approach described in Section 4.3.3 provides a good estimation of the column dimensions, feed flow rate, and feed time. However, there are many more process variables that can be adjusted, as, for instance, the recovery/regeneration protocols. The experimental optimization of a capture process with respect to the numerous relevant process variables is barely feasible in practice. In contrast, chromatographic models allow screening a large number of configurations, which helps to identify the conditions that optimize certain process performance parameters, for example, in terms of product recovery, buffer consumption, and productivity. In this section, we apply the basic concepts of process optimization presented in Section 2.6.2 to two illustrative examples.

Example 1

In the first example, a model-based optimization was applied to a mAb capture step performed with affinity chromatography using a three-column countercurrent process (BioSC) [254, 255]. A succinct presentation of the BioSC process is given in the following, while a detailed description is found in the original publication. Table 4.3 summarizes the different steps and the associated experimental conditions determined from a base case design performed on a single-column process. At the end of the sequence described in Table 4.3, the operations performed in each column are moved by one position to the right (i.e., $1 \mapsto 2$, $2 \mapsto 3$, and $3 \mapsto 1$). After another 9.4 min ($4 + 1.8 + 0.45 + 3.15$ min), the positions are changed again. At the end of this period (i.e., after a total of 3×9.4 min), the process is reinitialized to its starting configuration, which completes a cycle.

In the considered example, the objective function to be maximized was the productivity. The selected manipulated variables were the feed time, the interconnected and disconnected wash times, the time of the recovery phase as well as the number of columns and the column length. The constrained variables are the total pressure drop of the system, the purity, the product recovery and the column diameter. The different variables are summarized in Table 4.4.

The linear velocity was kept the same for all operations (i.e., load, wash, recovery, regeneration) and computed from the Kozeny–Kármán equation (see Equation 2.50) in order to reach a pressure drop of 2.5 bar. It follows that for each of the investigated

Table 4.3. Details of the sequence of operations in the BioSC process. The protein concentration in the feed was $C_{Ab}^{F} = 2.3$ mg/mL. The pH was set to 7.4 for the load, wash, and equilibration solutions, whereas it was set to 3 and 2 for the recovery and cleaning solutions, respectively. IC and DC stand for interconnected and disconnected.

Δt^j [min]	Position 1	Position 2	Position 3
4	IC load: 2 mL/min		recovery: 1 mL/min
1.8	IC load: 2 mL/min		cleaning: 2 mL/min
0.45	IC wash: 2 mL/min		equil: 1 mL/min
3.15	DC wash: 1 mL/min	DC load: 2 mL/min	equil: 1 mL/min

Table 4.4. Set of manipulated and constrained variables for the optimization of a multicolumn capture process in example 1.

Manipulated variables	Constrained variables
Δt^F	$\Delta P = 2.5$ bar
Δt^{ICwash}	$\vartheta_{Ab} = 0.95$
Δt^{DCwash}	$\chi_{Ab} = 0.98$
$\Delta t^{recovery}$	$D_{col} = 0.5$ cm
n^T	
L_{col}	

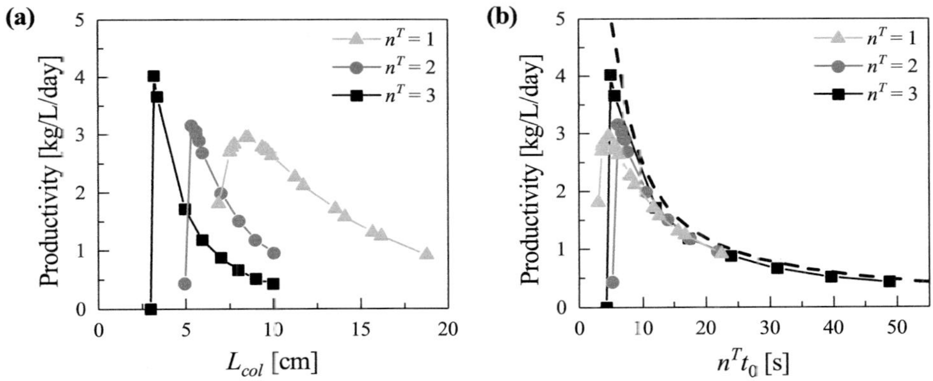

Figure 4.22 Optimal productivity as a function of the (a) column length and (b) total zero-retention time for different configurations: one-, two-, and three-column processes. The dashed line in panel (b) represents the maximal productivity $(\varepsilon_e C^F_{Ab})/t^T_0$ computed for $C^F_{Ab} = 1$ g/L and $\varepsilon_e = 0.3$. Experimental data reprinted from [255], copyright 2014, with permission from Elsevier.

configurations, the linear velocity is given by $7.5\ 10^{-3} \times n^T L_{col} u = 2.5$ where u is given in cm/min.

The optimization of the process performance requires testing many different configurations, i.e., different values of the manipulated variables to find the combinations of these variables delivering the highest productivity. Investigating systematically all the possible combinations of the manipulated variables is certainly not the most efficient approach to find out which configuration provides the best performance. For this kind of multi-variable optimization problems, evolutionary algorithms, among which genetic algorithms are certainly the most popular type, are often used. In such algorithms, the first combinations of variables are generated randomly to span the entire design space, while the subsequent iterations are created by combination of the best solutions of the previous ones. Genetic algorithms have been used to optimize the BioSC process under examination, and the results of the optimization are presented in Figure 4.22 for one-, two-, and three-column processes.

It is seen that the highest productivity is obtained with the three-column process for a column length of 3.2 cm. To this point corresponds a unique combination of the manipulated variables presented in Table 4.4, which can then be used to run the process. The data of Figure 4.22(a) are plotted as a function of the total zero-retention time $t_0^T = n^T t_0$ in Figure 4.22(b). It is observed that similar trends are observed as compared to those presented in Figure 4.16(b), even though many more manipulated variables have been investigated in this example. Notably, the productivity goes through a maximum and tends to the theoretical limit represented by the dashed line at high zero-retention times.

In this example, only the manipulated variables reported in Table 4.4 have been considered but many more could be part of the optimization procedure. Variables as the pH of the solution, the concentration of the feed solution, the different flow rates, etc. can be varied as long as their effect are accounted for in the chromatographic model.

Example 2

The second example deals with the optimization of the CaptureSMB process and its comparison with two other processes, the single-column process and a three-column process ($n^F = 2$, $n^R = 1$) [253]. As in the first example, the objective function is the productivity and the manipulated and constrained variables are summarized in Table 4.5. In particular, the minimum acceptable recovery was set to 95%, so that $CU \approx \beta$.

The results of the optimization are shown in Figure 4.23. For each process, the highest capacity utilization value is reported as a function of the productivity. It is seen that, for the case under investigation, the single-column process is slightly more productive than the multicolumn processes. However, the highest productivity in the single-column process is obtained at a very poor capacity utilization ($CU \approx 50\%$). This means that about half of the chromatographic bed is not involved in the protein capture. On the other hand, it appears that countercurrent multicolumn processes make a better use of the chromatographic medium for a given productivity. Interestingly, it is observed that the CaptureSMB behaves similarly as the three-column process, as already discussed in the context of Figure 4.21.

Table 4.5. Set of manipulated and constrained variables for the optimization of a multicolumn process for protein capture in example 2. For the CaptureSMB, both the interconnected and disconnected values of Δt^F and Q^F are manipulated variables.

Manipulated variables	Constrained variables
Δt^F	$Q^j \leq 1.7$ mL/min
Q^R	$\vartheta_{mAb} \geq 0.95$
Q^F	$\chi_{mAb} \geq 0.98$
	$\Delta t^R = n^R V_{col}/Q^R$

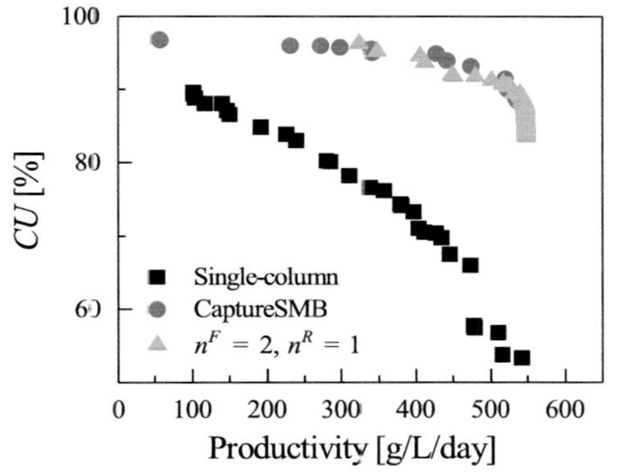

Figure 4.23 Pareto fronts of the single-column, the CaptureSMB and a three-column process showing the highest capacity utilization for a given productivity value. To each point in the graph corresponds a different combination of manipulated variables. Adapted from [253] by permission of John Wiley & Sons, Inc.

4.3.6 Process Control

Maintaining the constrained variables of multicolumn capture processes within specifications over time may be quite challenging. For example, the composition of the feed stream is likely to vary due to fluctuations in the bioreactor. In addition, the properties of the chromatographic media are subject to alterations caused by aging, fouling or leaching of the affinity ligands. Such disturbances may impact the process performance parameters and compromise the quality of the purified product. To reduce the influence of disturbances on the process, controllers can be implemented to take corrective actions on the manipulated variables and minimize deviations from the set point. Process control is a vast and complex field and the interested reader is referred to the specific literature for a detailed presentation of the topic [256–260]. Here, we confine ourselves to basic considerations and provide illustrative examples of process control in the context of protein capture.

To implement control strategies, it is necessary to quantify the response of the process to perturbations. This response can either be assessed directly from measurements performed at the process outlet (*feedback control*) or simulated with a suitable model feeded with measurements performed at the process inlet (*feedforward control*). In the case of feedback control, one can, for example, measure the protein concentration at the column outlet with UV, infrared or Raman spectroscopy or analytical chromatography. In the case of feedforward control, chromatographic models similar to those described throughout this book can be used to simulate the breakthrough curves based on the measured concentration at the column inlet. In both cases, the detection method should be selected so as to provide enough information in a sufficiently short time to rapidly correct the deviations from the set point. Once the process response has been quantified (measured or simulated), one has to decide which manipulated variables to act on. For multicolumn processes, these are typically the flow rates and the duration of the various phases.

A preliminary implementation of feedforward control was proposed in the literature for mAb capture by affinity chromatography [261], where the cycle time of the capture step was adjusted (manually) based on the at line measurement of the inlet concentration so as to maintain the process recovery constant. In the following, we present two examples of feedback control strategies based on the measurement of UV signals.

ΔUV Control

The first example relies on the comparison between the UV signal of the feed stream with the UV signal measured at the outlet of the first column of the load zone [262]. A schematic representation of these measurements is shown in Figure 4.24(a). The dashed line represents the UV signal of the feed stream (denoted UV^F), which arises from the presence of both the target protein and the impurities. On the other hand, the solid line represents the UV signal at the outlet of the first column of the load zone (denoted UV), which corresponds to the elution of the impurities (first plateau) and then of both impurities and target protein (second plateau). The UV signal of the first plateau defines the "baseline," i.e., the signal of the impurities only. Once this baseline is detected, the ratio $\Delta UV/\Delta UV^F = (UV - \text{baseline})/(UV^F - \text{baseline})$ is computed as a function of time.

As soon as the measured $\Delta UV/\Delta UV^F$ reaches a set value (e.g., 70%) the controller adjusts the load conditions, for instance, by reducing the feed flow rate or increasing the step time Δt^F. A proper threshold of the $\Delta UV/\Delta UV^F$ value should be selected to operate the process with a satisfactory recovery, while achieving a sufficiently high productivity.

In Figure 4.24(b) are shown the inlet (dashed line) and outlet (solid lines) UV signals of a three-column process (called 3C-PCC, with $n^F = 2$ and $n^R = 1$) equipped with a ΔUV controller. In this experiment, the inlet concentration of an antibody solution

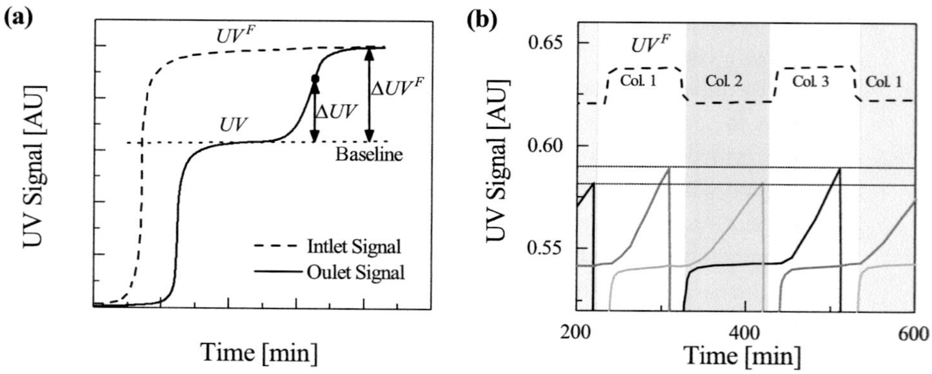

Figure 4.24 (a) Illustration of the ΔUV control strategy. (b) UV signal at the inlet (dashed) and outlet of the first column of the load zone (solid line) of a 3C-PCC process obtained when periodically switching the inlet protein concentration from 1.5 mg/mL (gray areas) to 1.8 mg/mL (white areas). Adapted from [262] © General Electric Company – Reproduced by permission of the owner.

was varied periodically to mimic fluctuations in the composition of the feed stream. It is seen that when the feed concentration is low (gray zone), the peak maximum of the outlet UV signal is lower than when the feed concentration is high. This indicates that the controller adjusted the volume of feed mixture loaded in the unit to reach the set value of $\Delta UV/\Delta UV^F$. In the end, similar quantities of protein were recovered at the process outlet with both feed concentrations.

Peak Area

Besides their main role, which is to cope with disturbances in the operating conditions (as feed composition and chromatographic medium aging), controllers can also be useful to optimize process performance parameters by adjusting the manipulated variables until reaching an optimal value of the objective function. Such strategy has been applied at the laboratory scale to optimize a single-column and a CaptureSMB process at fixed flow rate from a suboptimal starting point [263]. The CaptureSMB process was constituted of two columns identical to that of the single-column process. The principle was the following:

1. The process is initialized by loading a small amount of protein, i.e., a sufficiently small Δt^F is selected.
2. The UV signal is recorded at the column outlet and the area below the peak measured during the recovery phase is integrated. This area is proportional to the amount loaded into the column provided that no protein is lost during the load phase.
3. The time Δt^F is increased step-by-step, keeping the flow rates constant. As long as the amount of protein recovered is proportional to the loaded amount, the time Δt^F is increased.

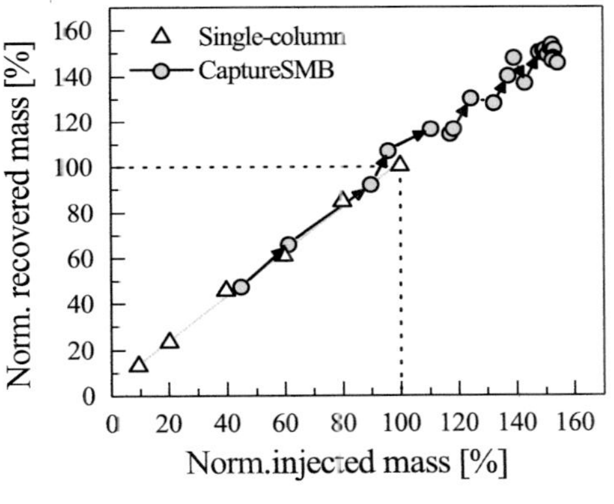

Figure 4.25 Recovered mass as a function of the injected mass of protein for a single-column and CaptureSMB processes equipped with a peak area controller. The injected and recovered masses have been normalized by the maximum mass that can be injected in the single-column process ensuring a recovery close to 100%. Successive actions of the controller can be seen following the arrows. Adapted from [264].

4. At some point, the protein starts breaking through the load zone. At this moment, the recovered amount is not proportional any longer to the amount injected. The controller then decreases the load time until no loss of protein is observed.
5. The controller continues to adjust the value of Δt^F until it finds a maximal value that ensures no protein loss (or rather less than a specification value).

The successive steps of the controller are illustrated in Figure 4.25, where the recovered mass of protein is plotted as a function of the corresponding injected mass. The values for both processes have been normalized by the maximum mass of protein that has been injected in the single-column process leading to a satisfactory recovery (close to 100%). It is seen that for both the single-column and CaptureSMB processes, the controller could considerably increase the injected mass of protein, which is related to an increase in productivity, when starting from a poorly selected operating point.

4.4 Conclusion

Protein capture usually involves a series of operations: load of the feed mixture, wash of the weakly interacting impurities, product recovery, and regeneration of the chromatographic medium. Operating conditions are selected in such a way that the target product is strongly interacting with the chromatographic medium during the load phase and almost not interacting during the recovery phase. In other words, the Henry coefficient changes by several orders of magnitude, from around 10^3 down to 1 between these two phases. The very strong protein adsorption during the load phase results in considerable internal mass transfer limitations, with characteristic times reaching several tens of minutes. In such cases, the contribution of the hydrodynamic dispersion to the overall broadening effect is negligible, so that the shape of the breakthrough curves is mainly imposed by the ratio between the characteristic time for internal mass transfer and the zero-retention time. Under recovery conditions, the adsorption strength is significantly reduced, which implies that the characteristic time for internal mass transfer is substantially shorter than during the load phase.

Most of the multicolumn processes for protein capture currently available are composed of two zones, namely a load zone and a recovery/regeneration zone. It is possible to design such units in terms of feed time, feed flow rate, and number of columns from few breakthrough experiments performed with a single-column process. This provides a quick method to obtain a first operating point to start with. However, a more detailed design (including column geometry) necessitates the use of chromatographic models. By performing model simulations, it was found that for acceptable recovery values, multicolumn processes allow reducing the total column volume for given specifications as compared to the single-column process. In addition, we have seen that the most productive multicolumn process strongly depends on the target specifications, especially the target protein recovery.

5 Countercurrent Chromatography for the Polishing Steps

5.1 Introduction

After the capture step, most of the impurities have been removed, except some process-related impurities from the previous steps (such as residual host cell proteins from the upstream and potential leachable affinity ligands), as well as product-related impurities (for example, due to posttranslational modifications, fragmentation, and aggregation). The role of polishing steps is to eliminate both types of impurities and to deliver the product at the target purity requirement.

We have seen in Chapter 4 that most of the situations encountered during the capture step share a number of common features among which are: very high selectivity (for a large part thanks to the use of affinity chromatographic media), binary separation product/impurities, noncompetitive adsorption, and successive step gradients for the recovery/regeneration phases. As a result, there are a lot of similarities between all the multicolumn processes dedicated to the capture step. In contrast, a wider diversity is observed among the polishing steps, varying in terms of chromatographic medium, nature and number of impurities, elution and collection strategies, etc. This is reflected in the broad variety of multicolumn processes designed for these steps.

Rather than drawing up an exhaustive list of conceivable scenarios and processes, we intend in this chapter to identify general concepts and to rationalize the design of countercurrent chromatographic processes for the polishing steps. In a first part, we analyze the relevant underlying physical phenomena with regard to the thermodynamics of fluid–solid equilibrium, hydrodynamics, and kinetics of mass transfer. In this context, we deepen the general concepts exposed in Chapter 2 to cope with multicomponent mixtures and competitive effects. In a second part, we elaborate on countercurrent chromatographic processes for polishing applications, starting from the classical SMB introduced in Chapter 3 and progressively moving toward processes that have been developed over the years to meet the needs of the biopharmaceutical industry.

5.2 Process Operations and Associated Physical Phenomena

5.2.1 Chromatographic Modes

Depending on the difficulty of the separation, single-column chromatographic processes are operated either in frontal or in elution mode. As described in Section 2.3.1, these two modes differ in terms of the volume of feed mixture injected into the column. While the injected volume is relatively small as compared to the retention volumes of the solutes in the case of elution chromatography, it is comparatively larger in the case of frontal chromatography.

Separation in Frontal Mode

Frontal chromatography is preferentially used when a large difference in terms of retention behavior is observed between the protein of interest and the impurities to be removed. This mode was already discussed in Chapter 4 in the context of protein capture processes, where the protein strongly interacts with the chromatographic medium, while the impurities flow through the column without being retained. During the polishing steps, the opposite situation is often encountered: the impurities strongly interact with the chromatographic medium, while the protein flows through the column without being retained. This type of frontal mode is often referred to as *flow-through chromatography* and is commonly well suited to the removal of process-related impurities, which significantly differ from the protein of interest, for example in terms of size, charge and hydrophobicity. Most of the downstream processes for mAb purification include at least one IEC step operated in frontal mode to remove HCP, DNA and leached Protein A ligands [265].

For illustrative purposes, Figure 5.1 shows a typical chromatogram obtained during the separation of a monoclonal antibody from HCP in frontal mode using an anion exchange chromatographic medium [266]. In this example, the composition of the eluent, in terms of solution pH and NaCl concentration, was selected so that the mAb does not interact with the chromatographic medium, thus resulting in a Henry coefficient value equal to 0. The sudden increase in the UV signal observed in Figure 5.1 soon after

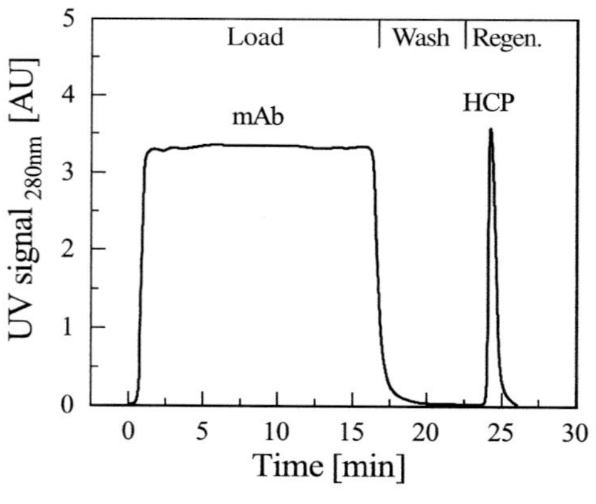

Figure 5.1 Chromatogram obtained in frontal mode for the purification of a mAb with IEC. The mAb is recovered in the flow-through fraction and during the wash step, while the HCP are strongly retained and elute only when the chromatographic medium is regenerated by changing the pH and the salt concentration of the eluent. Adapted from [266] by permission of John Wiley & Sons, Inc.

the injection indicates the breakthrough of the antibody. The elution goes on until around 15 min, after which the column is washed with the same eluent to recover the non-adsorbed mAb, thus leading to a decrease of the UV signal. Under the selected elution conditions, the HCP are negatively charged, which results in strong attractive interactions with the chromatographic medium. The HCP are only recovered after regenerating the column by changing the pH and the salt concentration of the eluent, as seen from the sharp peak at around 25 min in Figure 5.1.

Separation in Elution Mode

Elution chromatography is usually chosen for challenging separations characterized by a low selectivity. Such separations are, for example, encountered during the removal of product-related impurities, which may be very similar to the target protein, as charge variants or misconjugated proteins. Typically, the target product elutes between slightly less retained and slightly more retained impurities. This type of separation, involving impurities eluting on both sides of the product, is often referred to as *central-cut separation*.

Figure 5.2 presents an example of central-cut separation, where the product of interest, insulin-like growth factor (IGF), is separated from closely related impurities [267]. These impurities include misfolded proteins, aggregates, a variant of IGF containing a methionyl sulfoxide at the position 59 ($met^{59}O^{-}$) and a variant of IGF missing the N-terminal glycine and proline ($desGly^{1}Pro^{2}$). Figure 5.2(a) shows the preparative scale chromatogram obtained with a reversed-phase chromatographic medium using an aqueous solution of hexylene glycol as eluent. The collection window in which the IGF is

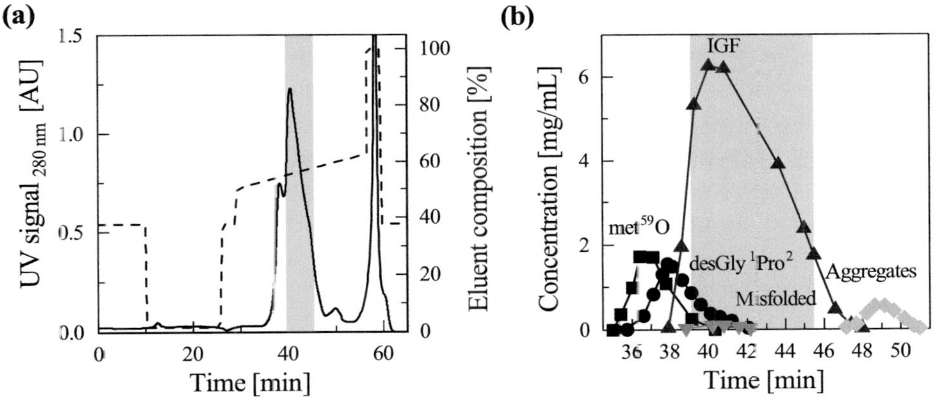

Figure 5.2 (a) Preparative scale purification of insulin-like growth factor on 6 × 25 cm reversed-phase C4 using a 100 mM K_2HPO_4 aqueous solution at pH 7 as eluent and hexylene glycol as modifier. The elution is performed by applying a linear gradient of hexylene glycol using a solution at 30% volume. The collection window is depicted in gray. (b) Results of the analytics form the VYDAC® assay performed on various collected fractions. Adapted from [267], copyright 1999, with permission from Elsevier.

collected with less than 1% of the most toxic impurity, i.e., met^{59}O^{-}, at the highest possible recovery is shaded on the chromatogram. The result of the analytics performed on collected fractions is reported in Figure 5.2(b), where it is seen that a nonnegligible loss of product is necessary to obtain the target purity. This is due to a quite low resolution between the target product and its closest impurities, in particular the misfolded protein and desGly^{1}Pro^2. A higher resolution could be obtained by reducing the slope of the modifier gradient, which would, however, lengthen the separation.

The presence of multiple impurities interacting at the surface of the chromatographic medium is expected to affect the retention behavior of the target protein. In the following, we deepen the concepts related to the modeling of protein chromatography introduced in Chapter 2 so as to better describe complex situations involving multicomponent systems.

5.2.2 Thermodynamics of the Fluid-Solid Equilibrium

Competitive Adsorption Isotherms

Modeling the thermodynamics of multicomponent solutions requires understanding the competitive effects between the various solutes at the surface of the chromatographic medium. In the absence of competition, for instance under linear adsorption conditions or if the solutes interact with different adsorption sites, modeling an n-component chromatographic system is nothing more than modeling n single-solute systems independently. However, difficulties arise when the presence of one species influences the behavior of the others. Although the competition of solutes for adsorption sites is certainly the most prominent manifestation of multicomponent phenomena, other effects may include a decrease in solubility, interactions of solutes with already bound species, as well as complexation and precipitation. In the following, we focus on competitive effects at the surface of the chromatographic medium.

In all generalities, a competitive adsorption isotherm links the concentration of a solute A bound to the chromatographic medium to the concentration of all the solutes present in solution. The use of multicomponent adsorption isotherms requires substantial experimental work to identify the appropriate model and to estimate the corresponding parameters. Indeed, it is necessary to perform experiments at different solute concentrations for various compositions of the multicomponent mixture. Langmuir adsorption isotherms (see Equation 2.21) are quite popular in protein chromatography due to their relative simplicity and their capability to represent many real systems, and are thus the only type of adsorption isotherms considered in this chapter. For other types of adsorption isotherms, such as the anti-Langmuir, Toth, Freundlich, and Moreau adsorption isotherms, the reader is referred to the following books [21, 158–161].

As an example, Figure 5.3 shows the results of batch uptake experiments obtained with a mixture containing γ-globulin (*glo*) and lysozyme (*lys*) in different proportions in contact with an ion exchange chromatographic medium [268]. The batch uptake method was described in Section 2.5 for a single-solute system and is extended here to a binary system. In the case under consideration, the experiments were performed at a total protein concentration of 8 g/L at different initial compositions (given in weight %

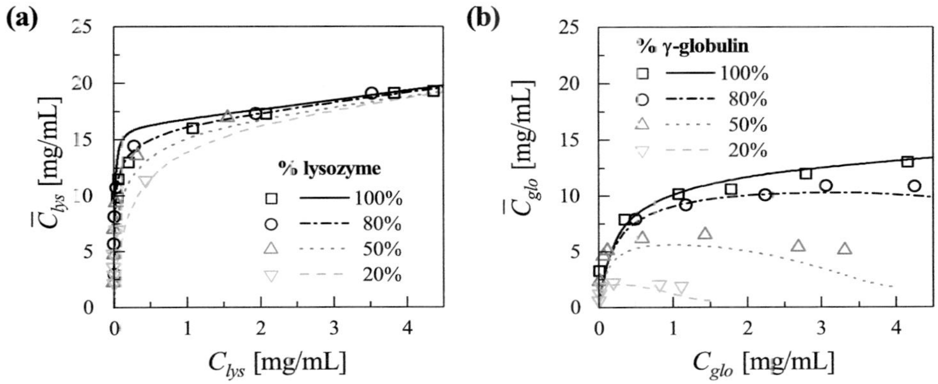

Figure 5.3 Lumped protein concentration in the particles as a function of the concentration in the surrounding liquid for (a) lysozyme and (b) γ-globulin. The batch uptake experiments were performed with an ion exchange chromatographic medium in a 100 mM sodium acetate buffer (pH 5) at a total initial protein concentration of 8 g/L and at different initial compositions, given in weight % in the legend. The adsorption isotherms of the pure solutes are also reported. Experimental data and simulations are represented by symbols and lines, respectively. The simulations were performed with a competitive binary Langmuir adsorption isotherm (Equation 2.21) using the following parameter values: $\tilde{K}_{glo} = 4$ L/g, $\overline{\overline{N}}_{glo} = 78$ g/L, $\tilde{K}_{lys} = 115$ L/g, $\overline{\overline{N}}_{lys} = 101$ g/L, $\varepsilon_i = 0.84$. Experimental data taken from [268] by permission of Taylor & Francis Ltd.

in the caption of Figure 5.3). The following mass balance equations were used to estimate $\overline{C}_{glo}$ and $\overline{C}_{lys}$ from the values of C_{glo} and C_{lys} measured at equilibrium:

$$\begin{aligned} VC^F_{glo} &= VC_{glo} + \overline{V}\overline{C}_{glo} \\ VC^F_{lys} &= VC_{lys} + \overline{V}\overline{C}_{lys} \end{aligned} \tag{5.1}$$

where C^F_{glo} and C^F_{lys} are the concentrations of γ-globulin and lysozyme initially present into the liquid phase. For each initial protein mixture, the volume of chromatographic medium $\overline{V}$ was kept constant to 0.25 mL, and the volume of liquid V was varied. The results are plotted in terms of $\overline{C}_{lys}$ versus C_{lys} and $\overline{C}_{glo}$ versus C_{glo} in Figures 5.3(a) and (b), respectively.

It can be seen that the binding behavior of the two proteins is very different. While the amount of bound lysozyme is almost unaffected when varying the protein ratio, the concentration of bound γ-globulin drops down when increasing the amount of lysozyme in solution. In a sense, γ-globulin, which is the least retained solute of the two, is losing the competition and is desorbed by lysozyme. One should pay attention that in the representation of Figure 5.3, the multicomponent experimental data are not adsorption isotherms strictly speaking because not only the concentration of the considered protein increases along the x-axis, but also the concentration of the second protein. This explains why some of the curves go through a maximum. If experiments were performed by varying only the concentration of one protein keeping the other constant, monotonic curves would be obtained, i.e., a given concentration in the particles would be associated with only one concentration in the surrounding liquid at thermodynamic equilibrium.

The experimental data were then fitted with a lumped binary Langmuir adsorption isotherm (see Equation 2.21) by solving the following set of equations:

$$\overline{C}_{glo} = \varepsilon_i C_{glo} + (1-\varepsilon_i)\frac{\tilde{K}_{glo}\overline{\overline{N}}_{glo}C_{glo}}{1+\tilde{K}_{glo}C_{glo}+\tilde{K}_{lys}C_{lys}}$$
$$\overline{C}_{lys} = \varepsilon_i C_{lys} + (1-\varepsilon_i)\frac{\tilde{K}_{lys}\overline{\overline{N}}_{lys}C_{lys}}{1+\tilde{K}_{glo}C_{glo}+\tilde{K}_{lys}C_{lys}} \qquad (5.2)$$

The first term in the preceding expressions corresponds to the concentration of protein in the liquid contained in the pores of the particles, while the second term corresponds to the concentration bound to the solid. The regression of the model parameters on the experimental data gave $\tilde{K}_{glo} = 4$ L/g, $\overline{\overline{N}}_{glo} = 78$ g/L, $\tilde{K}_{lys} = 115$ L/g, and $\overline{\overline{N}}_{lys} = 101$ g/L. This leads to lumped Henry coefficient values of $\overline{K}_{glo} = 50$ and $\overline{K}_{lys} = 193$, which correspond to the slopes of the adsorption isotherms of the pure components at infinitely low concentration. The large value of $\overline{K}_{lys}$ as compared to $\overline{K}_{glo}$ confirms the experimental observation that lysozyme displaces γ-globulin.

It is seen in Figure 5.3 that, although quite simple, the Langmuir adsorption model enables obtaining a reasonable description of the experimental data. In general, Langmuir adsorption isotherms provide a good description of experimental results obtained with ion exchange media provided that the modifier molar concentration is in excess as compared to the other solutes, which is the case in this study. A modification of this model accounting for the impact of the size difference between the two solutes on their adsorption behavior is also proposed in the original article [268].

In the absence of competitive experimental data, the ideal adsorption solution (IAS) theory allows computing the multicomponent adsorption isotherm based on the adsorption isotherms measured separately with the pure solutes [269]. The IAS was initially developed for gases [270] and then extended to dilute liquid solutions under the assumption that solute–solute interactions in both the liquid and adsorbed phases can be neglected [271]. Extensions to nonideal solutions encountered at higher concentrations require the introduction of activity coefficients to properly describe the thermodynamics of the adsorbed phase. This provides the basis of the real adsorbed solution (RAS) theory. Although very attractive from a conceptual point of view, the difficulty to predict activity coefficients makes experimental multicomponent equilibrium data necessary. For this reason, the RAS theory is seldom used for liquid chromatography, and in particular for protein applications.

Alternatively, some researchers derived colloidal models to compute adsorption isotherms based on the protein amino-acid sequence [272, 273]. The idea is to calculate electrostatic (attractive and repulsive) and Van der Walls forces between one protein in solution and the surface of the chromatographic medium as well as between two already adsorbed proteins. This allows estimating the concentration of proteins located nearby the surface of the chromatographic medium. Although the interaction potentials could in principle be computed a priori from the amino-acid sequence, the chemistry of the chromatographic medium and the composition of the eluent, some fitting parameters are in practice still necessary to properly describe the experimental chromatograms. This approach has been used with some success to model the elution behavior of mAbs on ion exchange media in the presence of sodium chloride.

Column Chromatography

The competition for the adsorption sites in multicomponent systems has a strong effect on the shape of the chromatograms. In the following, we consider first chromatograms obtained in frontal mode, and then in elution mode. Continuing the example of the binary system lysozyme/γ-globulin considered earlier, breakthrough curves obtained with the pure proteins and with a 50% mixture are shown in Figure 5.4(a) and (b), respectively. It can be seen that the breakthrough curves of lysozyme, which is the most strongly adsorbed species, are relatively similar in the presence and in the absence of γ-globulin. On the contrary, competitive effects considerably modify the breakthrough curves of γ-globulin as compared to the single-solute case. Lysozyme significantly displaces γ-globulin, which therefore breaks through the column at much shorter times compared to the case where lysozyme is absent and reaches a concentration higher than the feed concentration. Simulations using the previously determined parameters are also plotted as solid lines in Figure 5.4, showing satisfactory agreement with the experimental data.

Let us now consider the case of antibody charge variants to illustrate competitive effects in elution chromatography. Figure 5.5(a) shows an analytical chromatogram of a monoclonal antibody obtained by weak cation exchange chromatography [274]. It can be seen that the mAb solution under investigation contains a main variant, corresponding to about 70% of the total mass, and two groups of charge variants: one eluting before

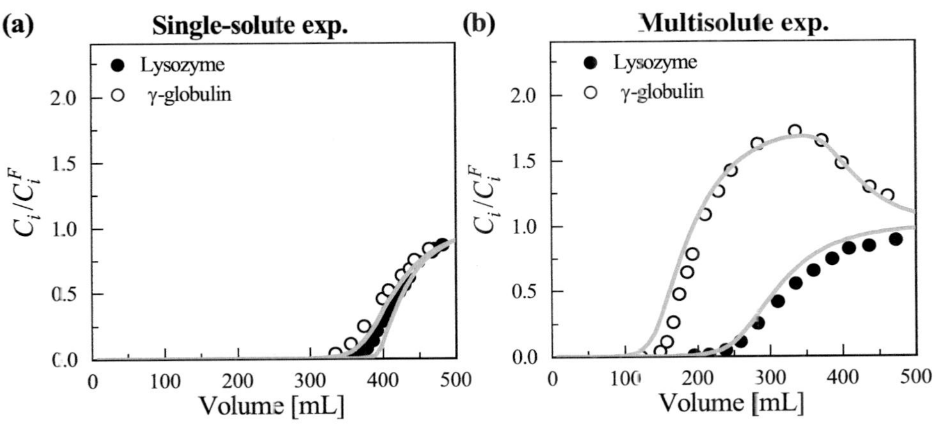

Figure 5.4 Frontal analysis of γ-globulin and lysozyme obtained for (a) the single-components and (b) considering an equimassic binary mixture. Experimental data taken from [268] by permission of Taylor & Francis Ltd. For the simulations, the following parameters have been used together with those reported in the caption of Figure 5.3: $\varepsilon_e = 0.33$, $C_A^F = C_B^F = 1$ g/L, $t_{i,glo} = 60$ min, and $t_{i,glo} = 53$ min.

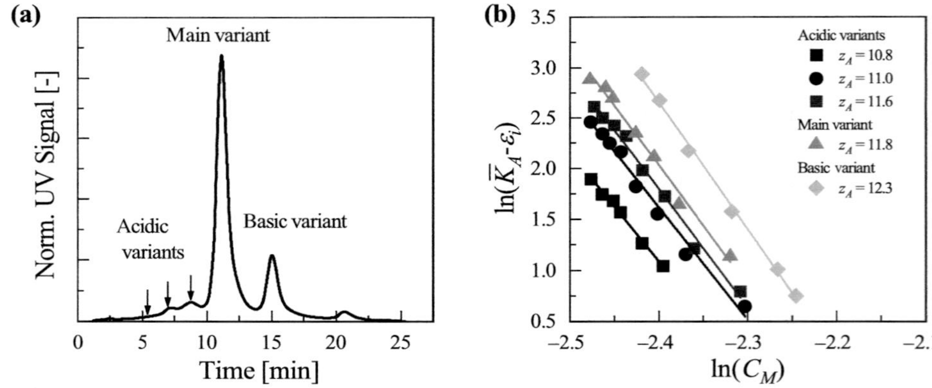

Figure 5.5 (a) Analytical chromatogram of monoclonal antibody charge variants separated by weak cation exchange chromatography at pH 5.5. Adapted from [274], copyright 2012, with permission from Elsevier. (b) Lumped Henry coefficient as a function of the modifier concentration (NaCl in mol/L) in the eluent. Straight lines correspond to fits with Equation 5.3. The slope is equal to z_A, the apparent charge of the protein interacting with the chromatographic medium.

the main peak (corresponding to more acidic variants) and one eluting after the main peak (corresponding to more basic variants). Acidic variants are typically formed by the addition of a carboxyl group or the disappearance of an amine, which results in lowering the net charge of the protein at a given pH value. On the contrary, basic variants are formed by the loss of a carboxyl group or the addition of an amino group, which leads to an increase of the protein net charge at the same pH value. While most of the commercially available monoclonal antibodies are known to be complex mixtures of numerous charge variants [67], some specific cases might require isolating one of the variants from the distribution. For instance, one variant could potentially be harmful or some variants could be less active than others, so that their removal would lead to the production of safer and more efficient drugs, respectively.

In order to quantify the effect of the apparent charge on the binding properties of the mAb under linear conditions, the lumped Henry coefficient of each charge variant was measured at various concentrations of sodium chloride in the eluent, as discussed in Section 2.5.2. According to Table 2.3, the lumped Henry coefficient is varying as a power function of the modifier concentration, with an exponent equal to the protein apparent charge z_A, as follows:

$$\overline{K}_A = \varepsilon_i + (1 - \varepsilon_i)\overline{\overline{K}}_{0,A} C_M^{-z_A} \tag{5.3}$$

This simple model is applicable as long as the modifier is in large excess as compared to the solutes to be separated, i.e., the molar concentration of the modifier in solution can be considered unaffected by the ion exchange mechanism occurring at the surface of the chromatographic medium.

In Figure 5.5(b) are presented the lumped Henry coefficients of the five charge variants identified in Figure 5.5(a) as a function of the modifier concentration. A regression

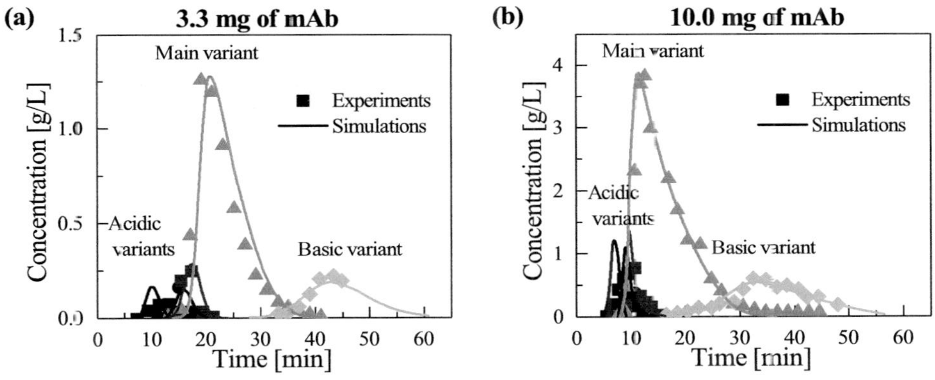

Figure 5.6 Chromatograms obtained under nonlinear conditions corresponding to the injection of (a) 3.3 mg and (b) 10 mg of mAb charge variants into a 5 × 85 mm column packed with a strong cation exchange resin. The elution was performed under isocratic conditions in 20 mM acetate buffer solution at pH 5.5 for a total ionic strength of 95 mM adjusted with sodium chloride. Experimental data and model simulations are represented by symbols and solid lines, respectively. Adapted from [274], copyright 2012, with permission from Elsevier. Details about the chromatographic model can be found in the original article.

with Equation 5.3 leads to an apparent charge z_A equal to 11.8 for the main variant, and varying from 10.8 for the most acidic charge variant to 12.3 for the most basic one.

To investigate the role of competitive effects on the shape of the chromatograms, experiments were performed at two different mAb concentrations under nonlinear conditions, corresponding to 3.3 and 10 mg of mAb loaded in the column. The resulting chromatograms are shown in Figure 5.6(a) and (b), respectively.

First, it can be seen that the five peaks distort to a more triangular shape whose front is moving toward shorter elution times when the amount of loaded mAb is increased. This behavior is typical for favorable Langmuir-like adsorption isotherms. Second, it can be seen that the early eluting acidic variants are compressed at the beginning of the main peak front. This indicates that the weakly bound acidic variants are displaced by the more strongly bound variants. Simulations have been performed in the original work with a PD model, using the LDF approximation and assuming Langmuir adsorption isotherms [274]. The simulation results shown by the solid lines in Figure 5.6 are in good agreement with the experimental data. This example illustrates how competitive effects impact the shape of chromatograms and how the retention times, of the weakly interacting solutes in particular, are modified as compared to noncompetitive cases.

5.2.3 Hydrodynamics and Kinetics of Mass Transfer

Lumping Strategies for Multicomponent Systems

In general, numerous and diverse impurities need to be considered to properly describe the protein polishing steps. There are different approaches to account for the various components present in complex feed mixtures, from the rigorous but unpractical (if not

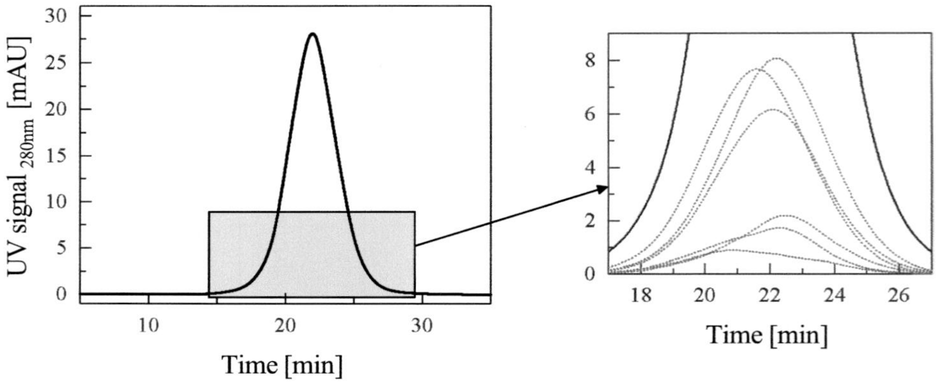

Figure 5.7 Experimental chromatogram of mono-PEGylated lysozyme injected in a column packed with a strong cation exchange chromatographic medium. The right-hand-side panel shows an enlargement of the gray area together with the results of the analytics obtained by pH-gradient IEC, which are represented by dashed lines. Each peak corresponds to a positional isomer of mono-PEGylated lysozyme.

unfeasible) description of each single solute, to lumping procedures grouping together multiple solutes in pseudospecies. For instance, common pseudospecies lump together all the weakly retained impurities on the one hand, and all the strongly retained ones on the other hand. In this section, we elaborate on two lumping methods: a discrete one, where solutes are grouped in a single pseudospecies, and a continuous one, where the solutes are described by a continuous distribution.

To illustrate the aforementioned approaches, we consider the example presented in Figure 5.7, which shows an experimental chromatogram obtained by injecting some mono-PEGylated lysozyme in a column packed with a strong cation exchange medium. At first sight, the chromatogram only shows one peak suggesting that the feed mixture contains a single solute. However, PEGylated proteins are known to be formed of multiple positional isomers, i.e., conjugates differing in the location where the PEG is grafted, as further explained in Chapter 6. A more detailed characterization of the collected fractions by analytical IEC revealed the presence of six positional isomers within this single peak, as shown in the right-hand-side panel of Figure 5.7. It appears from this example that the broadness of the experimental chromatogram arises from the combination of two contributions: on the one hand, the distribution of retention times of the six positional isomers, and on the other hand, dispersive effects, mainly due to mass transfer limitations.

Figure 5.8 illustrates three different approaches to model the nonresolved peak described earlier:

Detailed Approach A first modeling strategy consists in considering all of the six positional isomers with their respective parameters ($\overline{K}_j$, $D_{ax,j}$, and $t_{i,j}$, with j going from 1 to 6). This is shown in Figure 5.8(a), where the weight fractions of the different isomers are plotted as a function of the lumped Henry coefficient. This

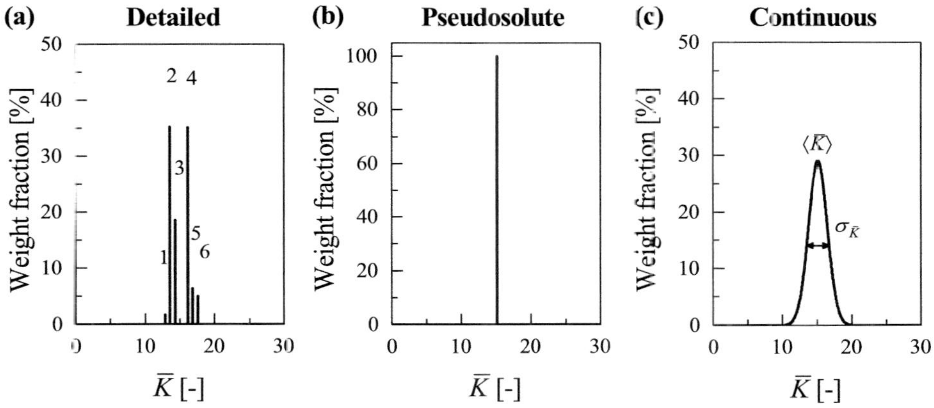

Figure 5.8 Approaches to describe a multicomponent system. (a) Detailed distribution of the weight fraction of the six positional isomers as a function of the lumped Henry coefficient. (b) Approximation of the mixture by one pseudosolute. (c) Approximation of the mixture by an infinite number of species following a Gaussian distribution of mean $\langle\overline{K}\rangle$ and standard deviation $\sigma_{\overline{K}}$.

approach is tedious and requires a tremendous amount of work to identify and characterize all the solutes, and then to estimate the associated model parameter values, especially when their number becomes very large.

Pseudosolute Approach As a more practical alternative, one can consider a single pseudosolute with average pseudoparameters ($\overline{K}^{pseudo}$, D_{ax}^{pseudo}, and t_i^{pseudo}), as shown in Figure 5.8(b). However, with this coarse-grained description, the range of validity of the model parameters is restricted to experimental conditions close to those where they have been determined.

Continuous Approach A third approach consists in approximating the composition of the feed mixture by an ensemble of solutes characterized by a continuous distribution of lumped Henry coefficients. For instance, the discrete solute distribution of Figure 5.8(a) is assimilated to a continuous Gaussian distribution in Figure 5.8(c). By doing so, only characteristic properties of the solute distribution, such as the mean lumped Henry coefficient $\langle\overline{K}\rangle$ and the standard deviation $\sigma_{\overline{K}}$, are necessary to model the behavior of a complex mixture of solutes. The mean and standard deviation associated with the continuous distribution are related to the concentration and lumped Henry coefficient of each solute as follows:

$$\langle\overline{K}\rangle = \frac{\sum_{i=1}^{n} C_i^F \overline{K}_i}{\sum_{i=1}^{n} C_i^F} \tag{5.4}$$

and

$$\sigma_{\overline{K}} = \sqrt{\frac{\sum_{i=1}^{n} C_i^F (\overline{K}_i - \langle\overline{K}\rangle)^2}{\sum_{i=1}^{n} C_i^F}} \tag{5.5}$$

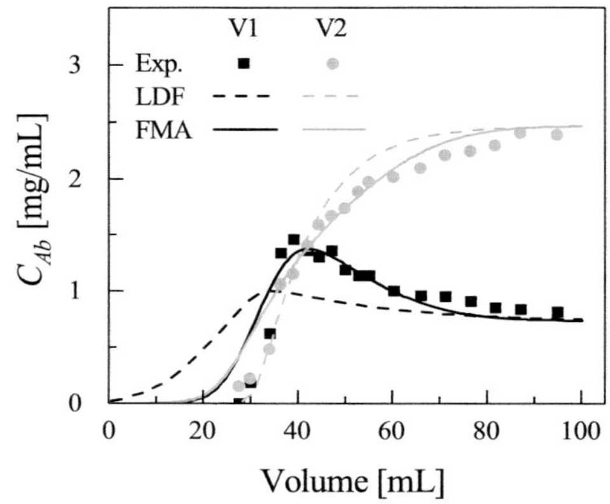

Figure 5.9 Breakthrough curves obtained with a polyclonal antibody composed of two main variants (23% wt. Variant 1 and 77% wt. Variant 2). The experiment was performed with a 5×43 mm column packed with a strong cation exchange medium at the flow rate of 0.12 mL/min. The buffer solution contained 20 mM of acetate adjusted to pH 5.0 and 70 mM NaCl. The fluid–solid equilibrium was modeled by a competitive Langmuir adsorption isotherm and the hydrodynamics by a MC model. The following parameter values were used: $\varepsilon_e = 0.39$, $J = 100$, $\overline{N} = 220$ g/L, $\overline{K}_{V1} = 220$, $\overline{K}_{V2} = 2180$, $t_{i,V1} = 39$ min, and $t_{i,V2} = 500$ min for the FMA and $t_{i,V1} = 163$ min and $t_{i,V2} = 80$ min for the LDF. Experimental data reproduced from [279], copyright 2008, with permission from Elsevier.

beginning of the front reaches a value higher than that in the feed. This phenomenon is typical for favorable competitive adsorption isotherms and suggests that Variant 1 is strongly displaced by Variant 2.

These experimental data were simulated using the classical LDF approach (Equation 2.58) and the FMA approach (Equation 5.10). The parameters of the adsorption isotherms were estimated independently from pulse injection experiments at different concentrations, whereas the characteristic times for mass transfer were fitted to describe the experimental breakthrough curves. The so-estimated parameters are reported in the caption of Figure 5.9. Considering the very different values of the lumped Henry coefficients (220 and 2180 for Variant 1 and Variant 2, respectively), it is expected from the equilibrium theory that the retention volume of Variant 1 is smaller than the one of Variant 2. Nevertheless, it is observed experimentally that the two variants start eluting at similar volumes. This can be explained by the large characteristic time for internal mass transfer of the strongly retained Variant 2, which flattens the elution profile, thus resulting in a shorter breakthrough time.

It can be seen in Figure 5.9 that the LDF approach underestimates the time at which the first variant breaks through and underestimates the maximum concentration value. Extensive simulations with different $t_{i,V1}$ and $t_{i,V2}$ values were performed, but no better fitting than the one presented in Figure 5.9 could be found. The breakthrough curves are instead correctly simulated when using the FMA approach, which better describes the coupling between solute diffusion and adsorption. In the case under investigation, it was found from the simulation results that the apparent characteristic time for internal mass transfer $t_{i,A} \times \frac{\Theta^* - \Theta}{\ln(\Theta^*/\Theta)}$ decreases by two orders of magnitude between the beginning of the elution (when the solute concentration is low) and the end of the elution (when the solute concentration reaches the feed concentration). This is in qualitative agreement with the discussion in Section 2.4.4, which comments on the expected decrease of the characteristic time for internal mass transfer with the protein concentration for favorable adsorption isotherms. As compared to the LDF approximation with variable characteristic time for internal mass transfer (see application in Section 4.2.3), the FMA model allows accounting for the competition between the solutes. Overall, this example illustrates that the LDF approximation reaches its limits in the case of strong and competitive adsorption, while the FMA better describes the kinetics of mass transfer in such situations.

5.3 Design of Countercurrent Chromatographic Processes

We have seen in Chapter 3 that the SMB process is extremely attractive for challenging separations that require efficient columns (e.g., due to a low selectivity) and/or for which it is difficult to reach a sufficient number of plates (e.g., due to a slow kinetics of mass transfer). There is thus a high potential for countercurrent technologies in the downstream processing of therapeutic proteins.

Nevertheless, it must be realized that the classical four-zone SMB is designed to perform binary separations only. Indeed, it splits the feed stream into two streams, extract and raffinate, characterized by different specifications. Some binary separations are of interest for the biopharmaceutical industry, such as desalting or aggregate removal. However, it is quite common that bio-applications require ternary separations, either to isolate one species eluting between more and less retained impurities (i.e., central-cut separations) or to recover three fractions with given specifications. Examples of ternary separations include the isolation of a protein from its fragments and aggregates or the purification of charge variants. In the following, we present some countercurrent chromatographic processes and their corresponding design procedures for binary separations first, and then for ternary separations. Finally, we briefly address some control considerations.

5.3.1 Binary Separations

Simulated Moving Bed (SMB)

Process Design

In this section, we show how to design a classical four-zone SMB process according to the approach described in Box 3.2. To do so, we continue the example of the separation of two antibody variants with IEC given in Section 5.2.3. We aim at recovering Variant 1 in the raffinate with a purity and recovery above 95% from a feed mixture containing only 23% of that variant. We consider a SMB process with two columns per zone and a sodium chloride concentration of 350 mM. At this salt concentration, the lumped Henry coefficients are equal to $\overline{K}_{V1} = 13$ and $\overline{K}_{V2} = 20$ for Variant 1 and Variant 2, respectively.

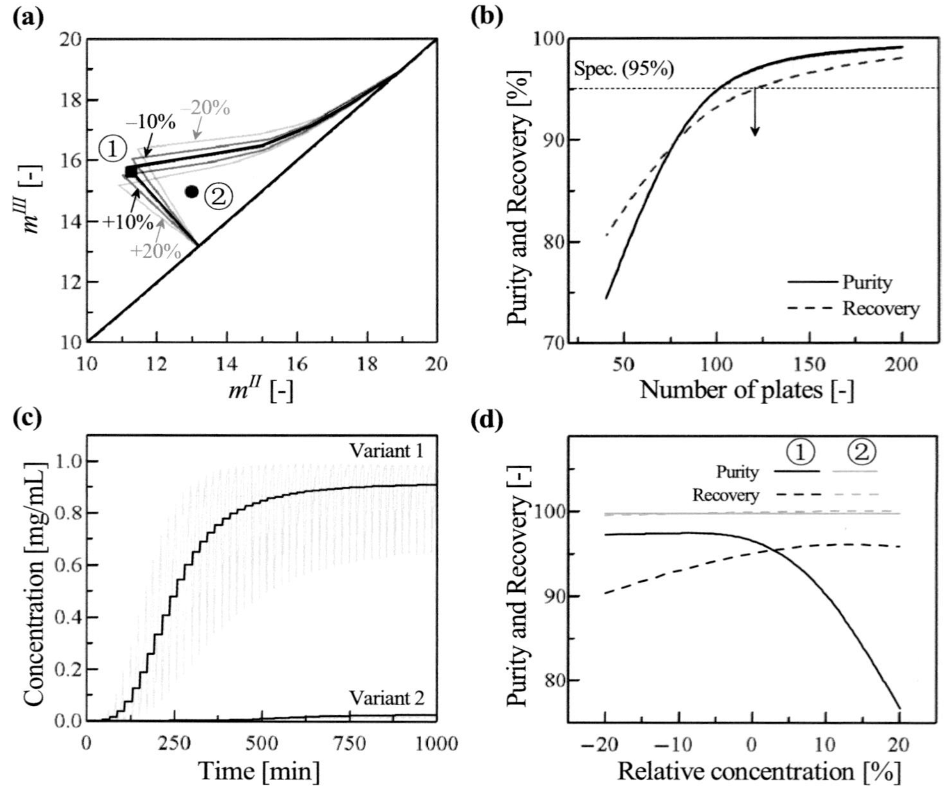

Figure 5.10 Design of a SMB process for the separation of two antibody variants. (a) Zone of complete separation in the m^{II}–m^{III} plane for several total protein concentrations in the feed around the nominal value. Two operating points are considered, denoted ① and ②, respectively. (b) Purity and recovery in the raffinate as a function of the number of plates for the operating point ① computed with the MC model and assuming thermodynamic equilibrium in each plate. (c) Simulated concentration profiles in the raffinate for the operating point ① using the MC model and the LDF approximation. Thin lines and thick lines represent instantaneous values and concentrations averaged over one cycle, respectively. (d) Purity and recovery in the raffinate for different values of the total protein concentration in the feed around the nominal value for the two operating points shown in panel (a). The following parameters were used: $\varepsilon_e = 0.39$, $J = 200$, $\overline{N} = 220$ g/L, $\overline{K}_{V1} = 13$, $\overline{K}_{V2} = 20$, $t_{i,V1} = t_{i,V2} = 1$ min, $m^I = 20$, $m^{II} = 11.3$, $m^{III} = 15.6$, $m^{IV} = 12.3$, $\Delta t = 21$ min, $Q^F = 1$ mL/min, $L_{col} = 4.95$ cm, $D_{col} = 1.44$ cm.

The zone of complete separation in the m^{II}–m^{III} plane computed using Langmuir adsorption isotherms with $\overline{N} = 220$ g/L is shown in Figure 5.10(a) with black lines. It is seen that the "triangle" is quite distorted, which reflects the nonlinearities of the adsorption equilibria. We select an operating point inside the complete separation region with a safety margin of 5%. This means that $m^F = m^{III} - m^{II}$ is 5% lower than the maximal value predicted by the equilibrium theory, and thus equal here to $4.5 \times 0.95 = 4.3$.

Any point located on the line $m^{III} = m^{II} + 4.3$ within the complete separation region satisfies the safety margin constraint. In this example, we consider the operating point that is equidistant from the boundaries of the complete separation region, which leads to $m^{II} = 11.3$ and $m^{III} = 15.6$. This point is represented by a black square in Figure 5.10(a) and denoted by the symbol ①. Similarly, we apply a safety margin on the selection of the m^{I} and m^{IV} parameters. Here, we select m^{I} and m^{IV} 5% larger and smaller than the maximal values predicted by the equilibrium theory, respectively, which leads to $m^{I} = 20$ and $m^{IV} = 12.3$.

Then, we perform simulations with the MC model assuming that equilibrium is reached instantaneously in each mixing cell. The simulated purity and recovery values of Variant 1 in the raffinate are shown in Figure 5.10(b) as a function of the total number of plates. It is seen that a minimum value of $N^{min} = 120$ plates is needed in order to meet the constraints in terms of purity and recovery.

Considering a maximum pressure drop of 0.3 bar in zone I and a value of $\delta_K = 0.05$ bar.min/cm^2 in the Kozeny–Kármán equation, Equations 2.92 and 2.93 lead to a column length of $L_{col} = 4.95$ cm and a fluid velocity of $u^{I} = 2.9$ cm/min. From the expression of m^{I} given in Table 3.2, we then obtain a switch time of $\Delta t = 21$ min. Finally, for a feed flow rate of $Q^{F} = 1$ mL/min, we estimate $D_{col} = 1.44$ cm with Equation 3.21 using Δt as the time interval.

Based on this design, we can now simulate a lab-scale SMB process using the MC model and the LDF approximation to describe the hydrodynamics and the kinetics of mass transfer, respectively (see Section 3.2.4 for the equations describing the SMB process). As compared to the situation reported in Figure 5.9 at 70 mM NaCl, the values of the lumped Henry coefficients at 350 mM NaCl are much lower, so that the classical LDF approximation provides satisfactory results. The simulated concentrations of the two variants in the raffinate are shown in Figure 5.10(c). Instantaneous values are represented with thin solid lines, while the thick solid lines correspond to the concentrations averaged over one cycle. It is seen that a periodic steady state is reached after around 10 h, i.e., approximately 30 cycles, and that only a residual amount of Variant 2 co-elutes with Variant 1.

Another example of SMB process will be presented in Chapter 7 as an interesting solution to eliminate mAb aggregates with size exclusion chromatography.

Robustness Analysis

Let us now investigate the sensitivity of the SMB process designed earlier upon a change in the feed concentration. To do so, we vary the total protein concentration in the feed by ±20% around the nominal value of 3.2 mg/mL and report the changes in purity and recovery in Figure 5.10(d) as black lines. It is observed that when the concentration increases above the nominal value, the purity decreases, whereas the recovery increases. This can be understood with Figure 5.10(a), where regions of complete separation are represented for different feed concentrations around the nominal value. When the feed concentration increases, the operating point ① (black square) falls above the region of complete separation, where the raffinate is polluted but the extract is pure (see Figures 3.5 and 3.10). This leads to a decrease of the purity of Variant 1 in the raffinate but to an increase of the purity of Variant 2 in the extract, which in turn results in an increase of the recovery of Variant 1 in the raffinate. A similar reasoning can be applied to rationalize the impact of the feed concentration below the nominal value. In this case, the operating point falls below the complete separation region, where the raffinate is pure but the extract polluted. This leads to an increase of the purity, but to a decrease of the recovery.

We select now the operating point ② shown with a black circle in Figure 5.10(a), which is located well inside the complete separation region for all the feed concentrations under investigation. Besides a change in m^{II} and m^{III}, all the other process variables previously determined are kept unchanged (L_{col}, D_{col}, $\Delta t \ldots$), and the same analysis is repeated. The results are shown with gray lines in Figure 5.10(d), where it is seen that both the purity and the recovery are very close to 100% independent of the feed concentration. This indicates that the process running at the operating point ② is very robust. However, the productivity at the nominal feed concentration dropped down by a factor of 3 as compared to the process obtained with the operating point ① (from 15 to 5 g/L/h). This reduced productivity is due to a lower value of m^F, or equivalently a lower value of the feed flow rate, since all the other process variables are kept unchanged. Overall, running the SMB process at operating point ① leads to a high productivity, but is risky because a slight change in the feed concentration may result in a product purity below the specifications. On the other hand, operating point ② is much less sensitive to variations in the feed concentration but leads to a low productivity.

This example illustrates how chromatographic models can be used to assess process robustness. The findings are relatively general, showing once more that designing a chromatographic process is always a matter of identifying the most appropriate compromise, between productivity and robustness in the present case.

Open-Loop Configuration

During the downstream processing of therapeutic proteins, the presence of impurities eluting close to the product of interest renders the range of process variables delivering the expected purity quite narrow. The risk to select conditions for which some impurities get recycled in the eluent stream and carried over from zone IV to I (or carried over with the solid phase from zone I to IV) is thus rather high. One possibility to prevent impurity recycling with the eluent flow is to lower m^{IV} below the value of the lumped Henry coefficient of the least retained impurity. For a nonadsorbing and completely excluded impurity, the lumped Henry coefficient is equal to 0, which leads to $m^{IV} = 0$. Such small value of m^{IV} results in a high solvent consumption and in the dilution of the raffinate. This is illustrated in Table 5.1, which summarizes the values of some process performance parameters obtained with the SMB process designed above at the operating point ①. When decreasing m^{IV} from 12.3 to 0 (keeping all the other process variables constant), the eluent consumption doubled and the concentration of Variant 1 in the raffinate dropped by a factor of 4.

Rather than decreasing m^{IV}, SMB processes are sometimes operated in an open-loop configuration, i.e., without recycling the eluent from zone IV to zone I. Such configuration is represented in Figure 5.11(a). As compared to the case where $m^{IV} = 0$, the open-loop configuration avoids diluting the raffinate stream for a similar eluent consumption, as shown in Table 5.1.

Table 5.1. Comparison of different SMB configurations in terms of eluent consumption productivity, and raffinate concentration. Values of the process variables and model parameters are reported in the caption of Figure 5.10. The operating point ① was selected.

Process	**Eluent consumption** [L/g]	**Productivity** [g/L/day]	**Raffinate conc.** [g/L]
Classical SMB	3.85	15.7	0.91
Classical SMB with $m^{IV} = 0$	7.27	16.3	0.20
Four-zone open-loop SMB	8.22	15.6	0.90
Three-zone open-loop SMB	7.42	21.5	0.19

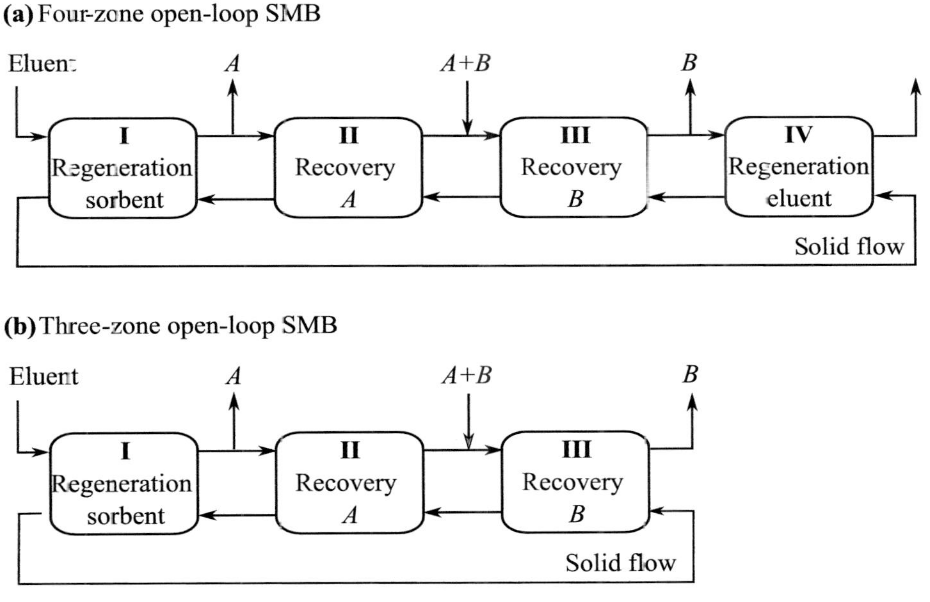

Figure 5.11 Scheme of the (a) four-zone open-loop SMB and (b) three-zone open-loop SMB.

In order to compensate for the cost associated with a larger eluent consumption, one may think of removing the fourth zone of the SMB to increase the productivity, as presented in Figure 5.11(b). Following our previous example, the productivity was increased by almost 40% when moving from a four-zone to a three-zone open-loop SMB, as indicated in Table 5.1.

Eluent consumption is a key component in the evaluation of large-scale chromatographic processes, so that open-loop configurations are economically viable only if large quantities of eluent can be treated (either regenerated to be reused or processed in the waste treatment unit to be discarded) at a relatively low cost. Lowering eluent consumption does not only allow decreasing the operational costs but also the investment costs by reducing the footprint of the system.

SMB with Step-Gradient

We have seen in Section 5.3.1 how to design SMB units based on the complete separation region determined in the frame of the equilibrium theory. Notably, the maximal normalized feed flow rate is given by $m^F = m^{III} - m^{II}$, which is equal to $\overline{K}_A - \overline{K}_B$ under linear conditions. This implies that the productivity of an isocratic process is intrinsically limited by the thermodynamics of the fluid–solid equilibrium under the selected conditions. To overtake this limitation, one could imagine applying different operating conditions in zones II and III so as to change the retention behavior of the solutes in the two zones, and thus to enlarge the complete separation region. Extending the analysis presented in Section 3.2, the boundary conditions for m^{II} and m^{III} leading to complete separation are given by:

$$\overline{K}_B^{II} \leq m^{II} \leq \overline{K}_A^{II} \tag{5.11}$$

$$\overline{K}_B^{III} \leq m^{III} \leq \overline{K}_A^{III} \tag{5.12}$$

where the indexes *II* and *III* indicate the zones where the lumped Henry coefficients are calculated. It follows that the maximal value of m^F under linear equilibrium conditions is $m^F = \overline{K}_A^{III} - \overline{K}_B^{II}$. By astutely selecting different operating conditions in zones II and III, the range of m^F values leading to complete separation can be enlarged, which in turn improves the productivity. Similar considerations apply for the normalized eluent flow rate given by $m^D = m^I - m^{IV}$, whose minimal value is equal to $\overline{K}_A^I - \overline{K}_B^{IV}$ under linear conditions. By selecting operating conditions in zones I and IV such that $\overline{K}_A^I \approx \overline{K}_B^{IV}$, the eluent consumption can be drastically reduced. Overall, selecting independently the operating conditions in the four zones would allow boosting the productivity and greatly diminishing the eluent consumption. For example, the solute retention behavior can be modulated by changing the temperature, pressure, or chemical composition of the eluent. This idea was first implemented for fatty acid purification using supercritical fluids as eluents and adjusting the pressure to modify the adsorption isotherms independently in the four zones of the SMB [280, 281].

In liquid chromatography, one can think of adjusting the eluent strength by changing the pH, ionic strength or concentration of organic solvent. In such cases, it would be quite challenging in practice to operate the four zones of the SMB under four different conditions. However, it is relatively easy to operate zones I/II and zones III/IV under different elution conditions by properly selecting the composition and flow rates of the feed and eluent streams. Such process is commonly referred to as the step-gradient SMB (SG-SMB) and is illustrated in Figure 5.12, where the height of the gray areas represent the concentration of modifier in each zone. Note that the modifier concentration in zones III and IV is necessarily lying in between the one of the feed stream and the one exiting zone II. Similarly, the modifier concentration in zones I and II is necessarily lying in between the one of the eluent stream and the one recycled from zone IV.

The SG-SMB is a rather complex process because it involves at least three solutes (two solutes to be separated and a modifier) interacting with the chromatographic medium. In the following, we present the process and its design for a simplistic case based on several assumptions: *(i)* the adsorption isotherms of the two solutes to be

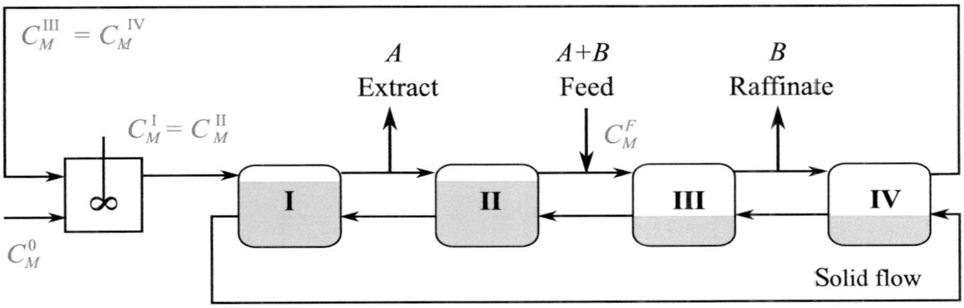

Figure 5.12 Schematic representation of the step-gradient SMB in the TMB-like representation. The height of the gray areas symbolizes the concentration of modifier in the different zones.

separated are linear, *(ii)* the modifier concentration is instantaneously uniform in a given zone, and *(iii)* the modifier does not adsorb at the surface of the chromatographic medium [282].

Let us assume first that C_M^{III} could be set independently of the internal flow rates (i.e., independently of the m values). In the case where the SMB is operated under isocratic conditions ($C_M^{II} = C_M^{III}$), the region of complete separation corresponds to the gray triangle represented in Figure 5.13(a). By playing with the modifier concentration in the feed stream, the values of $\overline{K}_A$ and $\overline{K}_B$ in zone III can be increased as compared to the values in zone II. This leads to an enlargement of the separation region, which progressively becomes rectangular as shown in Figure 5.13(b) and (c).

Although useful to understand the key idea behind the SG-SMB, the preceding considerations are certainly relying on an oversimplified view of the process. The modifier concentration in zone III is in fact dependent on the internal flow ratios and can be derived from a mass balance at the injection node to lead to the following expression:

$$C_M^{III} = \frac{m^{II} C_M^{II} + \left(m^{III} - m^{II}\right) C_M^F}{m^{III}} \tag{5.13}$$

For each value of m^{II}, the boundary conditions given by Equation 5.12 are computed by solving the following equations:

$$m^{III} - \overline{K}_A^{III} = 0 \tag{5.14}$$

$$m^{III} - \overline{K}_B^{III} = 0 \tag{5.15}$$

where the lumped Henry coefficients of A and B are computed as a function of C_M^{III} from the expressions given in Table 2.3. The corresponding zone of complete separation is shown in Figure 5.13(d). Note that the preceding equations are valid only under linear adsorption conditions. The reader interested in nonlinear isotherms, and in particular Langmuir adsorption isotherms, is referred to this work [283].

In the following, we report the example of an open-loop SG-SMB, which was implemented to purify the active dimeric form of morphogenic bone protein-2 (BMP-2) from a multicomponent renaturation broth [284]. As compared to the scheme shown in Figure 5.12, there was no zone IV in the considered implementation of the SG-SMB,

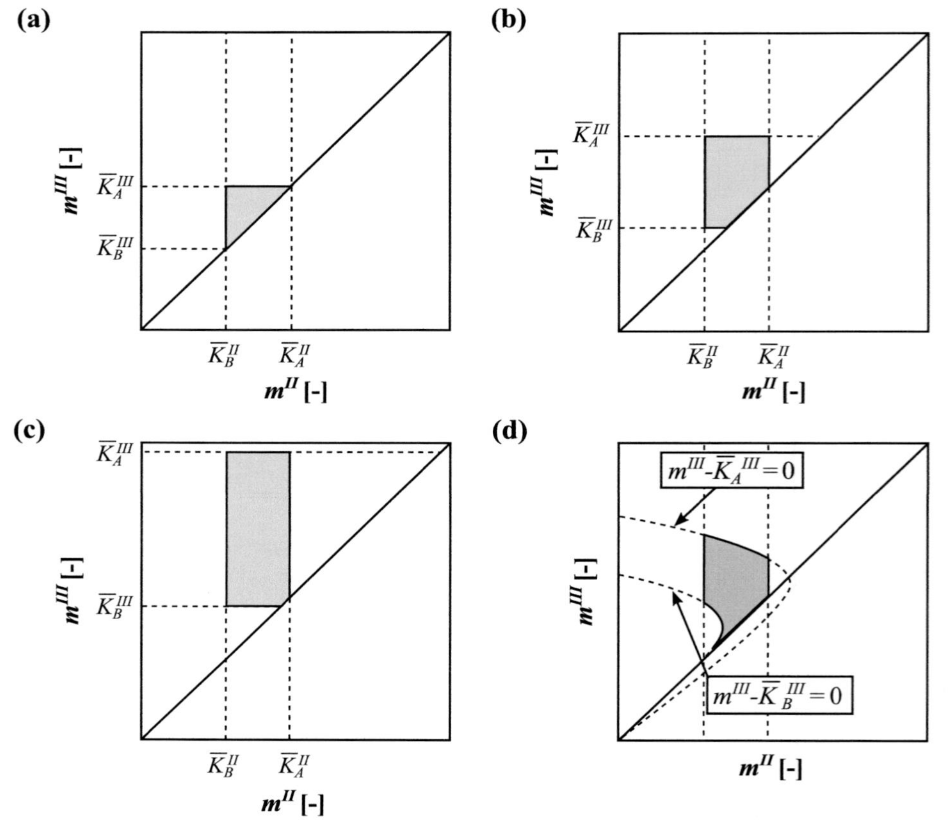

Figure 5.13 Zone of complete separation in a step-gradient SMB (gray areas). In panels (a)–(c) it is assumed that C_M^{III} is fixed independently of the values of the internal flow rates. Different values of C_M^{III} are considered in the three panels, thus leading to different regions of complete separation. In panel (d), the assumption of fixed C_M^{III} is relaxed. The curved boundaries of the complete separation region are computed by solving Equations 5.14 and 5.15.

but a zone 0 was introduced before zone I to regenerate the chromatographic medium. The purification was performed using a Heparin chromatographic medium and NaCl as a modifier. The salt concentration was set to 190 mM in the feed stream and to 600 mM in zones I and II. The region of complete separation for this SG-SMB process was determined based on the analysis presented earlier and is shown with a light gray line in Figure 5.14(a). At the feed concentration considered, the adsorption isotherms are linear and the expressions of the lumped Henry coefficient as a function of the salt concentration are given in the caption of Figure 5.14. The parameters m^{II} and m^{III} were selected inside the region of complete separation, as shown with the open square in Figure 5.14(a). Using the selected values of $m^{II} = 1.3$ and $m^{III} = 50.5$ to compute the modifier concentration in zone III with Equation 5.13 leads to $C_M^{III} = 200$ mM. Under these conditions, the lumped Henry coefficients of BMP-2 and of the impurities are

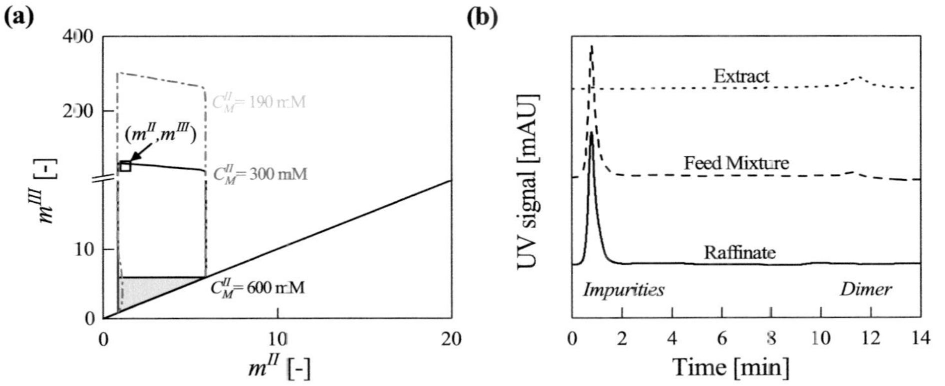

Figure 5.14 (a) Regions of complete separation in the m^{II}–m^{III} plane for the purification of BMP-2 with a SG-SMB. The process was implemented with one column per zone using a Heparin chromatographic medium and NaCl as a modifier. The following values of the process variables were used: $L_{col} = 5$ cm, $D_{col} = 0.7$ cm, $\varepsilon_e = 0.4$, $\varepsilon_i = 0.8$, $Q^F = 1$ mL/min, $Q^I = 0.7$ mL/min, $Q^{II} = 0.04$ mL/min, $Q^{III} = 1.04$ mL/min, $\Delta t = 59$ min. The lumped Henry coefficients of BMP-2 (denoted B) and of the impurities (I) vary with the modifier concentration as follows: $\overline{K}_B = 0.8 + 4.2 \times 10^{10} C_M^{-3.54}$ and $\overline{K}_I = 0.8 + 1.7 \times 10^{13} C_M^{-5.35}$, where C_M is in mM. The selected operating point is indicated with an open square. The concentration of NaCl in the feed was set to $C_M^F = 190$ mM. (b) Analytical chromatograms of the feed mixture, extract, and raffinate obtained on an Heparin column. Experimental chromatograms reproduced from [284], copyright 2007, with permission from Elsevier.

300 and 9, respectively, in zones I/II, whereas they are equal to 7 and 0.8, respectively, in zone III. The SG-SMB was operated with the conditions described earlier and the results of the analytics performed in the feed, extract and raffinate are presented in Figure 5.14(b). It is seen that a satisfactory separation between the active dimeric BMP-2 and the impurities is obtained.

To better understand the impact of the salt concentration in the SG-SMB process, Figure 5.14(a) shows the complete separation regions computed for different modifier concentrations in the feed from the selected value of $C_M^F = 190$ mM up to $C_M^F = C_M^{II} = 600$ mM. We consider here a constant value of C_M^{II}, but the latter is in practice tuned by playing with the modifier concentration in the eluent stream (i.e., C_M^0). It is seen that under isocratic conditions (i.e., $C_M^F = C_M^{II}$), the complete separation region is triangular, as discussed in Section 3.2.3. The lower C_M^F is as compared to C_M^{II}, and the larger is the complete separation region. In particular, much larger values of m^{III} can be used, which implies that the normalized feed flow rate m^F can be considerably increased as compared to the isocratic case. This is expected to lead to an increase in productivity.

Other examples of step-gradient SMB can be found in the literature for the purification of monoclonal antibody fragments [285], immunoglobulin Y [286] and various model proteins [284, 287].

Countercurrent Processes for Flow-Through Chromatography

We have seen in Chapter 4 how to implement multicolumn processes to perform binary separations in the context of protein capture, where the product of interest is much more retained than all the other impurities. Such processes are also well suited for flow-through chromatography, which, similar to protein capture, typically involves binary separations characterized by a high selectivity. In the case of flow-through chromatography though, the elution order is reversed, with the impurities being strongly retained as compared to the product of interest.

For illustrative purposes, we consider the removal of host cell proteins from an antibody solution, which has already been discussed in Section 5.2.1. Briefly, this separation was performed in frontal mode with IEC using NaCl as a modifier. The elution conditions were chosen such that the HCP strongly bind to the chromatographic medium, while the mAb is only weakly retained. In the following, we show how to design a multicolumn process capable of delivering the antibody solution with a 4-log reduction in HCP (i.e., $\log(\vartheta_{HCP}) = -4$ in the fraction containing the antibody) starting from a solution containing 5 g/L of antibody and 1 g/L of HCP. In addition, we compare this multicolumn process to the single-column process.

The protocol of the single-column process goes as follows: *(i)* load during a time Δt^F at the flow rate $Q^F = 10$ mL/min, *(ii)* wash with 0.5 CV of eluent at 0.1 CV/min to recover the antibody present in the liquid phase inside the column, *(iii)* regenerate the chromatographic medium by applying 5 CV of a 500 mM NaCl solution at 1 CV/min, and *(iv)* equilibrate the chromatographic medium with a buffer solution at low salt concentration with 5 CV of eluent at 1 CV/min.

To implement this separation in a continuous mode, we select a three-column process whose scheme is presented in Figure 5.15. In a first step of duration Δt^{s1}, two columns are connected in series in the load zone and the mAb is collected pure at the outlet of the second column, while the impurities are retained. Meanwhile, the third column is

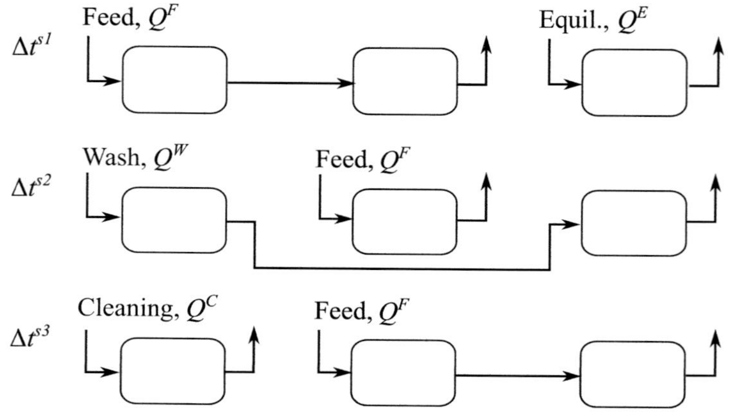

Figure 5.15 Scheme of a three-column process used for protein purification in flow-though chromatography. At the end of this sequence, the inlet and outlet streams move by one position to the right, and the three steps are repeated twice to complete a cycle.

re-equilibrated. When the impurities start breaking through the second column of the load zone, the first column is washed to recover the mAb. The outlet stream of this column is connected to the third column to capture weakly bound impurities. During this second step of duration Δt^{s2}, the load zone contains only one column at the outlet of which the mAb is also collected. Finally during a third step of duration Δt^{s3}, the first column is cleaned by changing the elution conditions to remove the strongly bound impurities, while the second and third columns are connected in series in the load zone. The feed time is given by the sum $\Delta t^{s1} + \Delta t^{s2} + \Delta t^{s3}$. At the end of this sequence, the inlet and outlet streams move by one position to the right, and the series of operations is repeated twice to complete a cycle.

In the following, we use the method given in Box 4.3 to design the single-column and multicolumn processes described earlier, i.e., to determine V_{col} and Δt^F. The first step of this shortcut method consists in defining the feed time from the maximal value predicted by the equilibrium theory. As an example, we keep a safety margin and target to load during 80% of the time derived from the equilibrium theory (i.e., $\beta = 0.8$), which leads to $\Delta t^F = 0.8 t_0 \left(1 + \phi_e \overline{C}^F_{HCP}/C^F_{HCP}\right)$. This value of 80% is arbitrary and should be adjusted when it comes to optimize a process. Considering the model parameters reported in the caption of Figure 5.16, we obtain $\Delta t^F = 121 t_0$.

Assuming that the contribution of the hydrodynamics to the overall peak broadening is negligible, designing this process consists in finding the value of $t_0 = \varepsilon_e V_{col}/Q^F$ that meets the specifications in terms of antibody purity (or equivalently in terms of HCP log reduction). Simulations of both the single-column and three-column processes

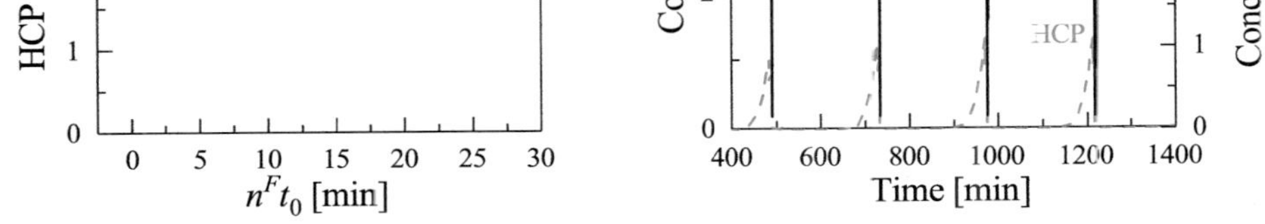

Figure 5.16 (a) Log reduction of the HCP concentration as a function of the zero-retention time for the single-column and three-column processes (considering the load zone only). (b) Simulated outlet concentrations at steady state of a mixture of mAb and HCP separated with a three-column process. A MC model was selected and the kinetics of mass transfer was described using the LDF approximation. The following values were used: $C^F_{mAb} = 5$ g/L, $C^F_{HCP} = 1$ g/L, $J = 20$, $\varepsilon_e = 0.4$, $\varepsilon_i = 0.5$, $L_{col} = 7.9$ cm, $D_{col} = 3.2$ cm, $\Delta t^{s1} = 718$ min, $\Delta t^{s2} = 5$ min, $\Delta t^{s3} = 5$ min, $Q^F = 1$ mL/min, $Q^E = Q^W = 1$ CV/min, $\overline{K}_{mAb} = 0.5$, $\overline{K}_{HCP} = 0.5 + 100 \times 10^{-0.1 C_M}$ with C_M in mM, $t^F_{i,mAb} = 4$ s, $t^F_{i,HCP} = 100$ s. Equation 4.13 was used to compute the characteristic times for internal mass transfer at high salt concentration.

Table 5.2. Comparison between a single-column and a three-column process for the removal of HCP from a mAb solution.

	Single-column	Three-Column
t_0 [min]	17.5	2
$n^T V_{col}$ [mL]	437.5	3 × 62.5
L_{col} [cm]	29.58	7.91
D_{col} [cm]	4.34	3.17
φ_{Ab} [g/L/day]	164	478

were performed at different t_0 values to identify the zero-retention time leading to the desired log reduction. The results are presented in Figure 5.16(a) as a function of $n^F t_0$ for the two processes, where n^F is the number of columns in the load zone. It can be seen that the multicolumn process is reaching higher log reduction values for shorter zero-retention times as compared to the single-column process. To obtain a 4-log reduction in the HCP concentration, a value of $n^F t_0 = 4$ min is necessary in the case of the multicolumn process, whereas more than 17 min are needed for the single-column process. Knowing the minimum value of t_0 delivering the product at the desired specifications allows computing the feed time for each process. Fixing $\Delta t^{s2} = \Delta t^{s3} = 5$ min according to the single-column process protocol leads to $\Delta t^{s1} = 718$ min.

Then, the required column volume is estimated from $V_{col} = Q^F t_0/\varepsilon_e$. Assuming a maximal pressure drop of 1 bar in the entire system and using the Kozeny–Kármán equation (see Equation 2.50) with $\delta_K = 0.05$ bar.min/cm^2, the column diameter and column length can be estimated. The values are reported in Table 5.2, where it is seen that a reduction of the total column volume by more than 2 could be obtained with the three-column process as compared to the single-column process. Such reduction of the system size has a positive impact on the productivity, which is almost tripled with the three-column process as compared to the single-column process.

For illustrative purposes, this process was simulated with the model parameter values given in the caption of Figure 5.16. The concentrations of mAb and HCP collected at the outlet of the process are represented by a black line and a gray line in Figure 5.16(b), respectively. It can be seen that the antibody is collected at a constant concentration equal to the feed concentration. The collection of antibody is stopped when the HCP concentration reaches the maximum acceptable value (corresponding to a 4-log reduction). The sudden decrease of both the antibody and HCP concentrations indicates the end of the collection period and the beginning of the wash step. Once the wash is completed, a novel step starts, indicated by the rise of the antibody concentration.

Note that a three-column process was selected here as an example. A study similar to the one reported in Section 4.3.3 in the context of protein capture could be performed to investigate the impact of the number of columns on process performances. Depending on the target recovery, two columns may be sufficient, or even more than three columns may be necessary to reach the maximum productivity. Besides productivity, other aspects such as the eluent consumption, the investment costs, process robustness, and process complexity should be considered in the final choice of the process.

5.3.2 Ternary Separations

Tandem SMB

In this section, we now focus on ternary separations. Several concepts have been developed over the years to extend the four-zone SMB process to ternary separations [288, 289]. One of the first ideas has been to use two SMB units, either merged within a single unit, i.e., an eight-zone SMB process [290], or using two SMB units in series [291]. In both cases, one outlet stream (either the raffinate or the extract) of the first SMB contains only one solute, while the other outlet stream contains the two remaining solutes, which are then separated in the second SMB.

The eight-zone SMB process has the disadvantage to impose the same column geometry for the different zones, while it is rather unlikely that both separations would require the same column dimensions. In particular, selecting the column length necessary to perform the most difficult separation implies that part of the system is oversized. Therefore, cascades of SMB in series are usually preferred due to their flexibility. As a matter of fact, the different SMB units can be better adapted (in terms of column dimensions, but also eluent composition, chromatographic medium, etc.) to each separation. In general, the easiest separation is performed in the first unit, while the second unit is reserved for the most challenging separation. In all cases, the tandem SMB process requires at least eight columns, that is one for each zone.

Figure 5.17 shows a tandem SMB process that has been used to purify insulin by separating zinc chloride salt and high molecular weight species from the monomeric insulin [291, 292]. As the three solutes exhibit quite different sizes, SEC was the technique of choice to perform both separations. The role of the first SMB unit was to separate the large aggregates (eluting in the raffinate) from insulin and salt (both eluting in the extract). The second unit was then loaded with the extract stream and used to separate insulin and the zinc salt.

We have seen in Section 2.4.2 that the fluid–solid equilibrium relations can generally be considered linear in the case of size exclusion chromatography. This implies that the zone of complete separation determined with the equilibrium theory is a triangle in the m^{II}–m^{III} plane (see Section 3.2). The following values of the lumped Henry coefficients were reported: $\overline{K}_{ZnCl_2} = 0.88$, $\overline{K}_{Ins} = 0.66$ and $\overline{K}_{Agg} = 0.17$ for the salt, monomeric insulin, and the aggregates, respectively. The zones of complete separation for each SMB unit computed with these parameter values are shown by gray areas in Figure 5.17(a). The black circles indicate the m parameters that were selected. In particular, the operating point of the first SMB was selected below the vertex of the triangle due to important mass transfer limitations. In other words, the normalized feed flow rate m^F was decreased as compared to the maximal feed flow rate in order to design a more robust process, ensuring that the extract stream was free of high molecular weight species. The experimental concentration profiles at steady state in the two SMB units are shown with symbols in Figure 5.17(b) and (c), respectively, while model simulations performed in the original work are shown with solid lines. It is seen that the aggregates are removed with the first SMB unit, whereas the second unit successfully separates the insulin/$ZnCl_2$ mixture.

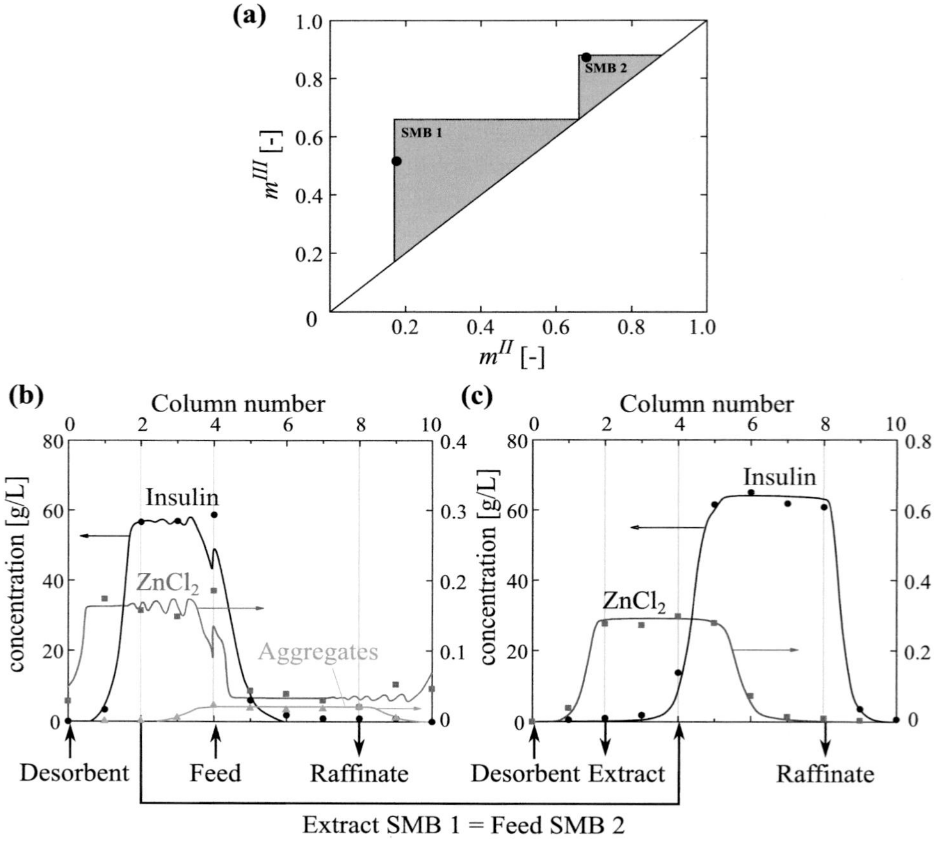

Figure 5.17 Tandem SMB for the separation of insulin, aggregates, and $ZnCl_2$ by SEC. Adapted from [292] by permission of John Wiley & Sons, Inc. (a) Design space representing the complete separation regions for both SMB units and the selected operating points. The first SMB separates the aggregates and the mixture insulin/$ZnCl_2$. The second SMB separates insulin and $ZnCl_2$. (b) Internal concentration profiles in the first SMB unit at steady state. (c) Internal concentration profiles in the second SMB unit at steady state. In panels (b) and (c), symbols and lines correspond to experimental data and model simulations performed in the original work, respectively. The concentrations for insulin and for the impurities are read on the left and on the right y-axes of the graphs, respectively.

SMB for Ternary Separations with a Strongly Bound Impurity

We have seen in Section 5.3.1 that different modifier concentrations can be applied in the different zones of the SMB process. This can be particularly useful in the presence of strongly retained solutes, which would require a large amount of eluent to be desorbed under isocratic conditions. If the regeneration of the chromatographic medium is not complete, some adsorbed solute may be recycled when the column in zone I enters zone IV. In order to avoid this situation, additional zones can be connected to the classical SMB to properly regenerate the chromatographic medium [293, 294].

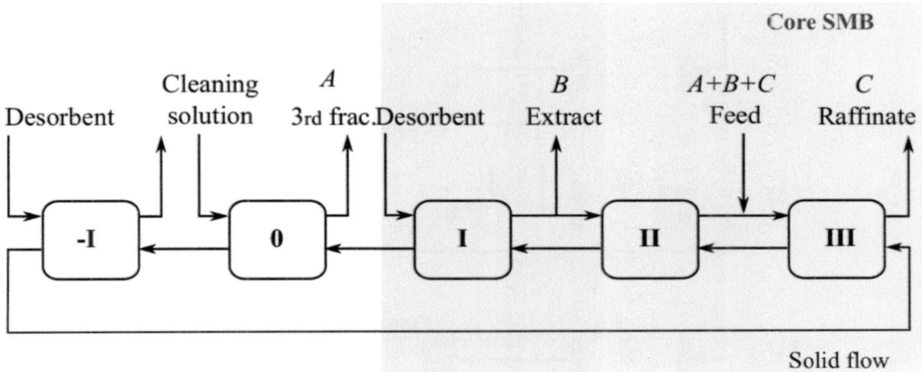

Figure 5.18 Scheme of the 3F-SMB process in the TMB-like representation. The gray zone indicates the three-zone open-loop SMB process, while the cleaning and equilibration zones are outside of the core SMB.

An example of such process is shown in Figure 5.18, where two zones have been added to a three-zone open-loop SMB:

- a zone 0 before zone I to implement the cleaning step, which typically involves eluents containing either sodium hydroxide, a high concentration of salt, or a high concentration of organic solvent, depending upon the selected type of chromatographic interactions;
- a zone −I before this cleaning zone to re-equilibrate the solid phase before it re-enters into the "core" SMB unit, which is highlighted with the gray area in Figure 5.18.

In addition to allowing the regeneration of the chromatographic medium, this process also opens the possibility to collect three pure fractions provided that the strongly adsorbed components are much more retained than the target protein. This process is thus referred to as the three-fraction SMB (3F-SMB) [295, 296]. Although useful to identify the role of the different zones, the TMB-like representation given in Figure 5.18 is not sufficient to understand the separation of three fractions. This necessitates representing the column configuration after each switch time, as it is done in Figure 5.19 with one column per zone. In the following, we show how to identify the complete separation region of the 3F-SMB to recover the three solutes pure. We rank the solutes A, B, and C such that A is the most strongly retained and C the least retained.

Let us first identify the complete separation region for the recovery of C for the binary separation AB/C. Considering the simplistic case of linear adsorption isotherms, the operating constraints on the internal flow rates are the following:

$$\overline{K}_C \leq m^{II} \leq m^{III} \leq \overline{K}_B \tag{5.16}$$

The constraints on m^{II} and m^{III} are identical to those obtained for the classical four-zone SMB in the frame of the equilibrium theory (see Section 3.2). Within the region

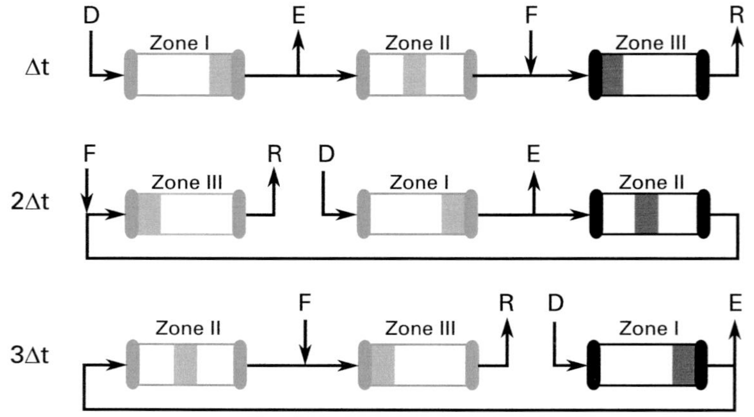

Figure 5.19 Representation of the zones I, II, and III in the 3F-SMB with one column per zone. The gray rectangles represent the bands of the strongly retained impurities inside the columns at different times. To better follow the time course of the elution of the impurities from zone III to zone I, only the column in which the feed mixture has been originally injected is highlighted.

defined by Equation 5.16, the least retained solute C can be recovered pure in the raffinate stream. However, if no particular precaution is taken, the extract will contain the most retained solutes A and B together. In order to collect A and B separately, the operating conditions should be selected so that the solute A does not leave the column it has been injected in before this column enters zone 0. This is illustrated in Figure 5.19, where the dark gray rectangles correspond to the eluting bands of the solute A at different times (Δt, $2\Delta t$, $3\Delta t$).

The total distance covered by A in the column it has been injected in is given by the sum of the distances covered in the zones III (z^{III}, with the flowrate Q^{III}), II (z^{II}, with the flowrate Q^{II}) and I (z^{I}, with the flowrate Q^{I}). In order to ensure that A does not elute before entering into zone 0, this covered distance should be less than the length of a column:

$$z^{III} + n^{II} z^{II} + z^{I} \leq L_{col} \tag{5.17}$$

where the covered distance in zone i is estimated from the equilibrium theory as $z^{i} = \Delta t \times (u^{i}/\varepsilon_e)/(1 + \phi_e \overline{K}_A)$ and n^{II} is the number of columns in zone II. Note that the preceding expression does not depend on the number of columns in zones I and III because it is sufficient to ensure that A does not leave the last column of zone I and the first column of zone III (as defined in the initial configuration). By introducing the definition of the m parameters, Equation 5.17 translates into:

$$\left(1 + \phi_e m^{I}\right) + n^{II}\left(1 + \phi_e m^{II}\right) + \left(1 + \phi_e m^{III}\right) \leq 1 + \phi_e \overline{K}_A \tag{5.18}$$

This inequality can be rearranged to provide an upper bound for m^{I}, which depends on m^{II} and m^{III}, as follows:

$$m^I \leq \overline{K}_A - \left(1 + n^{II}\right)/\phi_e - n^{II} m^{II} - m^{III} \tag{5.19}$$

Therefore, in addition to guaranteeing that the solute B does not leave zone I with the solid, which defines the lower bound of m^I as already discussed for the classical SMB (see Section 3.2), the parameter m^I should be such that the solute A does not leave zone I in the extract stream, which defines the upper bound of m^I. By setting m^I to a specific value, as for example its lower bound ($\overline{K}_B$ for linear adsorption conditions), a line can be drawn in the m^{II}–m^{III} plane below which the component A is not polluting the extract.

Overall, to determine the region of complete separation of a 3F-SMB process under linear adsorption conditions, it is first necessary to compute the triangle for the two least retained solutes (Equation 5.16) and then to draw the line below which the most retained solute does not pollute the extract (Equation 5.19). Note that a region of complete separation for the three solutes can be identified by properly selecting m^I only if $\overline{K}_A$ is sufficiently large with respect to $\overline{K}_B$ and $\overline{K}_C$ (see Equation 5.19). In other words, the three solutes can be recovered pure only if one of them is significantly more retained than the others.

An example of complete separation region obtained with a 3F-SMB is shown in Figure 5.20(a) [295]. This corresponds to the separation of three different nucleosides (A: 2′-deoxyadenosine, B: thymidine and C: 2′-deoxyguanosine) from an equimassic mixture using a reversed-phase medium and an aqueous solution of 1.5% vol. ethanol as eluent. Under these conditions, the Henry coefficients scale as $\overline{K}_A \geq \overline{K}_B \geq \overline{K}_C$ so

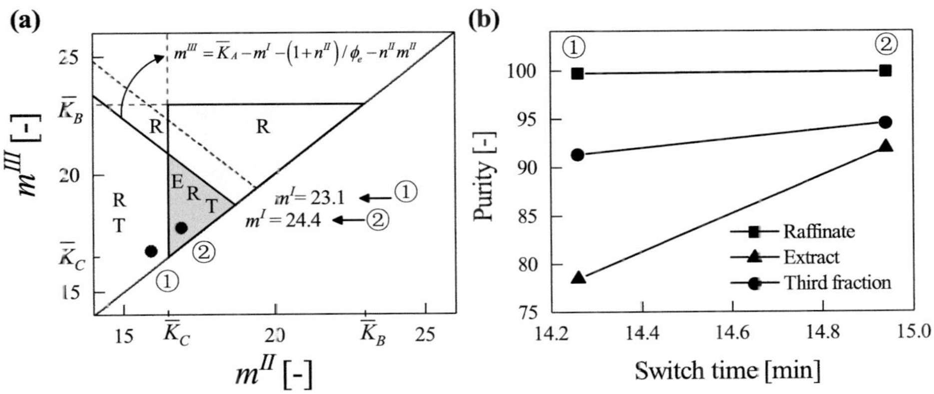

Figure 5.20 Example of application of the 3F-SMB process for the separation of three nucleosides (A: 2′-deoxyadenosine, B: thymidine, and C: 2′-deoxyguanosine) from a mixture containing 0.5 g/L of each nucleoside using reversed-phase chromatography. Adapted from [295], copyright 2004, with permission from Elsevier. (a) Complete separation region in the m^{II}–m^{III} plane. The combination of the light and dark gray areas correspond to the zone of complete separation for the binary separation AB/C, while the dark gray area corresponds to the zone of complete separation to recover the three solutes pure. The letters E, R, and T indicate the zones where the extract, raffinate and third fraction are pure, respectively. The points ① and ② represent experimental operating points. (b) Experimentally measured product purity as a function of the switch time (keeping the flow rates constant) for the operating points indicated in panel (a).

that A eluted from the cleaning zone, which was operated at the ethanol concentration of 12%, B was collected in the extract and C in the raffinate. The letters E, R, and T in Figure 5.20(a) indicate the zones where the extract, raffinate and third fraction are pure, respectively. The region where the three outlet streams are pure is shaded in dark gray. The upper boundary of this region was computed with $n^{II} = 1$, $\phi_e = 1.5$ and $m^I = 24.4$ (solid line).

Two operating points ① and ② were selected in the m^{II}–m^{III} plane, as represented in Figure 5.20(a). It was chosen to keep the fluid flow rates constant for these two points, and thus to move in the m^{II}–m^{III} plane by changing the switch time. This implies that ① and ② were characterized by different m^I values, namely 23.1 and 24.4, as indicated in Figure 5.20(a). The purity profiles obtained for the two selected operating points are shown in Figure 5.20(b), where it can be seen that the purity in the raffinate and in the third fraction was very high for both operating points, whereas the purity in the extract was lower for the operating point ① than for the operating point ②. These results are in agreement with the design space represented in Figure 5.20(a) because the operating point ② is located in the complete separation region, whereas the operating point ① is located in the region where the extract is polluted. This example confirms that the equilibrium theory, although neglecting hydrodynamic and kinetic effects, is a useful tool to explain the trends obtained experimentally with a multicolumn process.

Other variations of the classical SMB using several eluent streams to separate three solutes have been reported in the literature. For example, a three-zone open-loop SMB whose feed stream was periodically replaced by a second eluent stream was proposed [297, 298].

Processes with Internal Recycling of Nonresolved Fractions

Principle

As described in Section 5.2.1, central-cut separations are often encountered during polishing steps and the scheme of a single-column chromatogram corresponding to such separation is represented in Figure 5.21. The solute of interest B elutes in between a more retained impurity A and a less retained one C (where A and C represent either single solutes or a group of different impurities lumped together). Due to the large injected volumes, mass transfer limitations and low selectivity values, some overlapping between the product of interest and its closest impurities is often observed. In this section, we present a multicolumn process where the unresolved fractions $A + B$ and $B + C$ are retained in the unit in order to increase the recovery at a given purity of the target product. This process can accommodate either isocratic or gradient elution conditions and is referred to as the multicolumn solvent gradient purification (MCSGP) process [299, 300].

A central-cut separation can be decomposed into six phases occurring sequentially: *(i)* a load period, *(ii)* the elution of the pure solute C, *(iii)* the elution of an impure fraction containing both the solutes C and B, *(iv)* the elution of the pure solute B, *(v)* the elution of an impure fraction containing the solutes A and B, and finally *(vi)* the elution of the pure solute A. In order to transform this sequence into a continuous process,

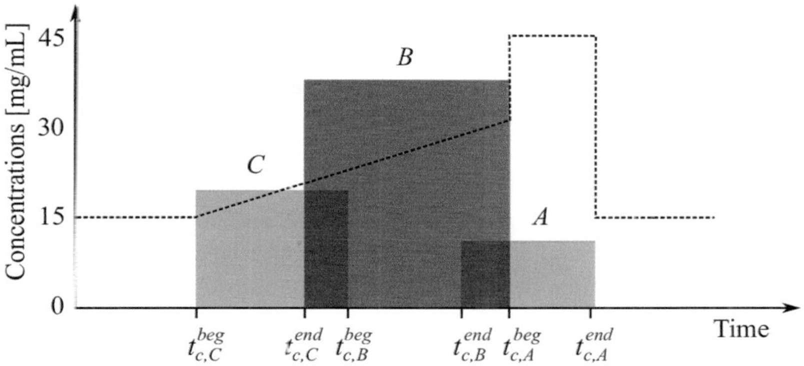

Figure 5.21 Scheme of a single-column chromatogram obtained for a ternary mixture presenting important overlapping between the different solutes. An illustrative modifier gradient profile is also shown with a dotted line.

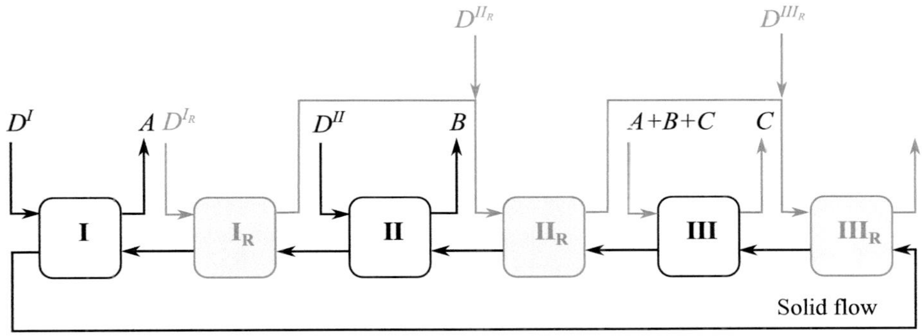

Figure 5.22 Scheme of the six-zone MCSGP process in the TMB-like representation. The protein of interest (B) is eluting between a strongly retained impurity (A) and a weakly adsorbed impurity (C).

operations that would be performed at different times in the single-column process are now performed at different locations, so that the question *when* is replaced by *where*. This leads to the scheme of the process shown in Figure 5.22. Let us take the simple case of a MCSGP process with one column per zone and follow the column initially in zone III (i.e., the zone where the feed solution is injected) through an entire cycle:

- While the feed mixture is loaded into this column, the solute C, with the weakest affinity for the solid phase, starts eluting at the outlet of zone III. This corresponds to the phases *(i)* and *(ii)* described earlier.
- Then, the positions of the inlet and outlet streams are moved by one step in the direction of the eluent. This corresponds to a movement of all the columns by one step to the left in Figure 5.22. This movement occurs just before the solute B

starts eluting so as to avoid losing the product in the stream containing the weakly bound impurity and thus maintain a high recovery. At this point, the column we are following enters zone II_R. A mixture of unresolved $B + C$ is now leaving the column (phase *(iii)*) and is recycled in zone III_R.

- Once this unresolved fraction is completely eluted, the positions of the inlet and outlet streams are switched again, and the column enters zone II, at the outlet of which B is collected pure (phase *(iv)*). The elution of the product continues until the strongly bound impurity breaks through at the column outlet.
- Then, the column becomes part of zone I_R, which is connected in series to the zones II_R and III_R. This allows recycling the second unresolved fraction $A + B$ (phase *(v)*) in order to increase the recovery of the product B.
- Once all the product B is eluted, the column enters zone I, in which the strongly adsorbed solute A is eluted (phase *(vi)*) and the column regenerated.

The MCSGP alternates between a process involving three single-columns in parallel, to collect the pure solutes, and a process involving three columns connected in series to recycle the unresolved fractions in a countercurrent configuration. The MCSGP therefore oscillates between an open-loop configuration (disconnected mode, for the collection of the products) and a recycling configuration (interconnected mode, for the recovery of the unresolved fractions). Different operating conditions (as flow rate, solvent composition, temperature …) can be used in the zones I, II and III because the three zones are disconnected, each one being operated as a single column. For instance, the operation of cleaning-in-place, needed in protein purification to regenerate the column after each cycle, can be integrated within this process [301].

Three important volumes can be identified for the MCSGP process, namely the collection volume of the target product and the two recycling volumes of the unresolved fractions. While the first one is mostly related to the productivity and product purity, the other two mainly affect the recovery. The countercurrent contact implemented to recycle internally the unresolved fractions is expected to improve process performance as compared to the single-column process, or even the single-column with recycling. Indeed, we have seen in the previous chapters that countercurrent processes better deal with mass transfer limitations, which allow using shorter columns and higher fluid velocities, which both have a positive impact on productivity.

The MCSGP process has been especially developed for ternary separations, i.e., either to perform a central-cut separation or to recover three different fractions. The concept can be further extended to recover four, five, and even more fractions from a multi-component feed mixture. For n fractions, the number of required zones is $2 \times n$, i.e., n zones for the pure fractions used during the disconnected phase, and n zones for the unresolved fractions used during the interconnected phase. For example, the eight-zone MCSGP process has been applied to the recovery of two antibody variants from a pool of four antibody charge variants [302].

Another process capable of performing central-cut separation with solvent gradient is known as the gradient with steady state recycle (GSSR) process and the interested reader is referred to the corresponding literature for more details [303].

Table 5.3. Definitions of the m parameters for the six-, four-, three-, and two-column MCSGP ensuring the analogy with the single-column process.

	Single-column	6C-MCSGP	4&3C-MCSGP	2C-MCSGP	
				Phase 1	Phase 2
m^I	$\frac{Q\left(t_{c,A}^{end}-t_{c,A}^{beg}\right)}{(1-\varepsilon_e)V_{col}}$	$\frac{Q^I\,\Delta t}{(1-\varepsilon_e)V_{col}}$	$\frac{Q^I\,\Delta t^{DC}}{(1-\varepsilon)V_{col}}$	–	$\frac{Q^I\,\Delta t^{DC,2}}{(1-\varepsilon)V_{col}}$
m^{II}	$\frac{Q\left(t_{c,B}^{end}-t_{c,B}^{beg}\right)}{(1-\varepsilon_e)V_{col}}$	$\frac{Q^{II}\,\Delta t}{(1-\varepsilon_e)V_{col}}$	$\frac{Q^{II}\,\Delta t^{DC}}{(1-\varepsilon)V_{col}}$	$\frac{Q^{II,1}\,\Delta t^{DC,1}}{(1-\varepsilon)V_{col}}$	$\frac{Q^{II,2}\,\Delta t^{DC,2}}{(1-\varepsilon)V_{col}}$
m^{III}	$\frac{Q\Delta t^F}{(1-\varepsilon_e)V_{col}}$	$\frac{Q^{III}\,\Delta t}{(1-\varepsilon_e)V_{col}}$	$\frac{Q^{III}\,\Delta t^{DC}}{(1-\varepsilon)V_{col}}$	$\frac{Q^{III}\,\Delta t^{DC,1}}{(1-\varepsilon)V_{col}}$	–
m^{I_R}	$\frac{Q\left(t_{c,A}^{beg}-t_{c,B}^{end}\right)}{(1-\varepsilon_e)V_{col}}$	$\frac{Q^{I_R}\,\Delta t}{(1-\varepsilon_e)V_{col}}$	$\frac{Q^{I_R}\,\Delta t^{IC}}{(1-\varepsilon)V_{col}}$	–	$\frac{Q^{I_R}\,\Delta t^{IC,2}}{(1-\varepsilon)V_{col}}$
m^{II_R}	$\frac{Q\left(t_{c,B}^{beg}-t_{c,C}^{end}\right)}{(1-\varepsilon_e)V_{col}}$	$\frac{Q_R^{II}\,\Delta t}{(1-\varepsilon_e)V_{col}}$	$\frac{Q^{II_R}\,\Delta t^{IC}}{(1-\varepsilon)V_{col}}$	$\frac{Q^{II_R,1}\,\Delta t^{IC,1}}{(1-\varepsilon)V_{col}}$	$\frac{Q^{II_R,2}\,\Delta t^{IC,2}}{(1-\varepsilon)V_{col}}$
m^{III_R}	–	$\frac{Q^{III_R}\,\Delta t}{(1-\varepsilon_e)V_{col}}$	$\frac{Q^{III_R}\,\Delta t^{IC}}{(1-\varepsilon)V_{col}}$	$\frac{Q^{III_R}\,\Delta t^{IC,1}}{(1-\varepsilon)V_{col}}$	–

Empirical Design

The alternation of the MCSGP between interconnected/disconnected phases and the potential variations of the Henry coefficient with the modifier composition makes its model-based design and its robustness analysis relatively complex [300, 304] even when using the equilibrium theory [305, 306]. Nevertheless, we have seen in Section 3.5.1 that the unified design approach offers a tool to rapidly estimate process variables of multicolumn chromatographic processes based on a single-column chromatogram. Let us recall that this empirical approach provides a reasonable operating point to start with, but by no means guarantees optimal performance of the process. The appropriate definitions of the m parameters ensuring the analogy between the MCSGP and the single-column processes for a given operating point are summarized in Table 5.3.

The unified design approach can help understand how to adjust the recovery and purity of the target product by relating the process variables of the MCSGP to the elution times read on the single-column chromatogram (Figure 5.21). On the one hand, the recovery of B can be increased by increasing the volume of the recycled fraction (i.e., by decreasing $t_{c,C}^{end}$ and increasing $t_{c,A}^{beg}$). On the other hand, the purity of B can be increased by reducing the product collection window (i.e., by increasing $t_{c,B}^{beg}$ and decreasing $t_{c,B}^{end}$). These four elution times are then used to compute the flow ratios m^{II}, m^{I_R}, and m^{II_R} of the MCSGP as shown in Table 5.3. It is worth noting that changing the values of $t_{c,B}^{beg}$ and $t_{c,B}^{end}$ also affects the value of the normalized feed flow rate (m^{III}) because the amount injected per switch time should be equal to the amount collected during this interval. In the implementation of the process proposed in this section, both the unresolved fraction containing B and C and the fraction containing C pure are recycled. In addition, the aforementioned recycling window starts immediately after the end of the loading phase (i.e., $t_{c,C}^{beg} = \Delta t^F$). The parameter m^{III_R} does not have an equivalent expression in the single-column process and is selected so that no solute elutes from zone III_R (i.e., C is only collected at the outlet of zone III).

Finally, the eluent flow rates can be computed from mass balance equations at the nodes. In the case of gradient elution conditions, the composition of the eluent at the inlet of the different zones can be adjusted dynamically to trigger the elution of the different solutes. For instance, one can mimic the modifier gradient profile used in single-column experiments by properly selecting the modifier concentration in the inlet streams. This presents no technical difficulty and simply requires writing mass balance equations at the nodes for the modifier.

Reduction of the Number of Columns

The initial implementation of the 6C-MCSGP presented earlier has been modified over the years toward a reduction of the number of columns. When the number of columns is getting smaller than the number of zones, each column has to endorse successively multiple tasks that were previously operated in different zones. This implies that the loading cannot always be performed continuously, so that intermittent loading phases may have to be implemented.

This is the case of the 3C-MCSGP process for which the columns are sequentially connected and disconnected within a switch time [307]. The phase during which the three columns are disconnected (and the feed mixture is injected into a column) is characterized by the time Δt^{DC}. Similarly, the phase during which the columns are interconnected (and the loading is interrupted) is characterized by the time Δt^{IC}. The switch time is then given by $\Delta t = \Delta t^{IC} + \Delta t^{DC}$. To allow the continuous loading of the feed mixture, it is possible to add a fourth column to the process described earlier. An example of a 4C-MCSGP process will be given for the separation of conjugated proteins in Chapter 6. The m parameters associated with the design of the 3C and 4C-MCSGP are given in Table 5.3.

The Two-Column MCSGP

Pushing further the reduction of the number of columns, even more tasks can be associated with a column within a time period, as it is the case for the 2C-MCSGP process. The schematic representation of the process is given in Figure 5.23, while the four phases are briefly presented below, indicating the specific tasks taken by each of the two columns:

- **First interconnected phase** ($\Delta t^{IC,1}$): the nonresolved fraction of the chromatogram containing the product co-eluting with the weakly adsorbed impurity is recycled from the first column into the second column. The two columns are attributed to the zones II_R and III_R of Figure 5.22. The time $\Delta t^{IC,1}$ has to be selected such that all the weakly adsorbed impurity C leaves the column of zone II_R, so as to prevent pollution of the product B, which is eluted in the next phase.
- **First disconnected phase** ($\Delta t^{DC,1}$): the purified product is collected at the outlet of the first column, while the second column is loaded with the feed solution. The two columns are attributed to the zones II and III of Figure 5.22. The time $\Delta t^{DC,1}$ has to be selected so that the strongly adsorbed impurity A does not break through the column of zone II, which would otherwise pollute the product B.

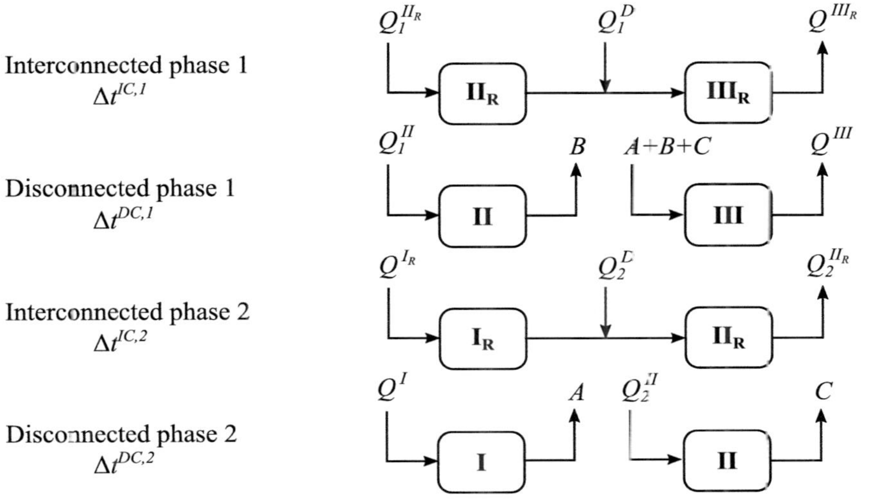

Figure 5.23 Schematic representation of the 2C-MCSGP process. The protein of interest (B) is eluting between strongly retained (A) and weakly retained (C) impurities.

- **Second interconnected phase** ($\Delta t^{IC,2}$): the nonresolved fraction containing the product and the strongly adsorbed impurity is recycled from the first to the second column. The two columns are attributed to the zones I_R and II_R of Figure 5.22. The time $\Delta t^{IC,2}$ has to be selected so as to elute completely the product B from the column in zone I_R, which would otherwise elute with the strongly retained solute A in the next phase, thus lowering the product recovery.
- **Second disconnected phase** ($\Delta t^{DC,2}$): the strongly adsorbing impurities are stripped out of the first column. Then the solid phase is cleaned and re-equilibrated before starting the new cycle. In parallel, the weakly adsorbed impurities elute from the second column. The two columns are attributed to the zones I and II of Figure 5.22. The time $\Delta t^{DC,2}$ has to be selected so that the product B does not leave the column of zone II with the weakly adsorbed solute C, which would lower the product recovery.

At the end of the last phase, the position of the columns are virtually interchanged through proper valve switching and a new sequence can start. Note that when using the MCSGP under gradient elution conditions, inline dilution has to be implemented during the interconnected phases (between zones I_R and II_R and between zones II_R and III_R) to ensure the proper binding of the recycled solutes.

The m parameters of the 2C-MCSGP are given in Table 5.3, while the reader is referred to the following work for more details regarding the experimental design procedure from a single-column chromatogram [308]. This procedure is simple and requires a minimum experimental effort, that is the acquisition of a single-column chromatogram showing a satisfactory separation. However, it certainly does not provide optimized conditions for the 2C-MCSGP, which in general requires some subsequent tuning. This is facilitated by the analysis of the unit outlet stream compositions, which indicate which variables should be modified in order to improve the recovery and purity. For instance, the presence of the impurities in the product stream means that the volume of the collection window has to be reduced at the expense of the productivity. On the other hand, the recycled volume should be increased to reduce the loss of product.

Examples of Separation with the MCSGP Process

A series of applications of the MCSGP have been discussed in the literature [83]. These include the isolation of a charge variant in three commercial antibodies [309], the separation of three charge variants [310], peptide purification [311], removal of aggregates and fragments [312], and bispecific mAbs purification [313]. In addition, two MCSGP processes in series have been used for the capture step (with a cation exchange medium) and the polishing step (with a multimodal medium) to purify a mAb solution [314].

In order to give an illustration of the 2C-MCSGP process, we consider the purification of mono-PEGylated lysozyme, conjugated to 5 kDa polyethylene glycol. The feed mixture contained some unreacted lysozyme ($C_0^F = 3$ g/L), two types of mono-PEGylated protein varying in the position at which the polymer is grafted, denoted P_1' and P_1'' ($C_{1'}^F = 0.68$ g/L and $C_{1''}^F = 0.60$ g/L), and a small amount of di-PEGylated protein ($C_2^F = 0.13$ g/L). The goal of this experiment was to isolate the positional isomer P_1' from the reaction mixture. To do so, the 2C-MCSGP process was designed based on the empirical approach using the single-column chromatogram shown in Figure 5.24.

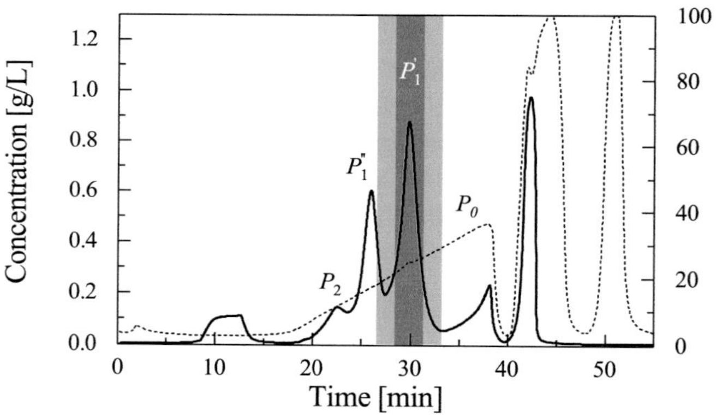

Figure 5.24 Single-column chromatogram for the purification of mono-PEGylated lysozyme performed with a 8 × 100 mm column packed with a strong cation exchange medium and using NaCl as a modifier. The following steps were performed at a flow rate of 2.5 mL/min: load (2 CV), wash with 20 mM phosphate buffer at pH 6.1 (3 CV), linear gradient elution in a 20 mM phosphate buffer at pH 6.1 from 50 to 400 mM NaCl (10 CV) and finally wash, cleaning, and equilibration with various solutions for a total of 10 CV. The composition of the eluent computed from the measured conductivity is represented by a dotted line. The collection and recycling fractions are shaded in dark gray and light gray, respectively. Data taken from [315].

Table 5.4. Parameter values of the 2C-MCSGP process for the purification of PEGylated lysozyme. Zones I, II, and III are disconnected, while zones I_R, II_R, and III_R are interconnected.

Zone	Q [mL/min]		Δt [min]	
	Phase 1	Phase 2	Phase 1	Phase 2
I	–	2.5	–	9.2
II	1.0	2.2	3	9.2
III	2.5	–	3	–
I_R	–	0.4	–	12
II_R	2.1	2.5	2.3	12
III_R	2.5	–	2.3	–

It can be seen that the first eluting solute is the di-PEGylated lysozyme P_2, then come the two types of mono-PEGylated lysozyme P'_1 and P''_1, and finally the unreacted protein P_0. Analytical ion exchange chromatography was performed to assess the purity of P'_1 during the elution. A target purity of 85% was considered in this example. The collection window delivering the product at the specified purity with the maximum recovery is shaded in dark gray in Figure 5.24. It is comprised between the times $t^{beg}_{c,P'_1} = 29$ min and $t^{end}_{c,P'_1} = 31$ min and leads to a recovery of 75%. The recycling windows are selected between $t^{end}_{c,P''_1} = 27$ min and t^{beg}_{c,P'_1} for the first unresolved fraction and between t^{end}_{c,P'_1} and $t^{beg}_{c,P_0} = 33$ min for the second unresolved fraction (shaded in light gray in Figure 5.24).

The 2C-MCSGP process was then designed with two columns identical to the one used in the single-column experiment. The flow rates and switch times were determined according to the empirical design approach described in [308] using the expression of the m parameters given in Table 5.3. The values of these parameters are reported in Table 5.4. The UV signals recorded at the outlet of each column are plotted in Figure 5.25 for the first six cycles. It is seen that the unit reaches steady state in a few cycles.

The stream containing the purified conjugate P'_1 at the outlet of zone II was analyzed by IEC. The result is presented with an open circle in Figure 5.26(a) in terms of protein recovery and purity. For comparison purposes, the results obtained with the single-column process used for the empirical design are also shown for various widths of the collection window. It can be seen that with the MCSGP, a higher recovery value can be reached while maintaining a high purity.

This is a rather general observation for the MCSGP process [83] and another example dealing with the purification of a monoclonal antibody charge variant with a 2C-MCSGP is presented in Figure 5.26(b) [308]. Here again, the open circle corresponds to the MCSGP and the black triangles to different collection windows of the single-column experiment used to perform the empirical design. In this example, 5-cm-long columns were used both for the single-column and MCSGP processes, which implies that the total bed length of the 2C-MCSGP was twice that of the single-column process. For the

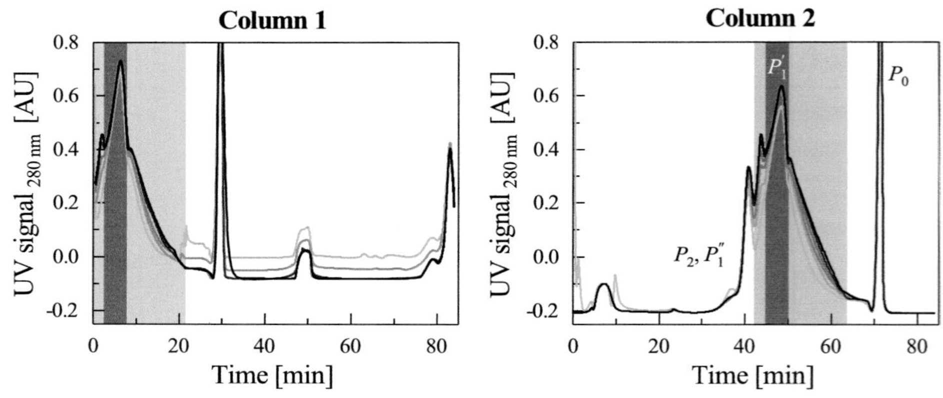

Figure 5.25 Purification of mono-PEGylated lysozyme with strong cation exchange chromatography. Overlay of the chromatograms measured at the outlet of the two columns of the 2C-MCSGP process for the first six cycles, with the darkest lines representing the highest cycle numbers. The dark zones correspond to the collection windows of the pure product, P'_1, while the light gray areas correspond to the recycled fractions. Data taken from [315].

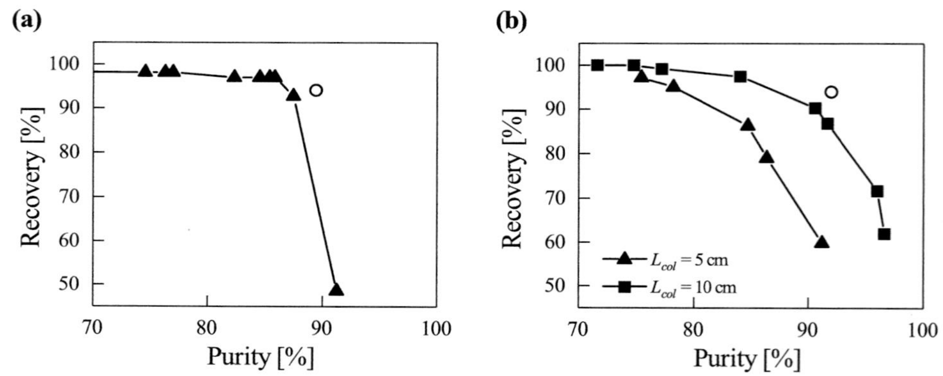

Figure 5.26 Product recovery and purity of the MCSGP process (open circle) designed based on the empirical approach and comparison with the corresponding single-column experiments (filled symbols). The different points for the single-column process were obtained by varying the product collection window (for a given injection). (a) Purification of mono-PEGylated lysozyme (positional isomer P'_1) with a 2C-MCSGP process. Data taken from [315]. (b) Purification of monoclonal antibody charge variants with a 2C-MCSGP process. The MCSGP is constituted of two columns of 5 cm bed height. For comparison, two different single-column processes with one and two columns of 5 cm in series were used. Data adapted from [308], copyright 2017, with permission from Elsevier.

sake of completeness, the results obtained with a single-column of the same bed length as the 2C-MCSGP (i.e., 10 cm) are shown with black squares. Overall, it can be seen that thanks to the countercurrent contact and the internal recycling of unresolved fractions, the MCSGP process allows reaching higher product recovery values for a given purity as compared to the single-column process used for its design. In particular, it is seen that for a purity of 92% the MCSGP leads to a yield of 94%, compared to 50% and 85% for the single-column processes with 5-cm and 10-cm column length, respectively. These conditions imply productivities of 2.6 g/L/h for the MCSGP, and 2.8 g/L/h and 1.8 g/L/h for the single-column processes with column length of 5 and 10 cm, respectively. These results are in line with those found in Chapter 4 in the context of protein capture, namely the single-column process is the most productive at the expense of a very low recovery. The experimental performances given in this example derive from the empirical design approach and therefore do not correspond to optimized processes. A more rigorous comparison between the processes in terms of productivity and eluent consumption for target purity and recovery values will be performed in the next section.

Comparison between Processes

In this section, we use the approach described in Section 3.5.3 to compare the processes presented earlier: the tandem SMB, the 3F-SMB and the MCSGP, which all allow performing ternary separations in a continuous manner but differ in their mode of operation. In particular, the 3F-SMB and the MCSGP both perform ternary separations within a single unit, whereas the tandem SMB consists of two coupled units, each performing a binary separation. In addition, the implementation of the 3F-SMB is restricted to cases where one of the components is strongly retained as compared to the others, whereas no such limitation exists for the tandem SMB and the MCSGP. Besides comparing these three multicolumn processes between each other, we also compare them to the single-column process. To do so, we take the example of a ternary feed mixture containing the solutes *A*, *B*, and *C* in equal proportions. We aim at isolating the product *B*, while the solutes *A* and *C* are treated as impurities. For the sake of simplicity, we consider linear adsorption isotherms and isocratic conditions. To make the comparison between the different processes more concrete, we examine two specific cases differing in terms of selectivity. In case *(i)*, $\overline{K}_A = 3$, $\overline{K}_B = 2$, $\overline{K}_C = 1$, whereas in case *(ii)* $\overline{K}_A = 9$, $\overline{K}_B = 2$, $\overline{K}_C = 1$. The two selected examples have only illustrative purposes with no generality.

Boxes 2.1 and 3.2 summarize the shortcut design approach of the single-column and of the classical four-zone SMB process, respectively. The latter can be adapted to other countercurrent chromatographic units as detailed in the following. Briefly, the efficacy of the different processes is estimated by computing the purification factor (defined as the average between the purity and recovery) as a function of the total number of plates. Then, the column dimensions and the flow rates are determined for a given separation factor (selected here equal to 95%) and a given production requirement (selected here to 10 L of feed solution treated per hour) for a maximal acceptable pressure drop (here $\Delta P^{max} = 5$ bar). The following parameters are used in the van Deemter and Kozeny–Kármán equations: $b_{VD} \approx 0$, $c_{VD} = 2.8 \times 10^{-5}$ h, and $\delta_K = 0.1$ bar.h/m^2.

Tandem SMB

The first step of the shortcut design consists in determining the *m* parameters that maximize the normalized feed flow rate while ensuring complete separation. As discussed in Section 3.2, this corresponds to the vertex of the triangle of the binary separation under

Table 5.5. *m* parameters for the tandem SMB. Case *(i)* corresponds to $\overline{K}_A = 3$, $\overline{K}_B = 2$, $\overline{K}_C = 1$, and case *(ii)* corresponds to $\overline{K}_A = 9$, $\overline{K}_B = 2$, $\overline{K}_C = 1$.

	Case *(i)*		Case *(ii)*	
	SMB1	SMB2	SMB1	SMB2
m^I	2.2	3	9	2
m^{II}	1	2	2	1
m^{III}	2	3	9	2
m^{IV}	1	2	2	1

consideration in the m^{II}–m^{III} plane. Considering that the separation with the largest selectivity is performed in the first unit, i.e., the removal of C in case *(i)* and the removal of A in case *(ii)*, the corresponding m^{II} and m^{III} parameters are summarized in Table 5.5 for the two cases under investigation. The selection of the m^I and m^{IV} parameters of the first SMB unit requires some discussion. In general, m^I is selected equal to the largest lumped Henry coefficient to avoid recycling strongly bound solutes with the solid phase. However, in case *(i)*, it may be advantageous to let the most retained solute A partly elute in the raffinate (i.e., $m^I \leq \overline{K}_A$) to ease the second separation and reduce the solvent consumption. The lower bound of m^I is $\overline{K}_B$ because a lower value would lead to a loss of B in the extract. Therefore, the parameter m^I can be selected somewhere between $\overline{K}_A$ and $\overline{K}_B$ to lead to the best compromise between purity and recovery. Based on simulations with the MC model, it was set here to 2.2. Similar considerations hold on the selection of m^{IV}, which is usually set to the lumped Henry coefficient of the least retained solute. However in case *(ii)*, it may be advantageous to let the weakly bound solute C leak in the extract stream by selecting a larger value of m^{IV}, which was set here to 2.

In a second step, the total number of plates required to obtain a target separation factor is estimated using the MC model assuming instantaneous equilibrium in each cell. As compared to the other processes considered in this section, the tandem SMB is quite flexible because different columns can be used in the two SMB units, which allows reducing the size of the unit performing the easiest separation. For each SMB, we select here as an example a 2–2–2–2 configuration (i.e., a total of eight columns per unit) and for each plate number used in the first unit, we explore a range of plate numbers to be used in the second unit. We then select the combinations leading to the best separation factor values for the smallest number of plates, which as a first approximation leads to the highest productivity. In most applications, the stream leaving the first SMB unit is concentrated before being reinjected in the second SMB unit. This is typically performed with a falling film evaporator in the case of thermally stable molecules and could be performed with a tangential flow filtration system for proteins. Here, we choose to concentrate the outlet stream of the first SMB to reach the same total concentration as in the feed stream. Because the adsorption equilibria are linear, this has no impact on the number of plates required to obtain a given separation, but this will have an impact

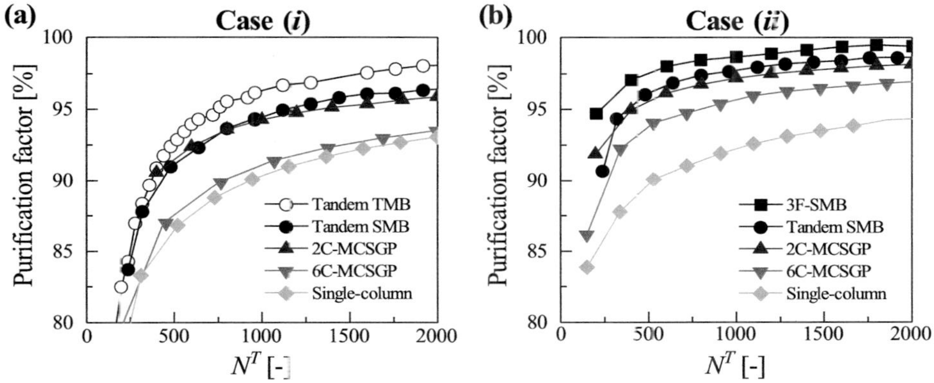

Figure 5.27 Separation factor (defined as the average between purity and recovery) as a function of the total number of plates for different chromatographic processes in (a) case (i) and (b) case (ii). Case (i) corresponds to $\overline{K}_A = 3$, $\overline{K}_B = 2$, $\overline{K}_C = 1$ and case (ii) corresponds to $\overline{K}_A = 9$, $\overline{K}_B = 2$, $\overline{K}_C = 1$.

on the eluent consumption and on the diameter of the columns of the second unit. The separation factor is plotted as a function of the total number of plates of the tandem SMB with black circles in Figure 5.27. It is seen that to obtain a purification factor of 95%, a total number of plates of 1280 (16 columns of 80 plates each) is required for case (i). For case (ii), 640 plates are sufficient (8 columns of 20 plates for the first unit and 8 columns of 60 plates for the second unit).

Equations 2.92 and 2.93 are applied on each SMB unit separately to estimate the average fluid velocity and the total length of the eight columns taken together. This allows computing the column diameters, the total volume of the tandem SMB, the productivity, and the eluent consumption. The results are summarized in Table 5.6.

MCSGP

We have seen in Section 5.3.2 that several implementations of the MCSGP are possible differing in terms of column numbers. We choose here to investigate the six-column and two-column configurations. The most straightforward method to determine the *m* parameters of the MCSGP consists of applying the unified design approach and using the analogy with the single-column process, for which the *m* parameters are summarized in Table 5.3. When the MCSGP is used in isocratic mode, few simplifications arise because the inline dilutions are not necessary, so that Q^{D,II_R} and Q^{D,III_R} are set to 0.

For the 6C-MCSGP process, the maximum pressure constraint applies on the recycling zone, i.e., on the part of the process where three columns are connected in series (corresponding to the succession of the zones I_R, II_R, and III_R). In practice, this means that the target number of plates used in Equations 2.92 and 2.93 is only half of that estimated from Figure 5.27. Moreover, the so-estimated length corresponds to the cumulative length of the three columns in series, while the velocity corresponds to the fluid velocity in the columns of zones I_R, II_R, and III_R. Knowing Q^{III} (10 L/h imposed by

Table 5.6. Comparison between various processes allowing to perform a ternary separation. Case *(i)* corresponds to $\overline{K}_A = 3$, $\overline{K}_B = 2$, $\overline{K}_C = 1$ and case *(ii)* corresponds to $\overline{K}_A = 9$, $\overline{K}_B = 2$, $\overline{K}_C = 1$. The variables u^{SC} and L_{col}^{SC} correspond to the average fluid velocity and column length used in the shortcut design approach (i.e., on the part of the system on which the pressure drop constraint applies).

Case *(i)*		**Single--column**	**tandem SMB** SMB1	SMB2	**2C-MCSGP**	**6C-MCSGP**	**3F-SMB**
N^T	[-]	4900	640	640	1400	3000	
u^{SC}	[m/h]	19	53	53	36	35	
L_{col}^{SC}	[cm]	261	94	94	139	144	
n^T	[-]	1	8	8	2	6	
L_{col}	[cm]	261	12	12	70	48	
Q^F	[L/h]	20	10	–	20	10	n/a
D_{col}	[cm]	5.8	3.1	2.7	4.9	3.9	
V_{col}^T	[L]	6.8	1.3		2.6	3.5	
φ_B	[g/L/day]	345	1447		900	660	
γ_B	[L/g]	0.51	0.32		0.35	0.34	

Case *(ii)*		**Single--column**	**tandem SMB** SMB1	SMB2	**2C-MCSGP**	**6C-MCSGP**	**3F-SMB**
N^T	[-]	2700	160	480	450	720	300
u^{SC}	[m/h]	26	106	61	63	71	84
L_{col}^{SC}	[cm]	194	47	82	79	71	60
n^T	[-]	1	8	8	2	6	7
L_{col}	[cm]	194	6	10	40	24	10
Q^F	[L/h]	80	10	–	80	10	10
D_{col}	[cm]	9.9	1.5	2.2	4.5	2.7	2.4
V_{col}^T	[L]	15	0.41		1.25	0.83	0.32
φ_B	[g/L/day]	158	5492		1821	2762	6340
γ_B	[L/g]	0.81	0.27		0.86	0.75	0.80

the production constraint) and m^{III} (from the analogy with the single-column under ideal conditions), the different flow rates can be estimated. The column diameter is then simply computed from the flow rate and velocity in the zone I_R. In order to reduce the solvent consumption, the value of m^I has been adjusted so that the most retained solute is not entirely eluted from the column located in zone I and is carried over to finally elute together with the least retained solute from zone III_R.

In the case of the 2C-MCSGP, four time steps should be estimated, namely $\Delta t^{IC,1}$, $\Delta t^{DC,1}$, $\Delta t^{IC,2}$, and $\Delta t^{DC,2}$, whose sum is equal to the cycle time Δt. The ratio $\Delta t^{IC,1}/\Delta t$, i.e., the fraction of time during which the feed solution is injected, is

Table 5.7. *m* parameters for the 2C-MCSGP and 6C-MCSGP. Case *(i)* corresponds to $\overline{K}_A = 3, \overline{K}_B = 2, \overline{K}_C = 1$, and case *(ii)* corresponds to $\overline{K}_A = 9, \overline{K}_B = 2, \overline{K}_C = 1$.

	Case *(i)*			Case *(ii)*		
	2C-MCSGP		6C-MCSGP	2C-MCSGP		6C-MCSGP
	Phase 1	Phase 2		Phase 1	Phase 2	
m^I	–	0	0	–	5	4
m^{II}	1	0.67	1	1	0.67	1
m^{III}	1	–	1	1	–	1
m^{I_R}	–	0.5	1.67	–	0.5	1.67
m^{II_R}	0.5	0.5	1.67	0.5	0.5	1.67
m^{III_R}	0.5	–	1.67	0.5	–	1.67

given by the equilibrium theory and is equal to the ratio between the feed time and the cycle time of the single-column process (0.5 in case *(i)* and 0.125 in case *(ii)*). As the average feed flow rate per cycle is fixed to 10 L/h, the feed flow rate of the 2C-MCSGP is 20 L/h in case *(i)* and 80 L/h in case *(ii)*. The rest of the switch time is left to the three other subperiods. To ease the design, it has been chosen to set the volumes of the two recycling fractions by imposing $Q^{IC,1} = Q^{IC,2}$ and $\Delta t^{IC,1} = \Delta t^{IC,2}$, which leads to $m^{I_R} = m^{III_R} = 0.5$. Finally, the pressure drop constraint applies during the interconnected phase, i.e., when two columns are connected in series.

The separation factor is plotted as a function of the total number of plates for each MCSGP process with gray triangles in Figure 5.27. It is seen that to obtain a purification factor of 95%, a total number of plates of 1400 and of 3000 are necessary in case *(i)* for the 2C-MCSGP and 6C-MCSGP, respectively. In case *(ii)*, only 450 and 720 plates are required for the same processes. The result of the calculations of the column geometry, productivity and eluent consumption are reported in Table 5.6.

3F-SMB

Based on the analysis carried out in Section 5.3.2, it is found that the 3F-SMB can only be implemented in case *(ii)*. Similarly to the tandem SMB, we select two columns per zone in the core SMB of the 3F-SMB (i.e., in zones I, II, and III). As compared to the scheme shown in Figure 5.18, there is no zone -I in the present implementation because *A* is recovered under isocratic conditions, so there is no need to re-equilibrate the column. We consider here a configuration with only one column in zone 0. In addition, the *m* parameters in zones I, II, and III are selected identical to the ones of the second unit of the tandem SMB process.

The simulated purification factor is plotted as a function of the number of plates in Figure 5.27. It can be seen that a total number of 300 plates distributed over seven columns is necessary to reach the desired value of 95%. In particular, 258 plates are allocated to the six columns of the core SMB. Because the pressure drop constraint applies on these six columns connected in series, this value of 258 plates is then used together with Equations 2.92 and 2.93 to estimate the average velocity and the total length of the core SMB. The results of the design are reported in Table 5.6.

Single-Column

The first step of the shortcut design approach consists of determining the normalized maximal injection volume. In other words, one has to find the ratio $CV^F = V^F/V_{col}$ leading to the so-called touching band situation delivering a product with 100% purity and 100% recovery in the frame of the equilibrium theory. For a ternary separation under linear conditions, CV^F is given by the minimum between $(1 - \varepsilon_e)(\overline{K}_A - \overline{K}_B)$ and $(1 - \varepsilon_e)(\overline{K}_B - \overline{K}_C)$. Here, it is found that $CV^F = 0.6$ in both cases *(i)* and *(ii)*. Then, one has to figure out at which frequency the successive injections can be performed. Because there are no constraints on the recovery of the impurities *A* and *C*, successive injections can be performed so that the solute *A* from a previous injection co-elutes with the solute *C* from the next injection. This overlapping between two injections enables increasing the productivity and reducing the solvent consumption. In the frame of the equilibrium theory one finds that the injection time corresponds to 50% and 12.5% of the cycle time in cases *(i)* and *(ii)*, respectively. Accordingly, the feed flow rate of the single-column process has to be set to 20 L/h in case *(i)* and to 80 L/h in case *(ii)* in order to match the production requirement of 10 L/h.

Simulations were then performed with the MC model to estimate the required number of plates to reach a purification factor of 95%. It was found that a total of 4900 plates is necessary in case *(i)* and only 2700 in case *(ii)*, as shown with gray diamonds in Figure 5.27. Then, based on Equations 2.92 and 2.93, the average fluid velocity and the total length were estimated. This allowed computing the column diameter, the volume of the chromatographic bed, the productivity and the eluent consumption. The results are summarized in Table 5.6.

Comparison and Discussion

The purification factor as a function of the total number of plates is plotted for all the processes under investigation in Figure 5.27(a) and (b) for cases *(i)* and *(ii)*, respectively. As a comparison, the results of the tandem TMB, which represents the theoretical limit of a SMB with an infinite number of columns, are also shown in Figure 5.27(a).

It is observed that all the multicolumn processes perform better than the single-column process for the two cases under investigation (i.e., they require a smaller number of plates to achieve a given purification factor). Considering a target purification factor of 95%, it is seen that in case *(i)*, the tandem SMB and the 2C-MCSGP are those that require the smallest number of plates, followed by the 6C-MCSGP. In case *(ii)*, the 3F-SMB requires the smallest number of plates, followed by the tandem SMB, the 2C-MCSGP and the 6C-MCSGP.

Figure 5.27 allows a first comparison between the different processes but does not provide any information on the column dimensions and thus neither on productivity nor eluent consumption. These results are summarized in Table 5.6, where it is seen that the different processes vary greatly in terms of column dimensions and performance parameters.

Some interesting comments can be made from these results. Let us first consider the two most efficient processes of case *(i)*: the tandem SMB and the 2C-MCSGP. It is worth noting that with only two columns, the 2C-MCSGP requires a similar number of plates than the tandem SMB, which is constituted of sixteen columns in the implementation under consideration. However, the two processes have significantly different column dimensions. In order to achieve a similar total number of plates, the columns of the 2C-MCSGP need to be longer than those of the tandem SMB (70 cm versus 12 cm in case *(i)*). The columns also differ in terms of diameter for two reasons. First, because the tandem SMB is constituted of two units, the pressure drop constraint applies on each unit separately and is thus less restrictive than for the 2C-MCSGP, where the total number of plates is considered in Equations 2.92 and 2.93. Second, due to periodic injections of the feed stream, the feed flow rate of the 2C-MCSGP process has to be increased in order to match the production requirement of an average feed flow rate of 10 L/h ($Q^F = 20$ L/h in case *(i)*). Overall, the 2C-MCSGP process requires a larger column diameter than the tandem SMB (about 5 cm versus 3 cm in case *(i)*). The larger column dimensions of the 2C-MCSGP lead to a lower productivity value as compared to the tandem SMB. This has to be balanced with the relative simplicity of the 2C-MCSGP process, which only requires two columns, with respect to the tandem SMB, which necessitates sixteen columns in the present implementation and a concentration device between the two units.

It comes out from the results shown in Table 5.6 that all the multicolumn processes under investigation are more productive than the single-column. This is striking in case *(ii)*, where the multicolumn processes are between 10 and 40 times more productive than the single-column. The latter is indeed penalized by the presence of a strongly retained impurity whose significant band broadening due to a high Henry coefficient pollutes the product of the next cycle, whereas the other processes better deal with the removal of strongly retained impurities. In order to improve the single-column process, one may think of using two columns in series so as to get rid of the strongly bound impurities in between the two columns.

Among the considered multicolumn processes, the tandem SMB is especially versatile because the two units can be designed separately to better handle each separation. In case *(i)*, where the separations A/B and B/C are of similar difficulty, the size of the two units is comparable. On the other hand in case *(ii)*, the size of the first unit, where the easiest separation is performed, is largely reduced as compared to the size of the second unit. This allows increasing the productivity and decreasing the eluent consumption. The performance parameters of the tandem SMB may potentially be further improved by considering different chromatographic media and eluent compositions between the two units. This is a general advantage of performing ternary separations in two consecutive steps rather than in one. In the case where one of the three solutes behaves significantly differently than the other two (as in case *(ii)*), the performances of the single-column process may be improved by performing the two separations in two successive steps. Finally, it is worth noticing that the 3F-SMB is very productive and even surpasses the tandem SMB with a smaller number of columns. However, it is not a general solution because it requires one species to be strongly adsorbed. In addition, in its present configuration, the 3F-SMB has a very high eluent consumption compared to the tandem SMB due to the open loop configuration.

The two cases discussed earlier have been selected to illustrate the effect of the selectivity on the choice of the most suitable process. However, a combination of these two cases is often observed in polishing applications. Indeed, the target protein is commonly recovered from a multitude of impurities, some of which elute very close on both sides of the target, while others are much more strongly retained. This may require to combine the high separation efficiency of countercurrent processes with chemical modulation, which is also needed for the column regeneration.

Although the conclusions drawn from this example cannot be regarded as general, the methodology presented in this section can be applied to other countercurrent chromatographic systems and separation tasks.

5.3.3 Process Control

As already stated in Section 4.3.6, we do not intend to provide a comprehensive view of process control, and the reader interested in a more complete presentation of the topic is referred to the existing literature [256–260].

Here, we simply provide an illustrative example, where a control strategy was implemented to optimize a 2C-MCSGP process at the laboratory scale [312]. The basic idea of this work was to use two controllers to adjust the collection and recycling windows of a product eluting between two impurities. The first controller corrected the position of the beginning of the collection window of B by adjusting the switch time of the first interconnected phase ($\Delta t^{IC,1}$ in Figure 5.23), whereas the second controller took care of the end of the collection window by adjusting the switch time of the first disconnected phase ($\Delta t^{DC,1}$). The corrective actions were based on the amount of impurities in the collected fraction containing the solute B measured by at line HPLC. The analytical method was designed to be shorter than a cycle time of the 2C-MCSGP process to allow readjusting the collection window cycle after cycle. In particular, $\Delta t^{IC,1}$ was increased to reduce the concentration of the weakly bound impurities, while $\Delta t^{DC,1}$ was decreased to reduce the concentration of the strongly bound impurities. A mAb mixture containing a main protein variant eluting between two impurities was used as a model system for this study. Figure 5.28 shows the purity and recovery values resulting from the successive steps taken by the double controller. It can be seen that starting from a poorly optimized point ($\vartheta_B = 0.75$, $\varphi_B = 0.83$), the controller delivered a far better operating point after a few iterations only.

We have seen so far that chromatographic models are useful to understand physical phenomena, design, and optimize processes, as well as evaluate their robustness. In modern control strategies, *model predictive controllers* (MPC) [260] can be implemented to estimate a new set of manipulated variables based on the predictions of the chromatographic model. It may also be interesting to couple MPC with a *repetitive control* (RC) strategy (together denoted as RMPC), which allows keeping in memory the actions taken in the previous cycle. The control of SMB processes with RMPC has been thoroughly investigated in the literature, and commonly uses inlet flow rates

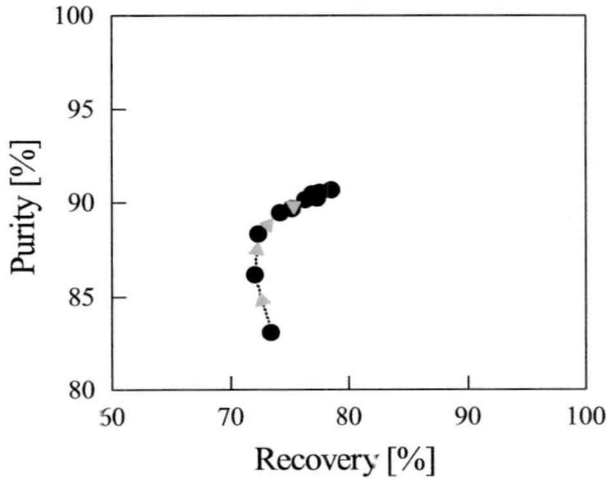

Figure 5.28 Time evolution of the recovery and purity of a mAb purified with a 2C-MCSGP process controlled by a double controller. Adapted from [312], copyright 2013, with permission from Elsevier.

and step time as manipulated variables [316, 317]. The potential of MPC has been well demonstrated for other processes such as the MCSGP process, for which additional manipulated variables as the modifier concentration and the gradient slope can be selected [318, 319].

5.4 Conclusion

In this chapter, we deepened the fundamental concepts of protein chromatography introduced in Chapter 2 to describe the physical phenomena involved in the polishing steps. As compared to the case of protein capture, where the separation protein/impurities can be regarded as binary and noncompetitive due to the high selectivity of the chromatographic medium, this approximation is rarely applicable in the case of the polishing steps. Most often at this stage, the feed mixture contains multiple solutes competing for adsorption at the surface of the chromatographic medium. This has a strong impact on the elution behavior, notably on the shape and position of the peaks/fronts. In particular, for competitive Langmuir adsorption isotherms, the weakly bound solute may elute much earlier than in the absence of competition. The kinetics of internal mass transfer may also be affected by the presence of multiple solutes, thus necessitating the revision of the LDF approximation in highly nonlinear multicomponent systems. In addition, we have seen how to decouple the contributions of the solute retention time distribution and of dispersive effects to the apparent broadness of an unresolved peak.

In a second part, we discussed the potential use of countercurrent chromatography for the polishing steps. The objective here is to meet strict product specifications, starting from the eluate of the capture step, which commonly contains a large variety of impurities eluting both before and after the target protein. This in principle requires the partitioning of the feed in three different streams. The classical four-zone SMB and the flow though processes (similar to those described in Chapter 4 in the context of protein capture) are limited to binary separations, seldom encountered in the downstream processing of therapeutic proteins. Nevertheless, these processes may be extremely valuable in a few applications, such as desalting or aggregate removal. One possibility to perform ternary separations is to use two binary separators in series, such as the tandem SMB. The main advantage of this strategy is that the two units can be designed independently to best cope with the separation task they have been assigned to, in terms of column dimensions but also chromatographic medium and eluent. This may be useful when orthogonal chromatographic methods are needed, e.g., IEC/SEC or IEC/HIC. Alternatively, many processes were developed over the years so as to cope with multifraction separations within a single unit, such as the 3F-SMB or the MCSGP. Considering two simple didactic examples, we performed model simulations to compare various processes performing ternary separations. It was found that all the multicolumn countercurrent processes under investigation outperformed the single-column process in terms of productivity and eluent consumption for high purity and recovery specifications. The improved performances of countercurrent systems lies in their ability to better cope with mass transfer limitations.

Throughout the previous chapters, we have seen two types of methods to design multicolumn processes. The first one is empirical and based on chromatograms obtained with single-column experiments. It allows rapidly determining process variables delivering the product with acceptable purity and recovery specifications for given column dimensions. If the so-obtained multicolumn process represents a good point to start with, it is generally not optimized in terms of productivity and eluent consumption. The second method is based on chromatographic models and allows a more complete design, notably in terms of column dimensions, with a view to optimizing process performance. Model-based approaches are not only very powerful to design chromatographic processes, but also to optimize process performance and investigate process robustness.

6 Protein Conjugation

6.1 Introduction

Protein-based drugs have emerged as a major class of pharmaceuticals in the 1980s due to their outstanding pharmacodynamics, selectivity, and targeting properties [320, 321]. However, therapeutic proteins often exhibit poor pharmacokinetics due to their rapid elimination by the organism and tend to suffer from a low solubility and a limited stability. There is a number of instances where it is desirable to improve the physicochemical properties of therapeutic proteins and extend their functionalities. This can be obtained with a conjugation reaction, which is the attachment of an add-on molecule to the protein [54, 322–325].

Common examples of add-on molecules are polymer chains, such as polyethylene glycol (PEG), which are widely used to increase the *in vivo* circulation half-life of the drug. Besides PEG, other polymers such as N-(2-hydroxypropyl)methacrylamide, polyglutamic acid, dextrin, hydroxyethylstarch, polyacetal Fleximer®, or polysialic acid can be employed [55]. Many successful examples are now available since the first PEGylated protein was approved in 1990 (Adagen®, PEG-bovine adenosine deaminase, ENZON Pharmaceuticals) [55]. Conjugation technologies also lead to major improvements in the treatment of various diseases and the most promising examples are certainly seen in oncology, where the attachment of cytotoxic drugs to cancer-cell-specific antibodies leads to highly potent conjugates often termed "magic bullet," "guided missile," or "armored antibody." For example, Adcetris® and Kadcyla® are two antibody-drug conjugates (ADCs) currently on the market [56]. Other types of therapeutic conjugated antibodies include antibody-enzyme, antibody-protein toxin, and antibody-radionuclide [56]. Besides numerous therapeutic applications, bioconjugation is also relevant for medical diagnosis and imaging. For example, conjugates derived from the attachment of fluorescent add-ons to proteins are used as biomarkers for the sensitive and selective detection of many different biological entities [326]. Finally, bioconjugation is also attractive in the field of chromatography [327]. Conjugation reactions are used, for example, to bind affinity ligands such as protein A to solid supports, or to attach specific tags, such as histidine groups, to proteins in order to purify them by IMAC. A summary of some of the most common protein conjugates is presented in Figure 6.1.

For any application, protein conjugation is raising challenges in the production process and in the characterization of the final product. Indeed, on top of protein

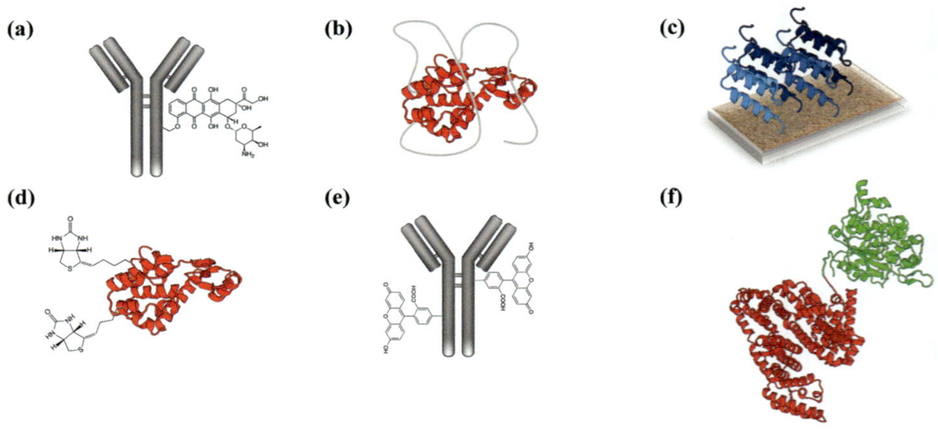

Figure 6.1 Common conjugates: (a) antibody-drug conjugate, (b) PEGylated protein, (c) Protein A attached to a solid surfaces, (d) biotinylated protein, (e) labeled antibody with fluorescent dye, and (f) a synthetic fusion protein.

heterogeneities (e.g., glycosylation, charge variants, and other possible mutations), the conjugation reaction generates a second layer of heterogeneities. This is because add-on molecules usually attach randomly to several residues across the protein surface, thus generating proteins conjugated to various numbers of add-on molecules and at different locations. Therefore, purification is an essential step to isolate the desired conjugate, which is generally a homogeneous product characterized by a given number of add-on molecules linked at well-defined positions into the protein sequence [328].

In this chapter, we review first various conjugation chemistries and present modeling strategies to describe the corresponding reaction kinetics. Then, we discuss the purification of the conjugates by single-column and multicolumn chromatographic processes. Finally, we analyze the possibility of coupling the conjugation reaction and the purification steps in an integrated process.

6.2 The Conjugation Reaction

6.2.1 Conjugation Chemistry

Most of the coupling strategies developed so far target functional groups naturally present in the protein sequence such as lysine and cysteine residues, disulphide bridges, as well as N-terminal amino groups and C-terminal carboxylic acids [55]. These functional groups are schematically shown in Figure 6.2 for a hypothetical protein. For each of these groups, typical reactive moieties of the corresponding add-on molecules are listed in Table 6.1.

Amino acid residues of the protein can also be pretreated in order to extend the range of possible conjugation reactions. For instance, the conversion of an amine to a sulfhydryl, of an aldehyde to a hydrazine or the functionalization of the C-terminal

Table 6.1. Most common conjugation chemistries: protein functional groups and associated reactive moieties of the add-on molecule. Adapted from [55], copyright 2014, with permission from Elsevier.

Protein	Add-on molecule
Carboxyl	Amine Hydrazide
Amine	NHS carboxylate Carbonylimidazole Acrylate Carboxylic acid Tosylate Dichlorotriazine Carbonate
N-terminal	Aldehyde Ortho-pyridylthioester
Thiol	Maleimide Ortho-pyridyldisulfide Thiol
Disulfide bridge	Propenyl sulfone

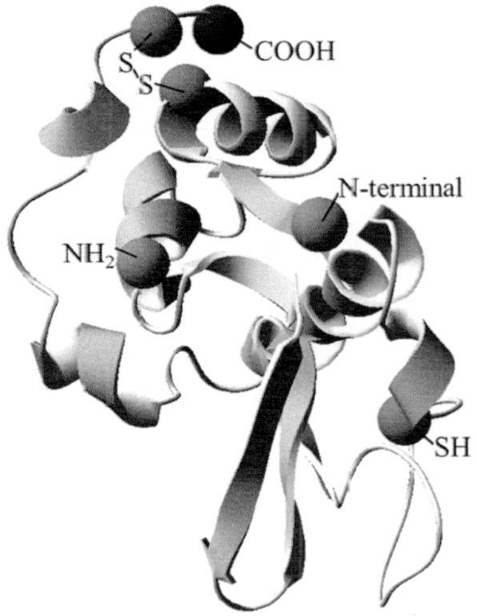

Figure 6.2 Common protein functional groups used for chemical conjugation reactions. Reprinted from [329], with permission from John Wiley & Sons.

carboxylic acid have been reported in the literature [327]. Some conjugation reactions may require the use of bifunctional linkers properly selected to best attach the add-on molecule to the protein. The use of a linker is particularly convenient in the case of conjugation sites hardly accessible to large or highly hydrophobic molecules, which have difficulties approaching the protein surface. The linker can also be tuned to keep the add-on molecule at a certain distance from the surface of the protein so as to facilitate its action by conferring it more flexibility. Finally, the linker is sometimes an important part of the mechanism of action of the conjugated protein, as for example in the case of cleavable linkers [330]. Such linkers are used for certain ADCs (Adcetris® and Mylotarg®) to make sure that the drug is efficiently released only after internalization in the cancer cell. Valine-citrulline (which contains hydrazone bonds cleaving under acidic conditions) or disulfide bridges (which are disrupted under the reducing conditions typically observed within cells) are typical examples of cleavable linkers.

It is worth noting that in all cases, the final utilization of the conjugate determines the choice of the biomolecule and of the add-on reagent because the product of the conjugation reaction exhibits the combined properties of its parts. For this, care must be taken to perform the conjugation reaction in such a way that both the protein and the add-on molecule retain their integrity. It is particularly important to ensure that the conjugation reaction does not affect the protein conformation and activity. For instance, it is essential that PEGylation, which aims at extending the *in vivo* circulation half-life, neither shields the protein active site(s) nor modifies its structure. The same applies to ADCs, whose antigen-binding domain should remain unaltered upon conjugation in order to guarantee the proper cell targeting capabilities in the organism. For these reasons, a careful selection of the conjugation strategy is crucial.

In general, the aforementioned chemical conjugation reactions suffer from a low selectivity. Indeed, most of the targeted sites are not unique within the protein sequence, and the conjugation reaction quite often leads to distributed products characterized by various degrees of conjugation (DoC) and various positional isomers. Considerable efforts are currently put forth to develop alternative methods and increase the selectivity of the reaction [331, 332]. For example, the genetic insertion of free cysteines or even unnatural amino acids has already led to some successful results [55, 333]. The great advantage of this kind of approaches is that the extent of modification and the position of the grafted molecules can be precisely controlled. Another strategy consists in reducing disulphide bridges and replacing them by three-carbon bridges to which the add-on molecule is covalently attached [334, 335]. Enzymatically catalyzed reactions are also receiving a growing interest for site-specific conjugation [330, 336]. Currently, more and more selective chemistries are under investigation to better control the conjugation reaction and ultimately deliver more homogeneous conjugates [330, 333]. However, selective conjugation reactions typically suffer from a lack of versatility as they require the development of tailor-made methods for a given protein/add-on pair, as compared to the chemical route which retains a wide generality.

6.2.2 Kinetics of the Conjugation Reaction

The Kinetic Model

The kinetic model proposed in this section describes the conjugation reaction between a multifunctional protein having n reactive sites and an add-on molecule A. It is assumed that the reactive sites at the protein surface react one after the other, so that the diconjugated protein can only be formed after the monoconjugated one, the triconjugated after the diconjugated, etc. As a first step, the following kinetic scheme can be proposed to describe the progressive conjugation of add-on molecules to the protein independently of the grafting position:

$$\begin{aligned} P_0 + \mathrm{A} &\rightarrow P_1 \\ P_1 + \mathrm{A} &\rightarrow P_2 \\ &\vdots \\ P_{n-1} + \mathrm{A} &\rightarrow P_n \end{aligned} \tag{6.1}$$

where P_0 refers to the protein and P_j to a protein conjugated to j add-on molecules. The number j corresponds to the *degree of conjugation* (DoC). In the simplistic case where only one add-on molecule could be conjugated to the protein, Equation 6.1 simplifies to its first line. The reaction rates describing the kinetics of protein consumption, $\mathcal{R}_0$, and of monoconjugated protein formation, $\mathcal{R}_1$, are then written as follows:

$$\begin{aligned} \mathcal{R}_0 &= -k_0 C_0 C_{\mathrm{A}} \\ \mathcal{R}_1 &= +k_0 C_0 C_{\mathrm{A}} \end{aligned} \tag{6.2}$$

where C_0 is the concentration of the nonconjugated protein, C_1 is the concentration of monoconjugated protein, C_{A} is the concentration of the add-on molecule and k_0 is the rate constant of the conjugation reaction between the amino acid residue (lysine, cysteine, N-terminal, S-S bridge ...) and the functionalized group of the add-on molecule.

However, the chemical modification of proteins is hardly ever so simple and often involves multiple reactive sites on the protein, each one being characterized by its own reactivity. This often leads to the formation of proteins conjugated to various numbers of molecules and at different positions. The number of possible positional isomers of a protein having n conjugation sites and conjugated to j add-on molecules is given by $n!/(n-j)!/j!$. Furthermore, the total number of species, i.e., species with different degrees of conjugation and their corresponding positional isomers, is 2^n. As an example, let us consider the reaction between interferon-α and a chemical reactant specifically targeting amino groups. Interferon-α has 12 possible reaction sites (11 lysine residues and one N-terminal), so a theoretical number of $2^{12} = 4096$ different products can be formed, among which 12 are monoconjugated, 66 diconjugated, 220 triconjugated, etc.

It clearly appears that the kinetic modeling of such complex reactions rapidly becomes very challenging. A mathematical method to deal with multifunctional molecules, originally developed for protein PEGylation, can be found in the literature [337]. Here, this model is extended to any kind of conjugation reactions. We present the rigorous and complete model, which allows a full characterization of the conjugated protein, that is the description of both the degree of conjugation and the position of the grafted add-on molecules. This involves a mathematical formalism that may look cryptic at first sight. A simplified version of this model, which addresses less details about the characterization of the product is presented in the frame of the iso-reactivity assumption. The reader only interested in describing the products of the reaction in terms of DoC (and not in terms of positional isomers) can refer directly to this section.

In the following, we illustrate this methodology with reference to a hypothetic protein having four conjugation sites. First, the idea is to rank the different protein isoforms according to a binary coding indexation where 0 corresponds to a free site and 1 to a conjugated site. For example:

> $[0, 0, 0, 0]$ for the nonconjugated protein,
> $[1, 0, 0, 0]$ for the monoconjugated protein with one add-on molecule in position 1,
> $[0, 1, 0, 0]$ for the monoconjugated protein with one add-on molecule in position 2...

The positions are arbitrarily selected and can, for example, simply be chosen as the order of appearance of the residues in the amino acid sequence.

The great advantage of this binary notation is that each vector characterizing the conjugation state of a protein can be associated with a single integer number i going from 0 to 2^{n-1} in the decimal notation. In other words, one simple integer number is sufficient to describe fully the complex state of a conjugated protein, both in terms of number and position of the grafted add-on molecules. For instance, the integer characterizing a protein having four conjugation sites already functionalized in the first and second positions is 3. Indeed, the decimal notation of $[1, 1, 0, 0]$ reads from left to right as $1 \times 2^0 + 1 \times 2^1 + 0 \times 2^2 + 0 \times 2^3 = 3$. For illustrative purposes, the binary codes of all the possible conjugates of a protein having four reactive sites are given in Table 6.2. Note that the sum of the ones for a given positional isomer corresponds to its DoC.

In the following, the concentration of the conjugated protein characterized by the integer i is noted c_i. This concentration should not be confused with C_i that denotes the total concentration of all the proteins conjugated to i add-on molecules. In other words, the capital letter is referring to the concentration of all the conjugated proteins having a given DoC value, while small letters refer to the concentration of the isoforms. With this notation, the total concentration of a monoconjugated protein having four reactive sites is $C_1 = c_1 + c_2 + c_4 + c_8$. Likewise, the total concentration of diconjugated protein is given by $C_2 = c_3 + c_5 + c_6 + c_9 + c_{10} + c_{12}$, etc. The same formalism is applied to the reaction rates, where r_i refers to the net rate of formation of the positional isomer indexed by the integer i, while $\mathcal{R}_j$ refers to the net rate of formation of the conjugate of DoC equal to j.

The conjugation rate constant of an add-on molecule to the functional group i of a protein already conjugated to j molecules is noted k_j^i. For example, the addition of an add-on molecule to the third position of a diconjugated protein defined by the vector $[1, 1, 0, 0]$ is k_2^3. It is assumed that the kinetic rate constant depends only on the position of the targeted residue and on the number of already conjugated molecules, but not on their location.

Let us now write the net reaction rates of formation of the positional isomers. As compared to the simple scheme presented in Equation 6.1, one needs to account for all the reactions occurring in parallel that can lead to the formation of the same products. Indeed, a single positional isomer can be formed in multiple ways depending on the order in which the sites at the protein surface react. For instance, the nonconjugated protein is consumed by the sum of all the possible reactions leading to the addition of one add-on molecule. Continuing our example with a protein having four reactive sites, the reaction rate of protein consumption is given by:

$$r_0 = -\left(k_0^1 + k_0^2 + k_0^3 + k_0^4\right)c_0 C_A \tag{6.3}$$

where $c_0 = C_0$ is the concentration of nonconjugated protein. If one wants to describe this reaction in the frame of the model presented in Equation 6.1, one would obtain $k_0 = \sum_i k_0^i$. Therefore, the model presented in Equation 6.2 corresponds to a simplified kinetic model, where the reactants and products involved are pseudospecies grouped according to their degree of conjugation.

Then, the net reaction rate of formation of the monoconjugated protein conjugated at its first position, i.e., indexed by $[1, 0, 0, 0] = 1$, is given by:

$$r_1 = k_0^1 c_0 C_A - \left(k_1^2 + k_1^3 + k_1^4\right)c_1 C_A \tag{6.4}$$

The term $k_0^1 c_0 C_A$ describes the formation of the considered monoconjugated protein due to the reaction of the nonconjugated protein of concentration c_0 with one add-on molecule at the first reactive site, while the term $(k_1^2 + k_1^3 + k_1^4)c_1 C_A$ describes the disappearance of the monoconjugated protein due to its reaction with one additional add-on molecule at the free positions 2, 3, and 4, respectively.

Table 6.2. Binary coding of the different isomers of a four-site protein Note that here the binary codes are written from left to right. Adapted from [329], with permission of John Wiley & Sons.

DoC	Binary code	Decimal notation
0	0 0 0 0	0
1	1 0 0 0	1
	0 1 0 0	2
	0 0 1 0	4
	0 0 0 1	8
2	1 1 0 0	3
	1 0 1 0	5
	0 1 1 0	6
	1 0 0 1	9
	0 1 0 1	10
	0 0 1 1	12
3	1 1 1 0	7
	1 1 0 1	11
	1 0 1 1	13
	0 1 1 1	14
4	1 1 1 1	15

Similarly, the net reaction rate r_2 can be written for the monoconjugated protein conjugated at its second site, i.e., indexed by $[0, 1, 0, 0] = 2$:

$$r_2 = k_0^2 c_0 C_A - (k_1^1 + k_1^3 + k_1^4) c_2 C_A \tag{6.5}$$

And for the protein indexed by $[1, 1, 0, 0] = 3$, one obtains:

$$r_3 = (k_1^2 c_1 + k_1^1 c_2) C_A - (k_2^3 + k_2^4) c_3 C_A \tag{6.6}$$

With the help of this formalism, the preceding equations can be generalized for any isoform i whose degree of conjugation is j, as follows:

$$r_i = \underbrace{(\boldsymbol{\delta_i} \cdot \boldsymbol{B}) \cdot (\boldsymbol{k_{j-1}} \circ \boldsymbol{c_i^\star}) \times C_A}_{\text{formation}} - \underbrace{(\boldsymbol{\delta_i} \cdot (\boldsymbol{1} - \boldsymbol{B})) \cdot \boldsymbol{k_j} \times c_i \times C_A}_{\text{consumption}} \tag{6.7}$$

where $\mathbf{B}$ is the matrix of size $2^n \times n$ of binary vectors ranging from 0 to $2^n - 1$. Continuing our example of a protein carrying four conjugation sites, $\mathbf{B}$ is given by:

$$\mathbf{B} = \begin{bmatrix} 0 & 0 & 0 & 0 \\ 1 & 0 & 0 & 0 \\ 0 & 1 & 0 & 0 \\ 1 & 1 & 0 & 0 \\ 0 & 0 & 1 & 0 \\ 1 & 0 & 1 & 0 \\ \vdots & \vdots & \vdots & \vdots \\ 1 & 1 & 1 & 1 \end{bmatrix} \tag{6.8}$$

In addition, $\boldsymbol{\delta_i}$ is the Kronecker row vector whose elements equal 0 except the $i + 1$ row which equals 1. Thus, the product $\boldsymbol{\delta_i} \cdot \mathbf{B}$ gives the row $i + 1$ of the matrix $\mathbf{B}$, which corresponds to the binary notation of i. Moreover, $\circ$ is the element-by-element product, j is the DoC of the positional isomer i and $\boldsymbol{k_j}$ is a column vector of n kinetic rate constants (i.e., the k_j^i with i going from 1 to n, for a given DoC equal to j). Finally, $\mathbf{1}$ is the $2^n \times n$ matrix whose elements equal 1 and $\mathbf{c}_i^\star$ is the column vector of n reactants involved in the formation of the component i (i.e., containing one conjugated site less than the conjugated protein i), which is computed as follows:

$$\mathbf{c}_i^\star = \begin{bmatrix} c_{i-1} \\ c_{i-2} \\ c_{i-4} \\ c_{i-8} \\ \vdots \\ c_{i-2^{n-1}} \end{bmatrix} \quad \text{if } i - 2^m \geq 0, \quad m \in [0, n-1] \text{ (0 otherwise)} \tag{6.9}$$

For example, the formation of the fully conjugated protein described by the vector [1111], derives from the conjugation of [0111], [1011], [1101] and [1110] (i.e., reactants whose DoC is equal to 3 and with a free site in position 1, 2, 3, or 4, respectively) with an add-on molecule. These vectors correspond to the subtraction of [1000], [0100], [0010], and [0001] to [1111]. In the decimal notation, this becomes 15 minus 1, 2, 4, and 8, respectively, i.e., 14, 13, 11, and 7. The elements of the vector $\boldsymbol{c}_{15}^{\star}$ are thus $\{c_{14}\ c_{13}\ c_{11}\ c_{7}\}$. If the result of the subtraction is negative, the corresponding element of the $\boldsymbol{c}_i^{\star}$ vector is set to 0.

With reference to a protein having four possible conjugation sites, the net reaction rate describing the formation of the conjugate 3 ($i = 3$, i.e., [1100] and $j = 2$) is written as follows:

$$r_3 = [1100] \cdot \left(\begin{bmatrix} k_1^1 \\ k_1^2 \\ k_1^3 \\ k_1^4 \end{bmatrix} \circ \begin{bmatrix} c_2 \\ c_1 \\ 0 \\ 0 \end{bmatrix} \right) \cdot C_{\mathrm{A}} - ([1111] - [1100]) \cdot \begin{bmatrix} k_2^1 \\ k_2^2 \\ k_2^3 \\ k_2^4 \end{bmatrix} \cdot c_3 C_{\mathrm{A}} \qquad (6.10)$$

On the other hand, the rate the of add-on molecule consumption, r_{A}, is given by the sum of all the possible conjugation events:

$$r_{\mathrm{A}} = -\sum_{i=0}^{2^n-1} (\boldsymbol{\delta}_i \cdot (\mathbf{1} - \mathbf{B})) \cdot \boldsymbol{k}_j \times c_i \times C_{\mathrm{A}} \qquad (6.11)$$

where the value of j corresponds to the degree of conjugation of the conjugate i. The kinetic model presented here requires a total of n^2 kinetic rate constants for a protein having n reactive sites. In the following, we propose an explicit expression of the reaction rate constant, which allows reducing the number of kinetic parameters down to $n + 2$.

The Kinetic Rate Constants

Conjugation reactions between proteins and large add-on molecules are sometimes limited by the diffusion of the reactants. This is likely to be the case with large therapeutic proteins conjugated with polymers, for instance. On the other hand, with fast-diffusing add-on molecules, the kinetics of reaction are rather likely to be controlled by the chemical event. An effective kinetic rate constant accounting for the diffusion of the reactants and the subsequent chemical conjugation reaction is given by [337]:

$$k_j^i = k^i \Psi_j \quad \text{with} \quad \Psi_j = \left[\frac{(R_{\mathrm{A}} + R_0)(1/R_{\mathrm{A}} + 1/R_j)}{1 + \kappa \frac{1/R_{\mathrm{A}} + 1/R_j}{R_{\mathrm{A}} + R_0}} \right] \exp(-j\alpha MW_{\mathrm{A}}) \qquad (6.12)$$

where i is the position of the conjugation site, j the number of already conjugated residues (i.e., the DoC value), k^i the intrinsic reaction rate constant between the site i and the add-on molecule, and Ψ_j a steric factor. The latter accounts for the diffusion of the reactants, their collision cross section, and the reactive surface available for conjugation. MW_{A} is the molecular weight of the add-on molecule, while R_{A}, R_0, and R_j are the hydrodynamic radii of the add-on molecule, of the nonconjugated protein and of the conjugated protein of DoC equal to j, respectively. In Equation 6.12, the impact of solute diffusion is reflected in the terms $1/R_{\mathrm{A}}$, $1/R_0$, and $1/R_j$ because $D \propto 1/R$ according to the Stokes–Einstein equation (see Equation 2.68). The hydrodynamic radii can either be estimated independently by SEC or light scattering measurements (see Chapter 7) or evaluated from literature correlations [338]. In addition, κ and α are two adjustable parameters of the kinetic model, as discussed in the following.

The parameter α describes the reduction of the number of accessible sites due to the coverage by the already grafted add-on molecules. This reduction of active surface is likely to be significant in the case of polymer-protein coupling reactions, like PEGylation, but much less significant for other reactions like acetylation of amine residues, where the size of the add-on molecule is negligible as compared to the size of the protein. On the other hand, the parameter κ depends upon the activation energy of the chemical reaction and on the type of interactions between the protein and the add-on molecules. The product $\kappa(1/R_A + 1/R_j)/(R_A + R_0)$ represents the ratio between the diffusion rate and the chemical reaction rate, which is commonly referred to as the inverse of the Thiele modulus [99]. Large values of this product indicate that the reaction rate is limited by the chemical reaction, while small values indicate that the reaction rate is limited by the diffusional approach of the two reactants toward each other. In the former case, Equation 6.12 simplifies to:

$$k_j^i = k^i(R_A + R_0)^2\kappa^{-1}\exp(-j\alpha MW_A) \tag{6.13}$$

In this equation, $k^i(R_A + R_0)^2\kappa^{-1}$ is the reaction rate constant of a given site i with the add-on molecule in the case where no other sites are conjugated.

Overall, Equation 6.12 includes a total of $n + 2$ adjustable parameters:

- n intrinsic kinetic rate constants k^i for the n target residues
- the two parameters α and κ

The benefit of Equation 6.12 lies in the reduction of the number of kinetic parameters from n^2 (i.e., considering all the k_j^i) to $n + 2$ (i.e., considering the k^i, κ, and α) by predicting the effect of the DoC on the reactivity of the protein. This in turn simplifies the model parameter estimation procedure, especially for proteins having a large number of reactive sites. To give an example, this means that modeling the PEGylation reaction of interferon-α at the amine residues and N terminal ($n = 12$) with the detailed model would require 144 parameters. With the proposed simplifications, the number of model parameters will be reduced to 14.

Iso-Reactivity Assumption

The determination of the conjugation rate constants k_j^i (where i refers to the position of the targeted residue and j to the number of already grafted molecules) necessitates identifying and quantifying the concentration of all the products of the reaction, not only according to their DoC but also according to the position of the grafted molecule. This is a rather cumbersome task, which requires the use of multiple and orthogonal techniques to resolve all the possible products. Analytical chromatography, capillary electrophoresis, coupled with peptide mapping and other reducing techniques have been used to this end [339–342].

For illustrative purposes, we report here the conjugation of a monoclonal antibody with a fluorescent dye specifically targeting free amines. The antibody was conjugated

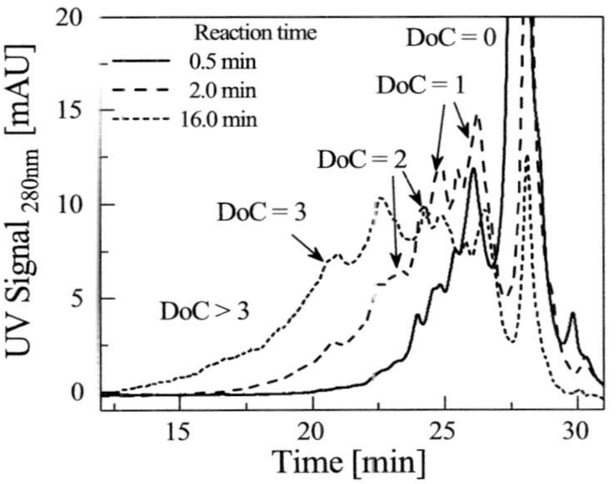

Figure 6.3 Kinetics of the conjugation between a monoclonal antibody and threefold molar excess of Atto488-NHS-dye. The products of the reaction were analyzed by weak cation exchange chromatography under LGE conditions (100 mM NaCl to 450 mM NaCl over 30 min at pH 5.5. The DoC of each peak was estimated according to the supplier's instructions [343]. Reprinted from [329], with permission from John Wiley & Sons.

with 3-molar excess of Atto488-NHS-dye, and the products of the reaction were analyzed by analytical weak cation exchange chromatography. Figure 6.3 shows the analytical chromatograms taken at different reaction times. The DoC of the proteins eluting in the different peaks was estimated by comparing the UV signals at 280 nm and 500 nm as suggested by the dye supplier [343]. The peak eluting at around 28 min corresponds to the unreacted protein, while all the other peaks eluting before correspond to conjugated proteins. One can see multiple peaks corresponding to the monoconjugated protein, diconjugated protein, the triconjugated protein and higher degrees of conjugation. For each degree of conjugation, multiple positional isomers are present, which leads to a heterogeneous family of mAb-dye conjugates getting more and more complex as the reaction time goes on.

In such cases, one is possibly no longer interested in modeling the entire distribution of isoforms, but only the distribution of degrees of conjugation. Accordingly, the species considered in the kinetic scheme are pseudospecies grouped according to their DoC, as shown in Equation 6.1. In this frame, a reasonable approximation consists in assuming that all the residues have the same reactivity, i.e., that the intrinsic rate constants k^i defined in Equation 6.12 are equal to k^{iso} for all the sites. Under this simplifying assumption, the kinetic rate constant only depends on the number of already grafted add-on molecules j through the steric factor Ψ_j so that $k_j^i = k^{iso}\Psi_j$ for all the conjugation sites. The overall reactivity of the pseudospecies with a DoC equal to j is given by the sum of all the corresponding positional isomers. For instance, for $j = 0$, the overall reactivity of the nonconjugated protein is given by $k_0 = \sum_i k_0^i = nk^{iso}\Psi_0$. This can be generalized to any DoC equal to j by using combinatorial calculus, which leads to:

$$k_j = (n - j)k^{iso}\Psi_j \tag{6.14}$$

where the expression of Ψ_j is given in Equation 6.12

This expression of the conjugation rate constant contains only three adjustable parameters: k^{iso}, α and κ. The iso-reactivity assumption therefore leads to a reduction of the number of model parameters from $n+2$ down to 3, which is particularly appreciable for large numbers of conjugation sites n.

In the frame of the iso-reactivity assumption, the kinetic rate constant k_j are apparent kinetic parameters that quantify the reactivity of the proteins characterized by a DoC equal to j, summing together the reactivity of all the positional isomers. The apparent kinetic rates, $\mathcal{R}_j$, can then be expressed as a function of C_j, the concentration of proteins with a DoC equal to j, as follows:

$$\mathcal{R}_j = k_{j-1}C_{j-1}C_{\mathrm{A}} - k_jC_jC_{\mathrm{A}} \tag{6.15}$$

For instance, the apparent rate of protein consumption is given by $\mathcal{R}_0 = -k_0C_0C_{\mathrm{A}}$, the apparent net rate of monoconjugate formation is given by $\mathcal{R}_1 = k_0C_0C_{\mathrm{A}} - k_1C_1C_{\mathrm{A}}$.

As an example, the PEGylation of α-lactalbumin on free amines using a 10 kDa methoxy-PEG succinimidyl acid is considered. The experiments were performed in a batch stirred tank reactor and the concentrations of the different PEGamers were measured by size exclusion chromatography at selected time points. The experimental results are shown in Figure 6.4 [344]. It is seen that these kinetic data confirm the general kinetic scheme presented in Equation 6.1. The protein is consumed by the addition of a first PEG molecule, thus forming a mono-PEGylated protein, which is consumed by the addition of a second PEG molecule, and so on until the complete consumption of the PEG or the complete conjugation of all the functional groups of the protein.

In the following, we estimate the reliability of the reaction rate expression shown in Equation 6.14, which was derived based on specific physical assumptions. For this, two different approaches are considered to simulate the experimental data:

- The kinetic rate constants are computed with Equation 6.14 and the three parameters k^{iso}, α, and κ are regressed on the experimental data (solid lines in Figure 6.4).
- The five k_j rate constants are regarded as independent parameters and regressed directly on the experimental data without using any of the previous expression of the reaction rate constants (dotted lines in Figure 6.4).

It can be seen that both models give satisfactory results and that the simulations using the kinetic constants computed with Equation 6.14 show a very good agreement with the experimental data although using only three fitting parameters (instead of five for the second approach). This supports the assumptions behind the iso-reactivity model.

Add-on Molecule Deactivation

A very common side reaction occurring during protein conjugation is the chemical deactivation of the functionalized add-on molecule, for example by hydrolysis. When water is in large excess, such reactions are commonly modeled as first-order kinetics so that the effective deactivation rate is computed as $\mathcal{R}_d = k_dC_{\mathrm{A}}$. Accordingly, Equation 6.11 has to be modified to account for this deactivation reaction. If one groups the different

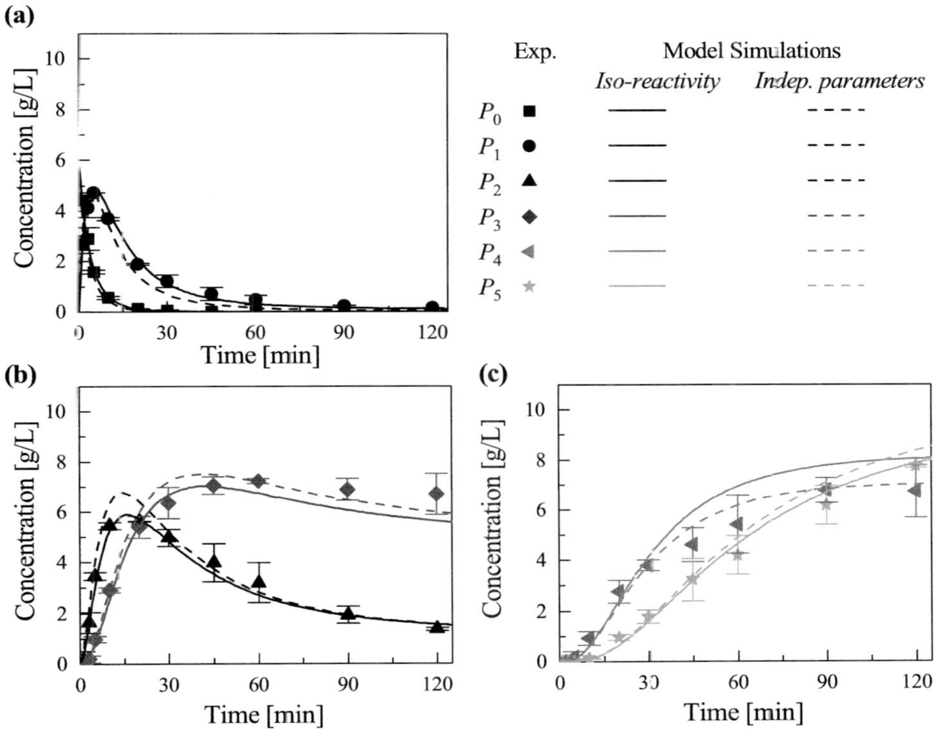

Figure 6.4 PEGylation of α-lactalbumin with a sixfold molar excess of 10 kDa mPEG-SPA. Comparison between experimental data (symbols) and simulated data with the kinetic rate constant given by Equation 6.14 and the following set of parameters: $n = 13$, $k^{iso} = 2.06$ mol/min/ mL, $\alpha = 1.8 \times 10^{-5}$ mol/g, and $\kappa = 790$ Å^2 (solid lines), or considering each kinetic rate constant k_j as an independent fitting parameter with the values $k_0 = 134$, $k_1 = 81$, $k_2 = 41$, $k_3 = 28$, and $k_4 = 27$ mL/mol/min (dashed lines). Adapted from [344], with permission from The Royal Society of Chemistry.

isoforms according to their degree of conjugation, the rate of add-on molecule consumption is given by:

$$\mathcal{R}_A = r_A = -k_d C_A - \sum_{j=0}^{n-1} k_j C_j C_A \tag{6.16}$$

As an example, we consider the deactivation of a 5 kDa mPEG-SPA at various pH values [345]. Briefly, the hydroxysuccinimide ester of the mPEG-SPA is undergoing a hydrolysis reaction releasing an N-hydroxysuccinimide molecule on the one hand and a methoxy-PEG propionic acid on the other hand. Figure 6.5(a) shows the UV absorbance recorded at 260 nm as a function of time at different pH values. Note that the absorbance data were log-linearized so that the slope of the lines shown in Figure 6.5(a) corresponds to the first-order reaction rate constant, k_d. The values of k_d are plotted as a function of the solution pH in Figure 6.5(b). It is seen that k_d is not very sensitive to the pH value below pH 6, but rapidly increases by two orders of magnitude when reaching pH 9.

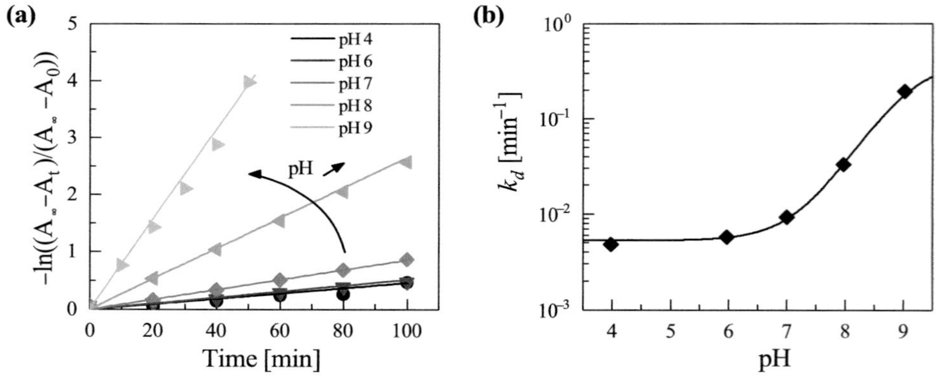

Figure 6.5 (a) Linearization of the kinetics of deactivation of the NHS ester of a 5 kDa mPEG-SPA measured by UV absorbance at 260 nm at different pH values. (b) Hydrolysis rate constant as a function of the pH value of the solution. Symbols represent experimental data, while the solid line corresponds to the physical model proposed in the original work [345]. Adapted from [329], with permission from John Wiley & Sons.

6.2.3 Conjugation Reactors

Definition of Conversion, Yield, and Selectivity

In the following, we give general definitions of the reactant conversion, product yield, and product selectivity, which are essential parameters to design conjugation reactors. We report first the corresponding definitions applied to a batch stirred tank reactor (BSTR) and then to a continuous stirred tank reactor (CSTR) and a plug flow reactor (PFR).

In a BSTR, the protein *conversion* is defined as the amount of protein reacted over the amount initially fed into the reactor. It is expressed as follows:

$$X_0 = \frac{n_0^F - n_0}{n_0^F} \tag{6.17}$$

where n_0 is the number of moles of the nonconjugated protein at a given time in the reactor, and n_0^F is its initial value. Similarly, it is convenient to define the conversion of the add-on molecule as:

$$X_A = \frac{n_A^F - n_A}{n_A^F} \tag{6.18}$$

where n_A is the total (activated and deactivated) number of moles of the add-on molecule at a given time in the reactor, and n_A^F its initial value.

The *yield* is defined as the number of moles of a given product over the number of moles of one of the reactants initially introduced in the batch reactor. In particular, for a conjugated protein with a DoC equal to j, the yield with respect to the nonconjugated protein is given by:

$$Y_j^0 = \frac{n_j}{n_0^F} \tag{6.19}$$

For example, Y_1^0 quantifies the fraction of the moles of nonconjugated proteins that reacted to form the monoconjugated protein. The yield could also be computed with respect to the add-on molecule in order to quantify which proportion of the loaded add-on molecules effectively lead to the desired product. For this, we can define Y_j^{A} by:

$$Y_j^{\mathrm{A}} = \frac{n_j}{n_{\mathrm{A}}^F} \tag{6.20}$$

The choice of the definition of the yield is particularly important for reactions in series when the conjugate of interest is an intermediate product.

The *selectivity* of a reaction is given by the ratio between the number of moles of desired product over the number of moles of converted reactant. In a BSTR, the selectivity of the reaction for the protein characterized by a DoC equal to j is computed as follows:

$$Z_j^0 = \frac{n_j}{n_0^F - n_0} \tag{6.21}$$

A selectivity of 1 implies that no side reactions occur and that the only product is the desired one. In the preceding equation, the selectivity is defined with respect to the nonconjugated protein, but it is also possible to define the selectivity with respect to the add-on molecule as:

$$Z_j^{\mathrm{A}} = \frac{n_j}{n_{\mathrm{A}}^F - n_{\mathrm{A}}} \tag{6.22}$$

It is worth noting that the yield can be expressed as a function of the conversion and selectivity as $Y_j^0 = X_0 Z_j^0$ and $Y_j^{\mathrm{A}} = X_{\mathrm{A}} Z_j^{\mathrm{A}}$.

Finally, for conjugation reactions, the initial molar ratio between the add-on molecule and the protein is particularly important. We therefore introduce ξ defined as:

$$\xi = \frac{n_{\mathrm{A}}^F}{n_0^F} \tag{6.23}$$

With this additional definition, one obtains $Y_j^0 = Y_j^{\mathrm{A}} \xi$.

Equations 6.17–6.23 are valid for batch reactors only. For continuous reactors, the definitions of the conversion, yield, selectivity, and ratios of reactants retain their physical meaning but have to be expressed in terms of molar flow rates, $\dot{n}_i = Q_i C_i$ instead of the molar quantities, n_i. In particular, $\dot{n}_i^F = Q_i^F C_i^F$ represents the molar feed flow rate of the reactant i.

Batch Reactors

The mole balance equations describing the time evolution of the different reactants and products in a BSTR are as follows:

$$\frac{\mathrm{d}C_j}{\mathrm{d}t} = \mathcal{R}_j = \begin{cases} -k_0 C_0 C_{\mathrm{A}} & \text{for } j = 0 \\ k_{j-1} C_{j-1} C_{\mathrm{A}} - k_j C_j C_{\mathrm{A}} & \text{for } 1 \le j \le n-1 \\ k_{n-1} C_{n-1} C_{\mathrm{A}} & \text{for } j = n \end{cases} \tag{6.24}$$

$$\frac{dC_A}{dt} = \mathcal{R}_A = -k_d C_A - \sum_{j=0}^{n-1} k_j C_j C_A \tag{6.25}$$

$$\frac{dC_{dA}}{dt} = \mathcal{R}_d = k_d C_A \tag{6.26}$$

where n is the number of conjugation sites and C_{dA} is the concentration of deactivated add-on molecule. These expressions are provided in terms of degree of conjugation for the sake of simplicity but they can be written for each isoform if one is interested in the formation of a selected positional isomer using the complete model discussed earlier in this section.

The system of ordinary differential Equations 6.24–6.26 can be integrated numerically to obtain the concentration profiles of the reactants and products, as it can be seen for example in Figure 6.4. This system of equations has also been solved analytically, originally for the PEGylation of recombinant hirudin, but the solutions proposed can be applied to any conjugation reaction that involves one, two, or three conjugation sites [346].

For most of the batch operations, designing the reactor consists in selecting the reaction time to obtain a target conversion X_0^* of the protein for a given composition of the feed mixture. Indeed, the longer the reactants are in the reactor, the more they are converted until reaching the complete consumption of one of the reactants. Introducing the definition of the conversion given in Equation 6.17, the design equation of the BSTR can be expressed as follows:

$$t^{BSTR} = C_0^F \int_0^{X_0^*} \frac{1}{-\mathcal{R}_0} dX_0 \tag{6.27}$$

Quite often, the desired conjugate is an intermediate product. In this case, finding the optimal reaction time does not only consist of optimizing the conversion of the protein, but also the yield for the target product. Low protein conversions necessarily lead to a low yield, but too high conversions also result in the consumption of the intermediate product to form overconjugated species, and thus a low yield too. This is seen in Figure 6.4, where the concentration of mono-PEGylated α-lactalbumin first increases with time and then decreases due to the formation of higher degrees of PEGylation.

In the case where the target product is the monoconjugated protein, the yield Y_1^0 can be expressed as a function of the protein conversion X_0 [98, 346]:

$$Y_1^0 = \frac{k_0}{k_0 - k_1}\left[(1 - X_0)^{k_1/k_0} - (1 - X_0)\right] \tag{6.28}$$

The conversion delivering the maximal yield is then given by [98, 346]:

$$X_0\left(Y_{1,\max}^0\right) = 1 - \left(\frac{k_0}{k_1}\right)^{\frac{k_0}{k_1 - k_0}} \tag{6.29}$$

For the system presented earlier, i.e., the PEGylation of α-lactalbumin, the numerical values of the kinetic rate constants reported in Figure 6.4 were used to evaluate Equation 6.28. The yield for the mono-PEGylated protein is reported as a function of

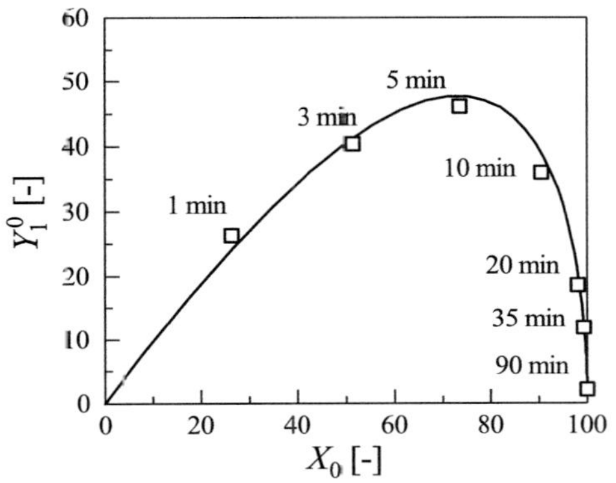

Figure 6.6 Yield for the mono-PEGylated α-lactalbumin as a function of the protein conversion. The solid line was computed with Equation 6.28, using Equation 6.14 and the parameters k^{iso}, α, and κ reported in Figure 6.4 to compute k_0 and k_1. The open squares correspond to experimental data obtained at various time points during the reaction [344].

the protein conversion in Figure 6.6 (solid line). Experimental data obtained at different time points are also included and represented by open squares [344]. It can be seen that the simple analytical expression given by Equation 6.28 describes well the experimental results. It is also observed that the maximum yield is equal to 47%. In other words, in the best case, less than half of the protein initially introduced in the reactor is used to produce the desired product. In this example, the maximum yield is obtained for a protein conversion of 73%.

Knowing the value of the optimal conversion from Equation 6.29, Equation 6.27 is then used to compute the optimal reaction time, t^{BSTR}. It is important to note that the value of the optimal reaction time depends on C_0^F, the initial protein concentration, and on ξ, the initial molar ratio between the add-on molecule and the protein. The simulated yield, Y_1^0, is reported in Figure 6.7 as a function of the reaction time for different values of ξ. It can be seen that for $\xi = 0.5$, the yield never reaches the theoretical maximum value computed with Equation 6.28. There is in fact a threshold value of ξ below which the reaction stops prematurely due to the complete consumption of the add-on molecule. In the present example, this value is 1.2, which is slightly above the stoichiometric value 1 because of the deactivation reaction, which consumes some of the add-on molecules before they can react with the protein. With $\xi = 1.2$, the optimal reaction time is theoretically infinite but in practice 99% of the maximum yield is obtained after 1 h. Then, it can be seen in Figure 6.7 that for ξ values above this threshold of 1.2, the maximum yield is reached at shorter and shorter reaction times when increasing ξ. This is because an increase of the concentration of add-on molecules at a given protein concentration leads to an increase in the conjugation rate. It is important to realize that for $\xi > 1.2$, the reaction must be quenched at t^{BSTR} before the monoconjugated product further reacts, which would result in a rapid decrease of the yield, Y_1^0.

The cost associated with the conjugation reaction is not only impacted by the cost of the protein, but also by the cost of the add-on molecule, which are sometimes

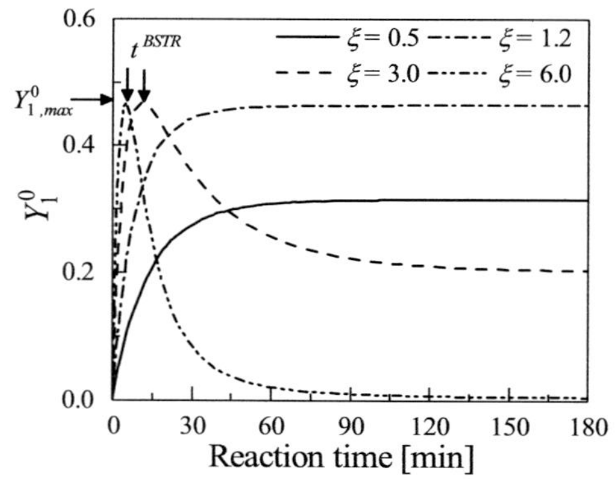

Figure 6.7 Simulated yield, Y_1^0, as a function of the reaction time for various add-on molecule/protein molar ratios, ξ, at the initial protein concentration $C_0^F = 6$ mg/mL. The simulations were obtained by integrating Equations 6.24–6.26. Arrows indicate the time at which the reaction should be stopped to obtain the maximum yield.

comparable. For instance, for PEGylated antibodies or antibody fragments (e.g., Cimzia®), the production cost of monoclonal antibodies is nowadays in the range of 100–200 \$/g for large-scale facilities, while monodispersed functionalized PEG borders 100 \$/g [265]. Therefore, besides the protein conversion and yield, the efficient incorporation of the add-on molecule into the desired product is also an aspect to be considered. In a BSTR, two main process variables can be tuned to improve the incorporation of the add-on molecule: the initial molar ratio add-on molecule/protein, ξ, and the initial protein concentration, C_0^F. The selected value of ξ is related to the stoichiometry of the reaction so that higher ξ values lead to higher DoC values. In addition, higher concentrations of protein favor the conjugation reaction with respect to the deactivation of the add-on molecule. This is because deactivation is a first-order reaction (with the characteristic time $1/k_d$), whereas conjugation is a second-order reaction (with the characteristic time $1/(k_0 C_0^F)$).

Continuing with the previous example, simulations were performed with different C_0^F and ξ values. The conversion of the PEG, X_A, and the yield of the mono-PEGylated α-lactalbumin with respect to the PEG, Y_1^A, computed at the time where Y_1^0 is maximal are shown in Figure 6.8(a) and (b), respectively. Because we are targeting a mono-PEGylated protein, $\xi > 1$ indicates that the PEG is in excess, whereas $\xi < 1$ indicates that the PEG is the limiting reactant. It is seen in Figure 6.8 that the best results in terms of conversion and yield of the PEG are obtained when the PEG is in default ($\xi = 0.5$ here). It is also observed that both X_A and Y_1^A increase when increasing the protein concentration. However, this increase is less pronounced when the PEG is in large excess ($\xi = 3$ in this example).

To conclude, the simulations presented in Figures 6.7 and 6.8 seem to indicate that, for the system under consideration, the reactor should be operated by taking the PEG as the limiting reactant (i.e., ξ slightly below 1.2) and the protein initial concentration at around 10 g/L.

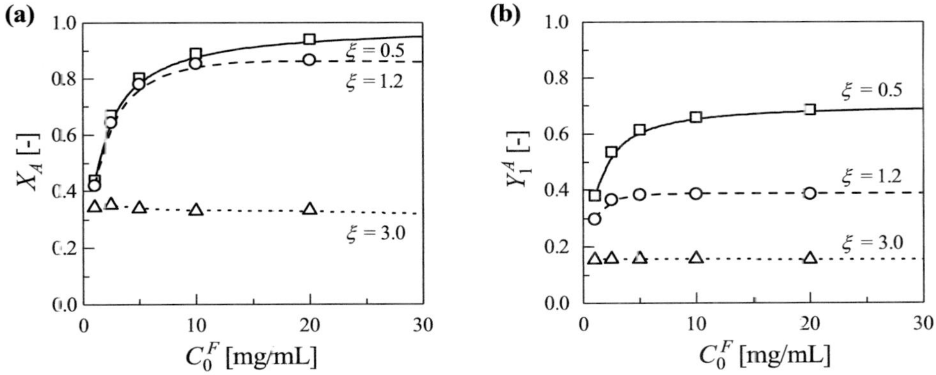

Figure 6.8 Model simulations of the (a) mPEG-SPA conversion, X_A and (b) the yield of the mono-PEGylated protein with respect to the polymer quantity, Y_1^A, as a function of the initial protein concentration, C_0^F, and for various values of the molar ratio ξ. These data were obtained at the theoretical maximum yield Y_1^0 corresponding to the optimal operating time t^{BSTR}.

Continuous Reactors

In this section, we discuss the design equations for continuous stirred tank reactors (CSTR) and plug flow reactors (PFR). As in the case of the BSTR, the mole balance equations are presented for pseudospecies of given DoC, lumping together all the positional isomers.

Continuous Stirred Tank Reactor (CSTR)

The general mole balance equations describing the kinetic scheme presented in Equation 6.1 in a CSTR of constant volume V^{CSTR} are the following:

$$\frac{dC_j}{dt} = \begin{cases} \frac{Q_0^F}{V^{CSTR}} C_0^F - \frac{Q}{V^{CSTR}} C_0 - k_0 C_0 C_A & \text{for } j = 0 \\ -\frac{Q}{V^{CSTR}} C_j + k_{j-1} C_{j-1} C_A - k_j C_j C_A & \text{for } 1 \leq j \leq n-1 \\ -\frac{Q}{V^{CSTR}} C_n + k_{n-1} C_{n-1} C_A & \text{for } j = n \end{cases} \quad (6.30)$$

$$\frac{dC_A}{dt} = \frac{Q_A^F}{V^{CSTR}} C_A^F - \frac{Q}{V^{CSTR}} C_A - \mathcal{R}_A \quad (6.31)$$

$$\frac{dC_{dA}}{dt} = -\frac{Q}{V^{CSTR}} C_{dA} + \mathcal{R}_d \quad (6.32)$$

where $\mathcal{R}_d = k_d C_A$ and $\mathcal{R}_A$ is given in Equation 6.16. In addition, Q_0^F and Q_A^F are the volumetric feed flow rates of protein and add-on molecule solutions, respectively, while C_0^F and C_A^F are the corresponding molar feed concentrations. The outlet volumetric flow rate, Q, is then simply given by:

$$Q = Q_0^F + Q_A^F \quad (6.33)$$

The design of a CSTR often consists in estimating the appropriate volume of reactor delivering a certain conversion at steady state ($dC/dt = 0$) for fixed inlet flow rates and feed concentrations of reactants. For this, the first line of Equation 6.30 can be rearranged to give the design equation of a CSTR:

$$\tau^{CSTR} = \frac{V^{CSTR}}{Q} = \frac{C_0^F (Q_0^F/Q) X_0}{-\mathcal{R}_0} \tag{6.34}$$

where $\mathcal{R}_0 = -k_0 C_0 C_A$ is the rate of protein consumption at steady state.

Plug Flow Reactor (PFR)

The mole balance equations of a continuous plug flow reactor without axial dispersion are given by:

$$\frac{\partial C_j}{\partial t} = -Q\frac{\partial C_j}{\partial V} + \mathcal{R}_j = \begin{cases} -Q\frac{\partial C_0}{\partial V} - k_0 C_0 C_A & \text{for } j = 0 \\ -Q\frac{\partial C_j}{\partial V} + k_{j-1} C_{j-1} C_A - k_j C_j C_A & \text{for } 1 \le j \le n-1 \\ -Q\frac{\partial C_n}{\partial V} + k_{n-1} C_{n-1} C_A & \text{for } j = n \end{cases} \tag{6.35}$$

$$\frac{\partial C_A}{\partial t} = -Q\frac{\partial C_A}{\partial V} + \mathcal{R}_A \tag{6.36}$$

$$\frac{\partial C_{dA}}{\partial t} = -Q\frac{\partial C_{dA}}{\partial V} + \mathcal{R}_d \tag{6.37}$$

with the following boundary conditions:

$$QC_0(t, V = 0) = Q_0^F C_0^F \tag{6.38}$$

$$QC_A(t, V = 0) = Q_A^F C_A^F \tag{6.39}$$

where Q is the total liquid flow rate given by Equation 6.33. Just like in the previous case, Equations 6.35–6.37 can be simplified at steady state because $\partial C/\partial t = 0$. In this case, these equations become identical to Equations 6.24–6.26 describing the BSTR if one replaces the time variable, t, by the residence time, $\tau^{PFR} = V^{PFR}/Q$. Similar to Equation 6.27, the design equation of the PFR is given by:

$$\tau^{PFR} = C_0^F \frac{Q_0^F}{Q} \int_0^{X_0^*} \frac{1}{-\mathcal{R}_0} \mathrm{d}X_0 \tag{6.40}$$

where the reaction rate $\mathcal{R}_0$ needs to be integrated along the reactor length.

Comparison between CSTR and PFR

In order to compare the performance of the two continuous reactors described earlier, it is interesting to analyze the volumes of the CSTR and PFR necessary to obtain a given conversion for the same reaction. To do so, let us assume that the composition and flow rate of the feed streams are the same for both reactors and plot the reciprocal of the reaction rate, $-\mathcal{R}_0$, as a function of the protein conversion X_0. This is illustrated in Figure 6.9(a) for the PEGylation of α-lactalbumin with 10 kDa mPEG-SPA and the process variables reported in the caption. Equations 6.34 and 6.40 can be used to estimate the volume of the CSTR and PFR, respectively, necessary to reach a given

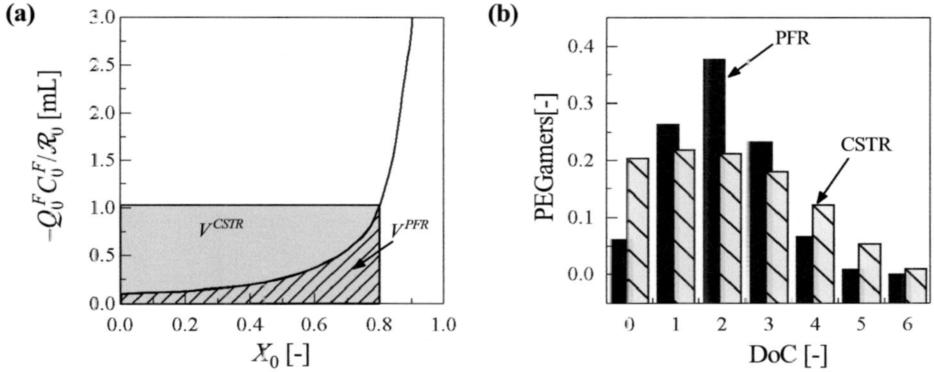

Figure 6.9 (a) Estimation of the volumes of continuous reactors (CSTR and PFR) for a given conversion of protein. This example corresponds to the PEGylation of α-lactalbumin with 10 kDa mPEG-SPA for $C_0^F = 6$ mg/mL, $\xi = 3$, $Q_0^F = 1$ mL/min. The solid line was computed with $\mathcal{R}_0 = -k_0 C_0 C_A$ and $k_0 = 134$ mL/mol/min (see Figure 6.4). (b) Distribution of DoC of PEGylated lysozyme centred on 2 for the CSTR and PFR.

protein conversion. The volume of the PFR corresponds to the integral below the curve (hatched area) while the volume of the CSTR corresponds to the area of the rectangle (gray area). For a conversion of 80%, we obtain $V^{PFR} = 0.25$ mL and $V^{CSTR} = 0.82$ mL. This shows that a PFR requires a smaller volume than a CSTR to achieve a given conversion. This result can be generalized to any positive reaction order [98, 99].

The two reactors do not only differ in terms of productivity, but they also lead to the formation of very different distributions of conjugates. To illustrate this point, Figure 6.9(b) presents the molar fractions of the products of the PEGylation reaction between lysozyme and mPEG-SPA performed in a CSTR and in a PFR. The reactor volume was chosen to produce distributions of conjugated products with average DoC equal to 2 in both reactors considering the same feed streams. It can be seen in Figure 6.9(b) that the shape of the distribution differs: the distribution obtained in the CSTR is much broader than the one obtained in the PFR. The difference comes from the residence time distributions in the two reactors: while all molecules spend exactly the same time inside the PFR, quite different residence times are observed for the different molecules in the CSTR, as shown in Figure 1.7. The proteins that spend a short time in the reactor are characterized by a low degree of conjugation, while the proteins staying longer exhibit a higher degree of conjugation. It is worth noting that the DoC distribution obtained after a given time in a BSTR is identical to the one observed at the outlet of a PFR with an equal value of the residence time.

Case Study: Continuous PEGylation of Lysozyme in a PFR

For illustrative purposes, we consider here the reaction of PEGylation of lysozyme in a plug flow reactor. The reactor is constituted of a microscale PTFE coil (0.5 mm internal diameter, 1.5 mL volume) connected to a HPLC pump. Lysozyme is solubilized in 25 mM phosphate buffer at pH 7.0 ($C_0^F = 10$ mg/mL) while the 5 kDa methoxy-poly(ethylene glycol) succinimidyl propionic acid is solubilized in DMSO

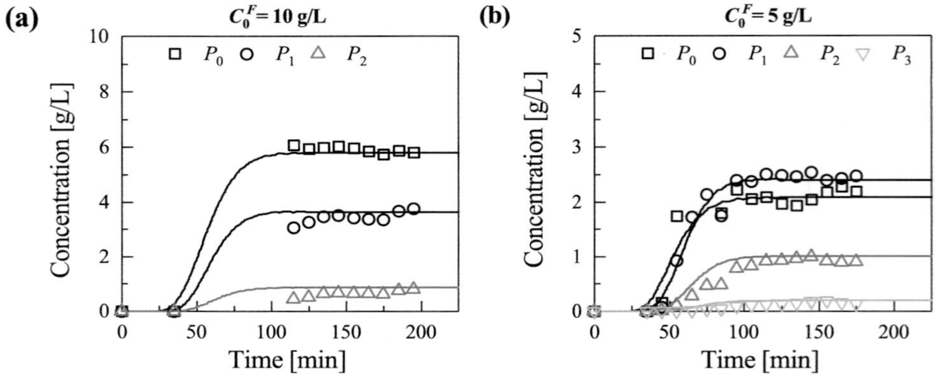

Figure 6.10 Concentration of unreacted lysozyme and PEGylated lysozyme as a function of time at the reactor outlet for two inlet lysozyme concentrations: (a) $C_0^F = 10$ g/L and (b) $C_0^F = 5$ g/L. The symbols correspond to fractions that were collected at the reactor outlet and analyzed by size exclusion chromatography. The lines correspond to model simulations performed with $k^{iso} = 10.7$ mL/mol/min, $\alpha = 3.1 \times 10^{-5}$ mol/g, and $\kappa = 0.32$ nm^2 [337].

($C_{PEG}^F = 30$ mg/mL) to prevent the hydrolysis of its ester group before entering the reactor. The volumetric flow rate of PEG solution was kept low, i.e., $Q_{PEG}^F/Q = 0.1$, to maintain a low amount of organic solvent in the reactor, thus avoiding protein denaturation. The molar flow rate of PEG was therefore controlled by adjusting the concentration C_{PEG}^F rather than the volumetric flow rate. The streams of reactants are mixed by means of a static mixer at the reactor inlet.

The kinetic constants for this system were determined previously using the iso-reactivity model: $k^{iso} = 10.7$ mL/mol/min, $\alpha = 3.1 \times 10^{-5}$ mol/g, and $\kappa = 0.32$ nm^2 [337], which with Equation 6.14 leads to $k_0 = 83.7$ mL/mol/min and $k_1 = 56.9$ mL/mol/min. At the protein feed concentration of 10 mg/mL, the ratio PEG/protein, $\xi = Q_{PEG}^F C_{PEG}^F/(Q_0^F C_0^F)$, is equal to 0.7. According to Equation 6.29, a maximum yield of 39% can be obtained. From Equation 6.28, the corresponding protein conversion is estimated to 54%. The required residence time to achieve this conversion is then computed from Equation 6.40, which gives $\tau^{PFR} = 50$ min for $C_0^F = 10$ mg/mL, or equivalently a flow rate $Q = 0.03$ mL/min. The inlet flow rates of reactants are therefore set to 0.027 and 0.003 mL/min for the protein and the PEG, respectively, to maintain the ratio $Q_{PEG}^F/Q = 0.1$.

The results of these experiments are presented in Figure 6.10(a), where the outlet concentrations of unreacted protein, P_0, mono-PEGylated protein, P_1, and di-PEGylated protein, P_2, are reported as a function of time. During the first 50 min, nothing is observed at the reactor outlet, which corresponds to the residence time of the plug flow reactor. After 50 min, the concentrations of the reactants and products of the conjugation reaction slowly increase until reaching steady state. Model simulations obtained by integrating the differential equations describing the PFR presented earlier are also shown. A second experiment at lower protein concentration, namely $C_0^F = 5$ mg/mL, is performed keeping the same concentration of PEG and the same volumetric ratio between the two inlet flow rates. The results of this second experiment are reported in Figure 6.10(b). It is seen that the model predictions are in very good agreement with the experimental data for both protein concentrations. This shows that the kinetic parameters estimated from BSTR experiments can be directly used to simulate continuous reactors. Indeed, the reaction kinetics is independent of the type of reactor where the reaction is taking place.

6.3 Purification of Conjugated Proteins

6.3.1 Separation Challenges

The purification of conjugated proteins implies removing the reaction by-products and the nonconverted reactants on the one hand, and isolating the target product among the various conjugates on the other hand. Depending on the applications, the desired product might be characterized by different degrees of homogeneity. It can be constituted of a single positional isomer (e.g., Cimzia®, mono-PEGylated Fab fragment conjugated to a site-specific engineered cysteine residue [347]), a distribution of positional isomers with a given DoC (e.g., PEG Intron®, a mono-PEGylated recombinant interferon α-2b containing 14 positional isomers [348]), or even a distribution of multiple degrees of conjugations (e.g., Kadcyla®, an ADC with DoC ranging from 0 to 8 [349]). Broad distributions of pharmacokinetic properties can be desirable, for example to enlarge the spectrum of action of the conjugated drug, or on the contrary, a single isoform may be requested, for instance if the other forms are toxic or show too little activity.

To illustrate the potential impact of the conjugation position on the drug pharmacodynamics and pharmacokinetics, Figure 6.11 shows experimental results obtained with

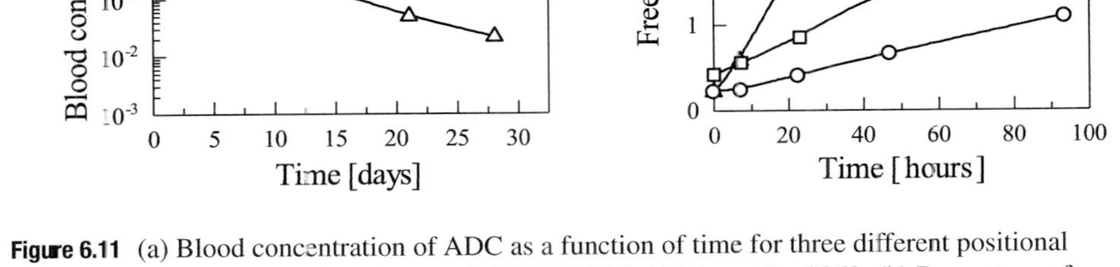

Figure 6.11 (a) Blood concentration of ADC as a function of time for three different positional isomers of thio-trastuzumab conjugates: LC-V205, HC-A114, and Fc-S369. (b) Percentage of MMAE released in plasma by the nonenzymatic cleavage of the linker. Reprinted by permission from Macmillan Publishers Ltd: Nature biotechnology [350], copyright 2012.

ADCs, namely three positional isomers of trastuzumab conjugated to monomethyl auristatin E (MMAE) at different thiol residues (LC-V205, HC-A114, and Fc-S369) [350]. Figure 6.11(a) illustrates the evolution of the ADC concentration in the blood for the three positional isomers. It is seen that different profiles are obtained depending on the position at which the cytotoxic group is covalently bound to the protein. Moreover, Figure 6.11(b) shows the fraction of cytotoxic drug (MMAE) released in plasma outside the tumor cells via the nonbenzymatic hydrolysis of the linker. It is seen that the rate of hydrolysis strongly depends on the location of the grafted drug. The nonenzymatic release of the drug is not desirable because it may reduce the action of the ADC against tumor cells and increase adverse side effects on healthy cells. Therefore, it may be necessary to exclude the positional isomers showing the largest MMAE release outside the tumor cells in the final drug.

The heterogeneity of the final product results from the conjugation chemistry on the one hand, and from the purification protocol on the other hand. Importantly, separation processes can compensate for the lack of selectivity of the conjugation reaction and improve the homogeneity of the drug, or alternatively create a well-defined distribution of conjugates [314] if necessary. In the following, we examine two techniques that are commonly employed in the purification of conjugated proteins, namely filtration and chromatography.

6.3.2 Filtration

Separation by filtration is primarily based on steric interactions between the solutes and the pores of the membrane, which allows only the smallest solutes to go through the membrane. Electrostatic repulsive interactions are also sometimes exploited to extend the capabilities of filtration membranes. Two types of filtration are commonly encountered in the purification of conjugated proteins, namely ultrafiltration and diafiltration. In both cases, the protein is retained on the retentate side, while buffer components flow through the membrane. However, the objectives differ from one case to the other: In ultrafiltration, the aim is to concentrate the protein solution, while in diafiltration the aim is to exchange the buffer solution.

Ultrafiltration and diafiltration are often used to separate the conjugated proteins from smaller molecules, such as the unreacted add-on molecules, reaction by-products and buffer solution components [351–354], which significantly differ in terms of charge and size from the protein. As a rule of thumb, a size difference by a factor 10 or more is often applied to consider filtration as a valid option [355].

The design of filtration units has been extensively reviewed, for instance by Zydney and coworkers, and the reader is referred to the literature for more details about the modeling of filtration processes [352, 353, 356]. Here, we only report concepts that facilitate the understanding of the general mechanisms involved. To do so, let us first introduce the sieving coefficient, $\mathcal{S}_A$, as the ratio between the concentration of solute A in the permeate, C_A^P, over the concentration of the same solute in the retentate, C_A^R, i.e., $\mathcal{S}_A = C_A^P/C_A^R$. The values of the sieving coefficient may vary in time so that the final and steady state values are often reported for batch and continuous processes, respectively.

High sieving coefficients, i.e., close to unity, indicate that the solute preferentially goes through the membrane and ends up in the permeate side. On the contrary, small values of $\mathcal{S}$ indicate that the solute is preferentially retained on the retentate side. The crucial parameter in a filtration process is the selectivity, which is defined as the ratio between the sieving coefficients of two solutes.

In the case of batch filtration units, the sieving coefficient can be related to the process operating conditions as discussed in the following. In diafiltration applications, the final concentration of the solute to be removed is given by $C_A^R = C_A^F \exp(-\mathcal{S}_A \times DV)$, where C_A^F is the feed concentration of A and DV is the number of diafiltration volumes. For example, the removal of small impurities (with $\mathcal{S}_A \approx 1$) up to 99.9% necessitates seven diafiltration volumes. For batch ultrafiltration processes, it is useful to define the volume concentration factor, VCF, as the ratio between initial and final volumes in the permeate. The recovery of A, ϑ_A, is related to the sieving coefficient measured at the end of the filtration step by $\vartheta_A = VCF^{-\mathcal{S}_A}$. In addition, the final solute concentration is given by $C_A^R = C_A^F \times \vartheta_A \times VCF$. If both ultrafiltration and diafiltration are performed simultaneously, the recovery of the filtration step is $\vartheta_A = \exp(-\mathcal{S}_A(DV + \ln(VCF)))$.

When the selectivity between solutes is too low due to a too small difference in size, membranes can be functionalized to improve their performances. In particular, charged membranes are highly interesting when dealing with proteins. For efficient filtration units, one has to select operating conditions, such as the pH of the solution, so that the charge at the membrane surface is of the same sign than the apparent charge of the protein. By doing so, the proteins are repelled from the surface of the membrane and remain on the retentate side, while neutral solutes (e.g., PEG) simply flow through.

For illustrative purposes, the sieving coefficient of α-lactalbumin, 5 kDa PEG, and mono-, di-, tri-, and tetra-PEGylated proteins filtered through a negatively charged membrane are reported in Figure 6.12 as a function of the concentration of sodium chloride [356]. At the selected pH value of 7.0, the protein and the conjugated proteins are negatively charged. It is seen that, at low ionic strength, the neutral and small 5 kDa

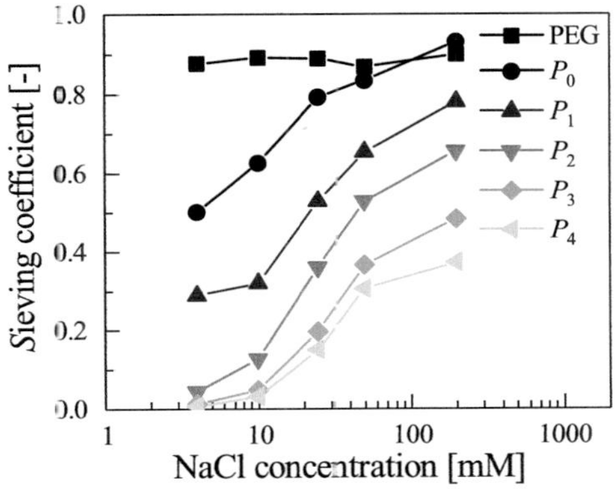

Figure 6.12 Sieving coefficients of 5 kDa PEG, α-lactalbumin (P_0), and mono- (P_1), di- (P_2), tri- (P_3), and tetra-PEGylated (P_4) α-lactalbumin at pH 7.0 as a function of the concentration of sodium chloride [356].

PEG exhibits the largest sieving coefficient and can therefore be easily removed from the PEGylation reaction mixture. The sieving coefficient of the other solutes go in the order α-lactalbumin, mono-, di-, tri-, and tetra-PEGylated protein. This is due to the increase in hydrodynamic radius with the extent of PEGylation, which makes it more difficult for the conjugates to cross the filter. On the other hand, an increase of the ionic strength leads to an increase of the sieving coefficient of the nonfunctionalized and PEGylated proteins. This is due to the presence of electrolytes that screen the charges at the protein and membrane surfaces, thus reducing the repulsive effect of the membrane. In Figure 6.12, this screening effect seems comparable for all the conjugates because the ranking in sieving coefficients remains the same at each ionic strength. However, the PEG is rather unaffected by the addition of salt because it is a neutral polymer. The consequence is a reduction of the selectivity at high ionic strength and therefore a poorer removal of the PEG from the conjugated proteins.

In general, ultrafiltration and diafiltration require a careful selection of the membrane material, pore size, charge of the membrane surface, pH, ionic strength, and flow rate through the membrane. A good selectivity can usually be obtained when it comes to removing by-products and unreacted add-on molecules from the reaction mixture. However, the separation of the different products of the conjugation reaction according to their DoC is more challenging and usually requires the use of chromatography, as described in the next section.

6.3.3 Chromatography

Chromatographic Media

Interaction Type

Chromatography is certainly the most versatile tool when it comes to isolating a given conjugated product in a mixture of conjugates varying in terms of DoC and position of the grafted add-on molecule. The choice of the type of chromatography (SEC, IEC, HIC, etc.) depends on the chemical nature of the various molecules to be separated. For instance, for biotinylated proteins, which are often used for detection purposes, affinity chromatography using avidin-based ligands is most often employed to isolate the target products. On the other hand, SEC is conveniently used when the conjugation reaction leads to an important increase of the hydrodynamic radius of the protein, as it is usually the case for PEGylation and other polymer-based conjugation reactions. In general, the chromatographic medium is selected so as to exploit at best the change in physicochemical properties of the protein caused by the conjugation of add-on molecules. In the following, we make use of the concepts introduced in Chapter 2 to characterize how the retention of conjugates varies with the DoC. We consider three examples with three different chromatographic media, namely SEC, HIC, and IEC.

In Figure 6.13(a) are shown the hydrodynamic radii of lysozyme conjugated to PEG chains of different molecular weights (5, 10, and 20 kDa) as a function of the DoC [55]. The hydrodynamic radii were estimated by comparing retention times from pulse injection experiments with suitable standards [338]. It is seen that the hydrodynamic radius of the conjugated proteins increases with increasing both the DoC and the PEG molecular

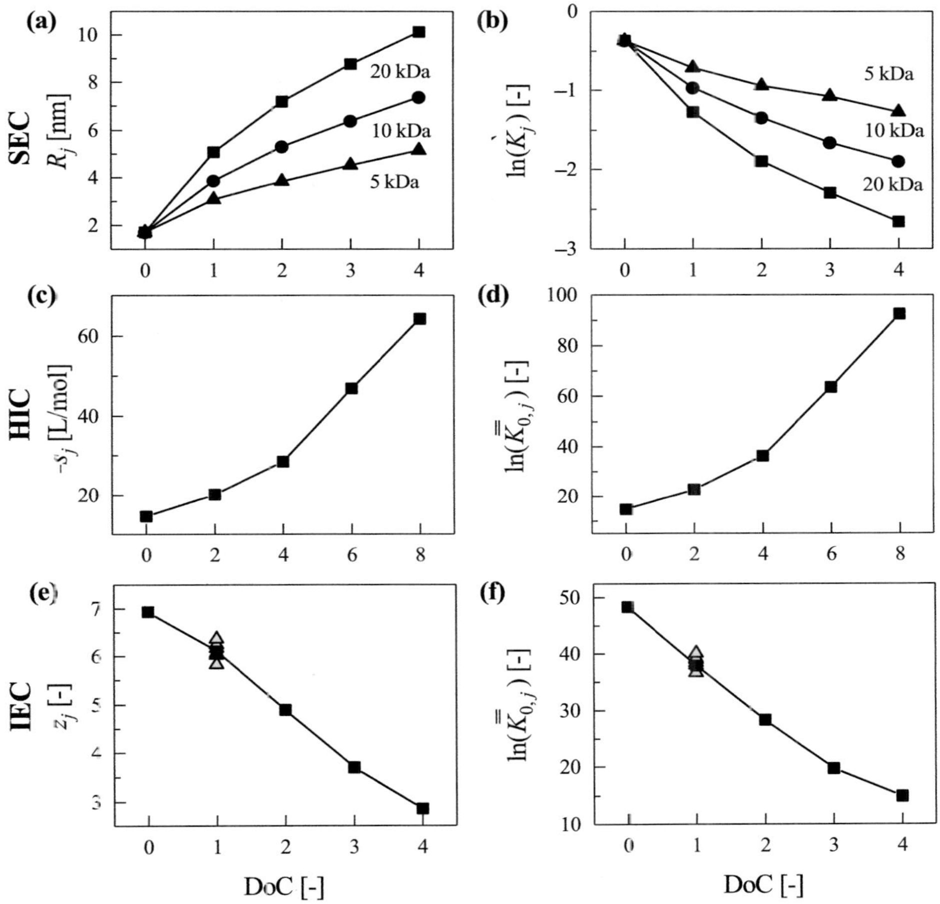

Figure 6.13 Variations of the parameters describing the Henry coefficient as a function of the DoC value for different chromatographic media and different conjugated proteins. *First row:* PEGylated lysozyme with 5, 10, and 20 kDa PEG analyzed by SEC, (a) hydrodynamic radius and (b) partition coefficient, $\grave{K}_j$ defined such that $\overline{K}_j = \varepsilon_i \grave{K}_j$ with $\varepsilon_i = 0.8$. Based on the experiments reported in [55]. *Second row:* brentuximab vedotin conjugated with a chemical drug analyzed by HIC by applying a linearly decreasing concentration of ammonium sulphate (c) sensitivity coefficient, s_j and (d) $\overline{\overline{K}}_{0,j}$ defined such that $\overline{K}_j = \varepsilon_i + (1 - \varepsilon_i)\overline{\overline{K}}_{0,j}\exp(-s_j C_M)$. Based on the experiments reported in [357]. *Third row:* lysozyme PEGylated with 5 kDa PEG in IEC (e) apparent charge, z_j and (f) $\overline{\overline{K}}_{0,j}$ defined such that $\overline{K}_j = \varepsilon_i + (1 - \varepsilon_i)\overline{\overline{K}}_{0,j} C_M^{-z_j}$. Gray triangles correspond to positional isomers of mono-PEGylated lysozyme. Based on the experiments reported in [358].

weight. In Figure 6.13(b), the same data are plotted in terms of the partition coefficients, $\grave{K}_j$, as defined in Chapter 2. It is seen that $\grave{K}_j$ decreases with increasing both the DoC and the PEG molecular weight, which is qualitatively in-line with the theoretical expression $\grave{K}_j = (1 - R_j/r_{pore})^2$ previously presented in Equation 2.15. It can also be seen that the difference between two consecutive values of $\grave{K}$ (i.e., between solutes of DoC

j and $j+1$) decreases with increasing the DoC value. This indicates that the separation by SEC of solutes with different DoC is easier for the smallest degrees of conjugation and becomes more and more challenging as the DoC increases.

For the purification of ADCs, HIC is particularly well suited because the add-on drugs are generally more hydrophobic than the antibody. This results in an overall increase of the hydrophobicity with the DoC, as confirmed by the elution ranking from the unconjugated to the most conjugated protein. The hydrophobicity of the add-on drug can even sometimes be problematic because it may promote aggregation of the conjugated product [359]. Hydrophobic proteins indeed tend to aggregate so as to reduce their contact area with the solvent, as discussed in Chapter 7. In order to estimate the chromatographic behavior of ADCs of various DoC (also referred to as drug-antibody-ratio in the case of ADCs) on HIC media, pulse injection experiments were performed under linear gradient elution conditions [357]. Various gradient slopes were used, and the parameters describing the evolution of the Henry coefficient with the concentration of ammonium sulphate were estimated as detailed in Section 2.5. In particular, the sensitivity coefficient, s_j, and the prefactor, $\overline{\overline{K}}_{0,j}$ (defined such that $\overline{\overline{K}}_j = \overline{\overline{K}}_{0,j}\exp(-s_jC_M)$, where C_M is the ammonium sulphate concentration) of brentuximab vedotin conjugates were evaluated for DoC values ranging from 0 to 8. The obtained values of s_j and $\overline{\overline{K}}_{0,j}$ are shown in Figure 6.13(c) and (d), respectively. It should be considered that because the parameter s_j is negative in HIC, the higher the value of $-s_j$ and the more hydrophobic is the solute. Figure 6.13(c) and (d) illustrate that the grafting of the MMAE increases considerably the hydrophobicity of the monoclonal antibody.

Finally, IEC is quite often used to separate the different products of protein/polymer conjugation reactions. Indeed, the addition of a neutral polymer is likely to shield the protein charges, thus reducing the strength of the binding with the charged chromatographic medium [360]. Similar to what has been done earlier for the ADCs, pulse injection experiments under linear gradient elution conditions on a strong cation exchange resin were performed. PEGylated lysozyme with various degrees of PEGylation was selected as an example [358]. Different gradients of sodium chloride were implemented, and the parameters describing the evolution of the Henry coefficient with the concentration of salt were estimated for the different isoforms. Briefly, the apparent charge, z_j, and the prefactor, $\overline{\overline{K}}_{0,j}$ (defined such that $\overline{\overline{K}}_j = \overline{\overline{K}}_{0,j}C_M^{-z_j}$ where C_M is the salt concentration), were determined as discussed in Section 2.5.2. The parameters z_j and $\overline{\overline{K}}_{0,j}$ are reported as a function of the DoC value in Figure 6.13(e) and (f), respectively. It can be seen that both the apparent charge, z_j, and the prefactor, $\overline{\overline{K}}_{0,j}$, decrease with increasing DoC values. In this case, the reduction of the number of charges is due to the consumption of one charged amine per conjugation event and to the presence of the PEG chain, which covers partially the surface of the protein. This shielding effect is also affecting the equilibrium constant of the ion exchange mechanism. Data corresponding to different positional isomers of mono-PEGylated lysozyme are also included in Figure 6.13(e) and (f) (gray triangles). It can be seen that both the apparent charge z_j and the prefactor $\overline{\overline{K}}_{0,j}$ differ slightly for the various positional isomers of DoC equal to 1. This is because PEG molecules at different positions shield regions that are characterized by different surface properties. However, these differences are rather subtle as compared to those caused by different degrees of conjugation. It is therefore usually much easier to separate the products of the conjugation reaction according to their DoC than to the position of the grafted add-on molecule.

Particle Size

In this section, we discuss the impact of the particle size on the separation of conjugated proteins. As an example, we consider a protein mixture obtained from the PEGylation reaction of lysozyme with 5 kDa PEG separated with different strong cation exchange chromatographic media whose particle size ranges between 2.5 and 75 μm [361]. The corresponding chromatograms, obtained under the same elution conditions, are shown in Figure 6.14. Figure 6.14(a) shows the analytical scale chromatogram obtained on TSKgel SP-NPR ($d_p = 2.5$ μm). It is seen that at least three positional isomers of mono-PEGylated lysozyme and two of di-PEGylated lysozyme can be detected.

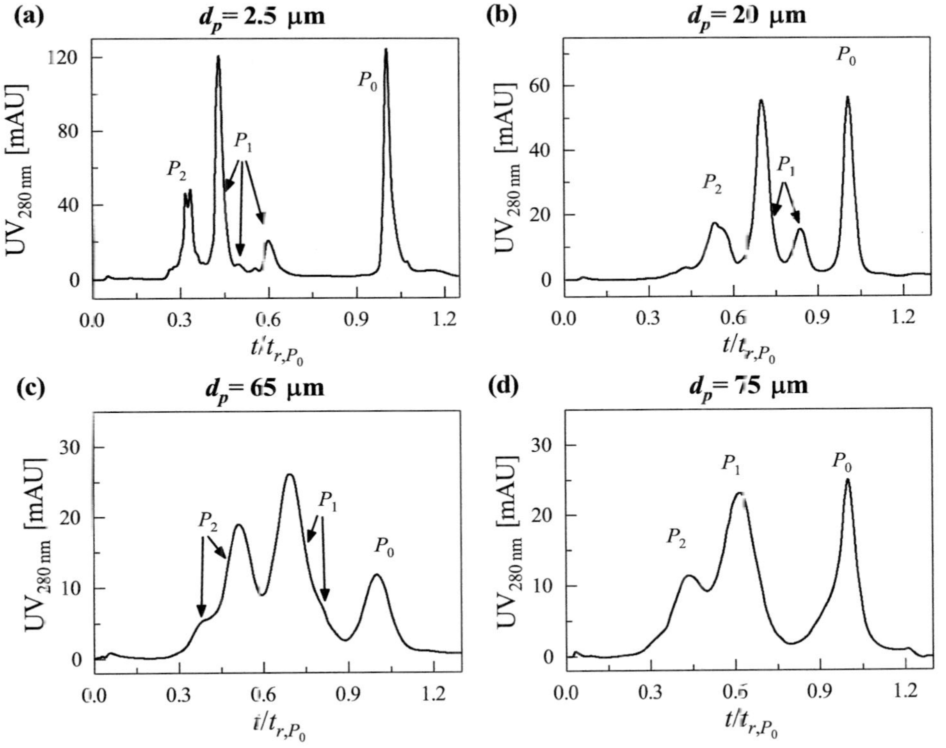

Figure 6.14 Separation of PEGylated lysozyme on strong cation exchange chromatographic media with different particle sizes. The feed mixture contains some unreacted lysozyme, mono-PEGylated proteins, and di-PEGylated proteins. (a) column TSKgel SP-NPR, ($d_p = 2.5$ μm), (b) column ToyoPearl SP-5PW ($d_p = 20$ μm), (c) column ToyoPearl SP-650M ($d_p = 65$ μm), (d) column ToyoPearl GigaCap S-650M ($d_p = 75$ μm). The DoC of the eluting species was determined by SEC. Adapted from [361], copyright 2010, with permission from Elsevier.

In Figure 6.14(b) to (d) chromatographic media with particle sizes increasing from 20 μm to 75 μm have been used. It can be observed that the resolution between the mono-PEGylated isoforms decreases with increasing d_p until only one single peak is detected per DoC. These experimental results illustrate how the peak broadening increases with the particle diameter, which was theoretically discussed in Section 2.4.

Single-Column Process

The separation and identification of all the possible products formed during the conjugation reaction is a major challenge, especially when the number of conjugation sites n is large. The complete resolution of the positional isomers sometimes requires the combination of orthogonal techniques, a first one dedicated to the fractionation of the conjugates according to their DoC and a subsequent one for the separation of the positional isomers of a given DoC. An example is reported in Figure 6.15, where a mixture of PEGylated lysozyme with degrees of PEGylation ranging from 0 to 3 is analyzed. In panel (a), the SEC chromatogram of the crude reaction mixture containing the unconverted protein, as well as the mono-, di-, and tri-PEGylated proteins, is shown. Solutes with a high DoC are eluting first due to the increase in the hydrodynamic radius upon

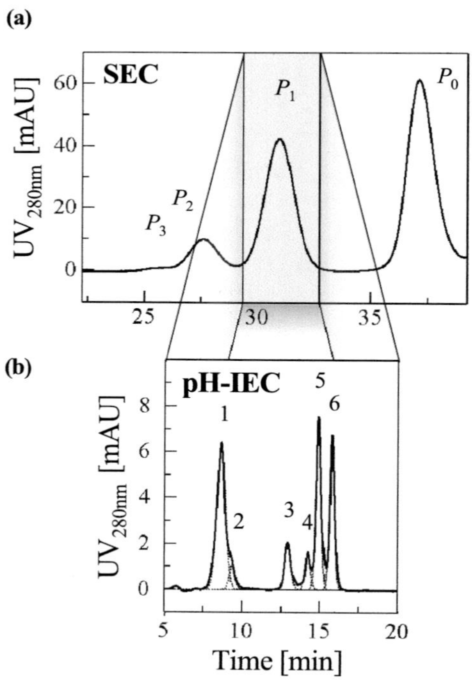

Figure 6.15 (a) Separation by SEC of a mixture of lysozyme, mono-, di-, and tri-PEGylated lysozyme (with 5 kDa mPEG-ester) according to the DoC. (b) Chromatogram of the mono-PEGylated positional isomers obtained by pH-gradient analytical IEC. Six different isoforms composing the mono-PEGylated lysozyme could be identified. The experimental conditions are reported in [337].

conjugation, as discussed earlier. In panel (b), the analytical IEC chromatogram of the fraction containing the mono-PEGylated lysozyme is shown. Six positional isomers were resolved by pH gradient ion exchange chromatography on an analytical strong cation exchange chromatographic column. In the present case, the different solutes eluted between pH 10 and 11, indicating that the grafted PEG affects the net charge of the protein differently depending on the position of the conjugation site.

Let us now design a single-column process for the purification of mono-PEGylated lysozyme from the reaction mixture. We consider reaction conditions similar to those described in the case study of Section 6.2.3, i.e., the PEGylation of lysozyme is performed in a batch reactor with $C_0^F = 9$ mg/mL, $\xi = 1.2$, $V^{BSTR} = 1.5$ mL, and $t^{BSTR} = 50$ min. Under these conditions, the solution at the end of the reaction is composed of 4.7 mg/mL of unreacted lysozyme, 4.5 mg/mL of mono-PEGylated lysozyme, 1.6 mg/mL of di-PEGylated lysozyme, and 0.4 mg/mL of unreacted and deactivated PEG.

The goal of the separation step is to recover the mono-PEGylated protein from the reaction mixture with a recovery superior to 92% and at a purity above 95%. According to the literature, strong cation exchange chromatography media are well suited for this separation and in particular, the Toyopearl GigaCap S-650M resin is often reported for this application [342, 345, 361]. The elution is performed under LGE conditions at pH 3.0 by linearly increasing the concentration of sodium chloride in the eluent from 0 to 1 M. At the end of the LGE, the high salt concentration is maintained for three column volumes, and the chromatography medium is re-equilibrated for 3 CV. The load volume is selected equal to the reactor volume ($V^F = V^{BSTR} = 1.5$ mL).

Having defined the load volume, the design of this chromatographic process reduces to selecting the column length, fluid velocity, and slope of the linear modifier gradient so as to obtain the highest productivity while fulfilling the recovery (>92%) and purity (>95%) constraints. We set the column diameter to 5 mm and vary the column length in the range 15 to 30 cm. Prior experiments showed that the pressure drop is related to the fluid velocity according to $\Delta P = 1.5 \times 10^{-4} u L_{col}$ with u in cm/h, L_{col} in cm and ΔP in bar. Considering a maximum acceptable pressure drop of 1 bar, the former relation provides the value of the fluid velocity for a given column. Finally, linear modifier gradients performed over various numbers of column volumes, ranging from 15 to 50 CV are investigated.

The results of the simulations are shown in Figure 6.16. The productivity, in milligrams of mono-PEGylated lysozyme per mL of column and hour, is plotted as a function of the column length for different gradient slopes in Figure 6.16(a). The productivity was estimated by considering the largest possible collection window delivering the mono-PEGylated protein at a purity above 95%. First, it can be seen that for a given LGE protocol, the productivity is decreasing when increasing the column length. Second, for a given productivity value, steeper gradients require longer columns to compensate the decrease in the resolution between consecutive peaks. The gray area in Figure 6.16(a) represents the region in which the system fails at delivering a protein recovery above 92% at a purity of at least 95%, no matter the selected collection window. The design of the single-column process is therefore restricted to the zone below the gray area.

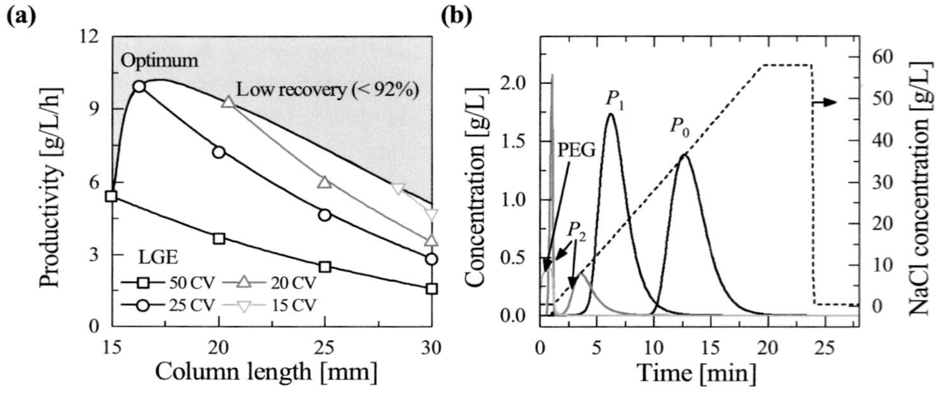

Figure 6.16 Design and optimization of the single-column chromatographic process for the purification of mono-PEGylated lysozyme from the reaction mixture. (a) Productivity of the mono-PEGylated lysozyme as function of the column length for different LGE protocols. The modifier gradient is implemented from 0 to 1 M NaCl over 15, 20, 25 and 50 CV (b) Chromatogram obtained with the selected optimal design ($L_{col} = 16$ mm, $Q = 1.4$ mL/min and gradient of 25 CV). The simulations were performed with the following model parameters [358]: $\overline{K}_M = 0.5$ and $t_{i,M} = 0.3$ s for NaCl; $\overline{K}_{PEG} = 0.4$ and $t_{i,PEG} = 10$ s for the PEG; $\varepsilon_i \grave{K}_j = 0.28$, 0.06 and 0.01, $z_j = 6.94$, 6.11 and 4.89, $(1 - \varepsilon_i)\overline{\overline{K}}_{0,j} = 1.18 \times 10^{12}$, 3.49×10^{8} and 7.05×10^{5} $(\text{mL/mg})^{-z_j}$, $\overline{N}_j = 150$, 120 and 80, and finally $t_{i,j} = 60$ s for the three proteins with $j = P_0, P_1$ and P_2, respectively.

Among the tested conditions, it seems that a column of 16 mm length, a flow rate of 1.4 mL/min and a modifier gradient performed over 25 CV are the conditions delivering the highest productivity. The chromatogram corresponding to these conditions is presented in Figure 6.16(b). It can be seen that the PEG is not retained and elutes first at the outlet of the column. Moreover, part of the di-PEGylated lysozyme breaks through the column during the load phase, which stops just before the mono-PEGylated protein also starts breaking through, i.e., at about 1 min. Then, a good separation between the remaining di-PEGylated protein and the mono-PEGylated lysozyme on the one hand, and the mono-PEGylated protein and the unreacted lysozyme on the other hand is obtained with the selected LGE protocol.

Continuous Countercurrent Multicolumn Processes

The purification of conjugated proteins is quite often a central-cut separation task where the product of interest, i.e., with the desired DoC or a given DoC distribution, elutes between underconjugated and overconjugated proteins. We have seen in Chapter 5 several countercurrent processes conceived to tackle central-cut separations.

In this section, we continue the example presented in Figure 6.16 and design a 4C-MCSGP process for the purification of mono-PEGylated lysozyme from the reaction mixture. The MCSGP process is described in Section 5.3.2, and more details about the 4C-MCSGP can be found in the literature [314]. Figure 6.17 shows a schematic representation of the two phases of the 4C-MCSGP process: the interconnected phase

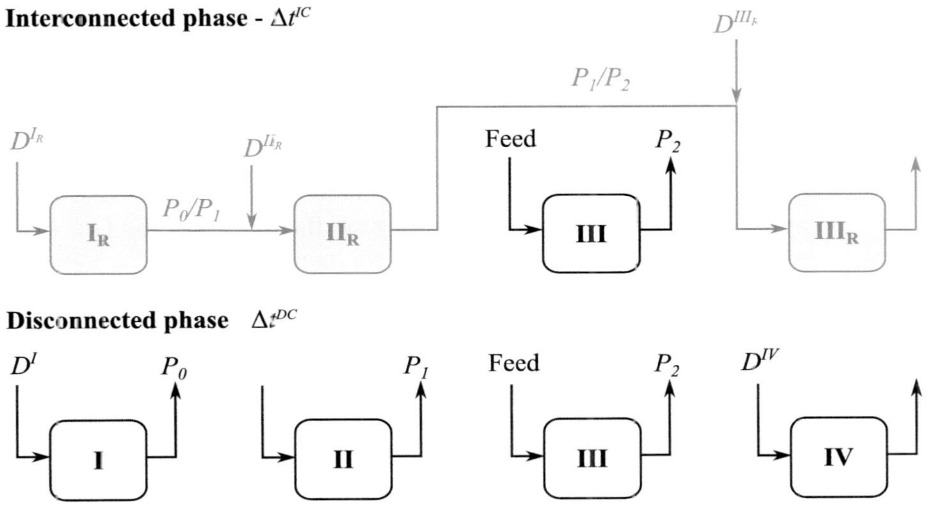

Figure 6.17 Simplified flowsheet of the 4C-MCSGP process. The first row corresponds to the interconnected phase while the second row represents the disconnected phase.

(top) and the disconnected phase (bottom). Designing such process is challenging due the large number of process variables to be selected. In order to simplify the approach, the multicolumn process will be first designed based on the single-column process given in Section 6.3.3. In a second step, we will show how the so-designed process can be optimized based on model simulations.

Let us start with the practical design approach, which is summarized in Box 3.1. To do so, four columns of 16 × 5 mm packed with the GigaCap S-650M resin are used, and the modifier concentration gradient is performed over 25 CV as described in Figure 6.16(b).

Due to the future integration with the PFR mentioned earlier, the feed flow rate of the MCSGP, Q^{III}, is set equal to the reactor flow rate, i.e., $Q^{III} = 0.03$ mL (see the case study in Section 6.2.3). The rest of the design consists in determining proper collection and recycling windows based on the chromatogram shown in Figure 6.16(b). In order to define the different collection windows in this chromatogram, we denote by $t_{c,A}^{beg}$ and $t_{c,A}^{end}$ the beginning and the end of the time interval during which the generic solute A is collected at the desired purity. Therefore, $t_{c,P_2}^{end} - t_{c,P_2}^{beg}$ is the collection window of the di-PEGylated protein P_2, $t_{c,P_1}^{end} - t_{c,P_1}^{beg}$ is the collection window of the mono-PEGylated protein P_1, and $t_{c,P_0}^{end} - t_{c,P_0}^{beg}$ is the collection window of the unreacted protein P_0. Consequently, the first unresolved fraction containing both P_1 and P_2 is comprised between the times t_{c,P_2}^{end} and t_{c,P_1}^{beg} while the second unresolved fraction containing both P_1 and P_0 is comprised between the times t_{c,P_1}^{end} and t_{c,P_0}^{beg}. The times t_{c,P_2}^{end} and t_{c,P_0}^{beg} are selected to maximize the product recovery. Considering the single-column chromatogram shown in Figure 6.16(b), we selected $t_{c,P_2}^{end} = 1.1$ min and $t_{c,P_0}^{beg} = 12.5$ min.

Then, the times t_{c,P_1}^{beg} and t_{c,P_1}^{end} between which the product is collected are selected within this interval so as to deliver the product at a purity above 95%. In addition, these times must be compatible with the constraints on the maximal internal flow rates and minimum modifier concentration in the eluent during the interconnected phase. In particular, with steep modifier gradients, the two times t_{c,P_1}^{beg} and t_{c,P_1}^{end} should be close enough to make sure the inline dilutions in the zones II_R and III_R are feasible. In the present case, the internal flow rate in zone II_R was set to the flow rate used in the single-column experiment, namely 1.4 mL/min. In addition, the initial concentration of modifier in the desorbent stream entering zones II_R (D^{II_R} in Figure 6.17) was set to 0 mM. Finally, we select $t_{c,P_1}^{beg} = 8$ min and $t_{c,P_1}^{end} = 11$ min for the collection of the mono-PEGylated lysozyme. With such a narrow collection window, the purity is slightly above 98%, but the product recovery does not exceed 28% with the single-column process.

After defining the collection and recycling windows, one has to determine the switch time. For this, it should be remembered that once steady state is reached, the amount of product loaded inside the column per cycle must be equal to the amount of mono-PEGylated lysozyme exiting the column, which in the present case corresponds to 28% of the quantity loaded in the single-column process. Therefore, we want to have $C_{P_1}^F Q^{III} \Delta t = 0.28 \times 1.50 \times C_{P_1}^F$ mL, where 1.50 mL is the loaded volume in the single-column. This relation leads to $\Delta t = \Delta t^{IC} + \Delta t^{DC} = 14.1$ min with $\Delta t^{IC} = \Delta t^{DC}$.

Knowing the collection and recycling windows from the elution chromatogram (Figure 6.16(b)), the feed flow rate, the switch times, the shape of the modifier gradient, the m parameters, the flow rates, and the composition of the eluent in the different zones can be computed as shown in Section 5.3.2 for the 4C-MCSGP process. In the end, two flow rates cannot be determined with this empirical approach, namely $Q_D^{III_R}$ and Q^{IV}. For this, the inlet flow rate of eluent in the zone III_R was selected to ensure a sufficient dilution of the recycled fraction to reduce the modifier concentration. In particular, with $Q_D^{III_R} = 2.0$ mL/min, the modifier concentration inside the column during this phase did not exceeded 10 mg/mL. This value is sufficient to ensure that the target protein is retained inside the column. After receiving this unresolved fraction, the chromatographic medium is washed in zone IV during the disconnected phase at 0.5 mL/min using a modifier concentration of 0.5 mg/mL to ensure the proper binding of the recycled solutes. After this washing step, the column is ready to receive the feed mixture in zone III. The obtained experimental results are summarized in Table 6.3.

The 4C-MCSGP process was simulated with the model parameters given in the caption of Figure 6.16 and the operating conditions given in Table 6.3. The results are shown in Figure 6.18. Figure 6.18(a) shows the steady state chromatogram simulated at the end of one of the four columns as a function of the eluted volume. It can be seen that, even though different flow rates are used in the different zones of the MCSGP, the gradient profile used in the single-column process is perfectly reproduced at the column outlet. In this figure are also shown the two unresolved fractions (light gray areas) and the target product collection window (dark gray area). With this process, a purity of 98% and a recovery of 92% are obtained. This shows that the practical design strategy presented here allows to rapidly obtain a multicolumn process delivering the product at the target specifications. However, this process is not optimized, and we show in the

Table 6.3. Process variables for the 4C-MCSGP with $\Delta t^{IC} = \Delta t^{DC} = 7.05$ min.

Elution time [min]	**Zone** [−]	m [−]	**Internal** Q [mL/min]	Q_D [mL/min]	**Inlet** C_M [mg/mL]		
					$t = 0$	Δt^{DC}	Δt^{IC}
$t^{end}_{c,P_0} = 28$	I	112.3	3.1		58.0	0.5	
$t^{beg}_{c,P_0} = 12.5$	II	21.2	0.6		21.8	31.21	
$t^{end}_{c,P_1} = 11$	III	1.6	0.03		0.5	0.5	
$t^{beg}_{c,P_1} = 8$	$\mathrm{I_R}$	10.3	0.3	0.3	31.21		35.90
$t^{end}_{c,P_2} = 1.1$	$\mathrm{II_R}$	49.6	1.4	1.1	0.0		18.10
$\Delta t^F = 1.1$	$\mathrm{III_R}$	124.1	3.4	2.0	0.0		0.0

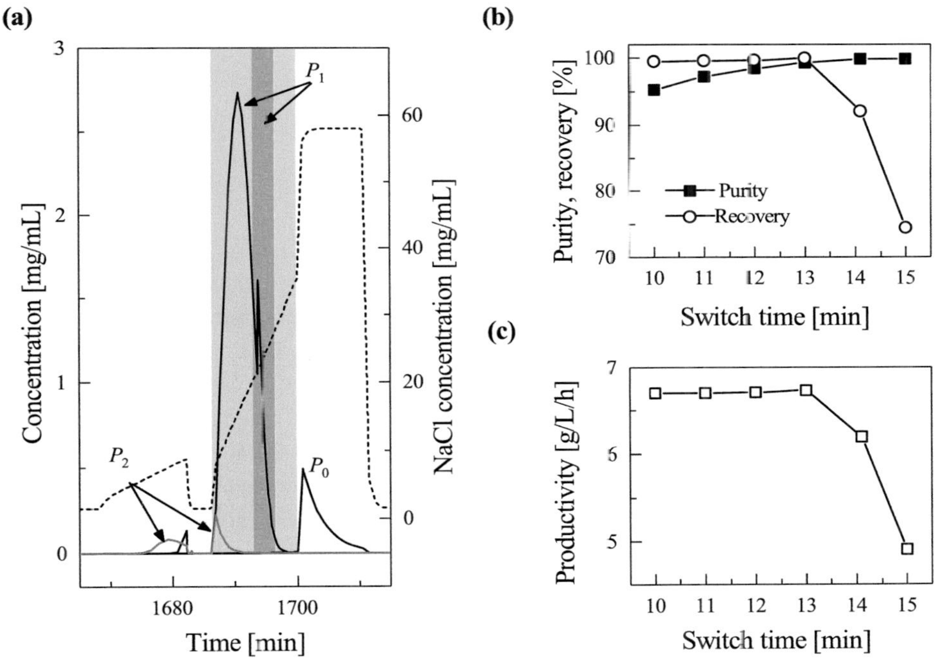

Figure 6.18 (a) Simulated steady state chromatogram at the outlet of the fourth column of the 4C-MCSGP computed using the values given in Table 6.3 and a switch time of 14.1 min. (b) Target product protein recovery and purity as a function of the switch time, keeping the flow rates and eluent composition unchanged. (c) Productivity as a function of the switch time.

following how to improve process performances by fine-tuning the previously selected process variables based on model simulations.

Here, we choose to vary the switch time while keeping the flow rates and eluent composition unchanged, but the same procedure could be used for any other process variables. The interconnected time is kept equal to the disconnected time, and their value is varied between 5 and 7.5 min, thus leading to a total switch time Δt ranging from 10 to 15 min. The purity of the mono-PEGylated lysozyme, its recovery and the associated productivity values are reported in Figure 6.18(b) and (c) as a function of Δt. In Figure 6.18(b), it can be seen that the purity of the mono-PEGylated protein slightly increases when increasing the switch time. When keeping everything else constant, an increase of the switch time leads to shallower gradients, which facilitate the separation between the di-PEGylated and the mono-PEGylated protein, thus resulting in a higher purity. On the other hand, the product recovery drops because of the prolonged load phase during which part of the product breaks though the column and is irremediably lost. It can be seen in Figure 6.18(c) that productivity follows a similar trend as recovery, going through a maximum for $\Delta t = 13$ min before dropping down for longer switch times.

In conclusion, keeping the values of the flow rates given in Table 6.3, a switch time of 13 min seems to deliver the highest productivity for a purity of 99% and a product recovery above 99%. This multicolumn chromatographic process will be further considered in the next section, in the context of the integrated reaction/purification continuous process.

6.4 Process Integration

6.4.1 Motivations

Process integration requires considering the entire reaction/purification process as a whole rather than designing and optimizing the different units (e.g., conjugation reactor, filtration, and chromatographic units) separately. By doing so, the synergy between the different unit operations can be better exploited. In the following, we discuss integrated processes designed for the production of conjugated proteins of a given DoC. The optimal DoC is usually fixed at a value exhibiting the best trade-off between the upgraded properties brought by the add-on molecule and adverse effects (e.g., loss of specific activity, increase in toxicity). For instance, PegIntron®, Pegasys®, and Mircera® are mono-PEGylated therapeutic proteins, which implies that the formation of multi-PEGylated conjugates should be avoided during their production. Consequently, the reaction has to be stopped prematurely before the target conjugate further reacts to form multiconjugated products so as to obtain the highest selectivity for the desired DoC. This is however not compatible with the high conversion of the protein. There is therefore a trade-off between conversion and selectivity, as discussed in Section 6.2.3.

This trade-off can be alleviated by taking advantage of the separation unit and recycling the unreacted protein (or conjugates with a DoC lower than the desired one). The schematic diagram of a general integrated process is presented in Figure 6.19. The system comprises a conjugation reactor, a separation unit and an ultrafiltration/diafiltration unit aiming at recycling the protein solution under the conditions used in the reactor (in terms of protein concentration and buffer solution composition). In principle, each step can be run in batch or in continuous mode, and we have seen in the previous sections that multiple technologies are already available to perform

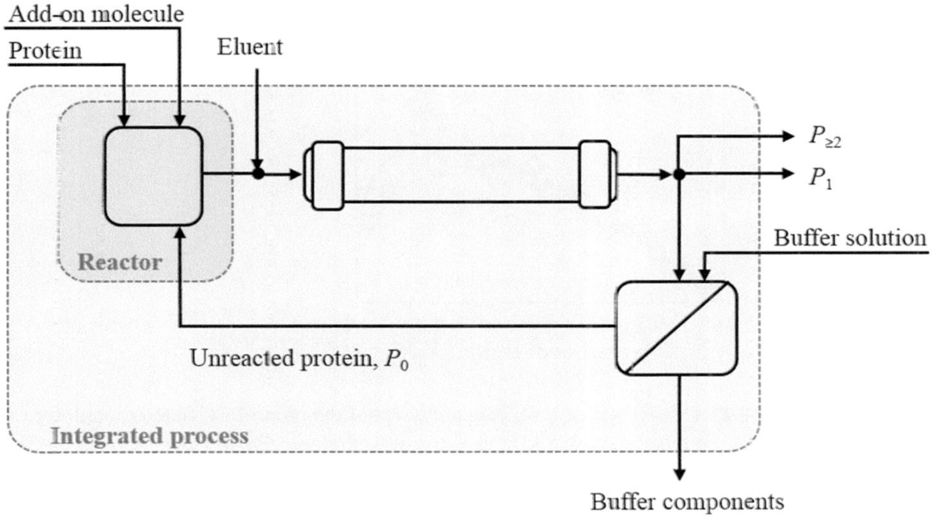

Figure 6.19 Schematic representation of the integrated process for the production of protein conjugates including a conjugation reactor, a separation unit (e.g., a chromatographic column), an ultrafiltration/diafiltration unit, and a recycling loop. Boundaries of the reactor and of the integrated process are indicated with dashed lines. This general scheme applies for both batch and continuous processes. Adapted from [345], with permission from John Wiley & Sons.

the reaction and purification steps in a continuous manner. Due to the recycling, one has to be careful when defining performance parameters such as conversion (see Equation 6.17) because its value depends on the considered boundaries of the system. In particular, one can define a conversion for the reactor (also referred to as conversion per-pass) considering the three inlet streams (including the recycling stream) and one outlet stream. Alternatively, one can define an overall conversion computed on the entire system considering only two inlet streams and the three outlet streams from the chromatography and filtration units (excluding the inert stream of eluent and buffer solutions as well as the internal recycle stream).

6.4.2 Protein Recycling and Overall Conversion

To illustrate qualitatively the impact of recycling on the conversion/yield trade-off, we continue the example of the PEGylation of lysozyme with 5 kDa methoxy PEG, which was already presented in Sections 6.2.3 and 6.3.3 for the reaction and purification aspects, respectively. A number of studies have investigated this PEGylation reaction in a BSTR under various conditions, including various solution pH, temperature, functionalized PEG reagents, reaction times [342, 345, 362–364]. The results of the different studies are summarized in terms of yield of the mono-PEGylated protein, Y_1^0, and overall conversion of the lysozyme, X_0, in Figure 6.20. This figure gathers PEGylation data obtained using different chemistries and processes based on three main strategies:

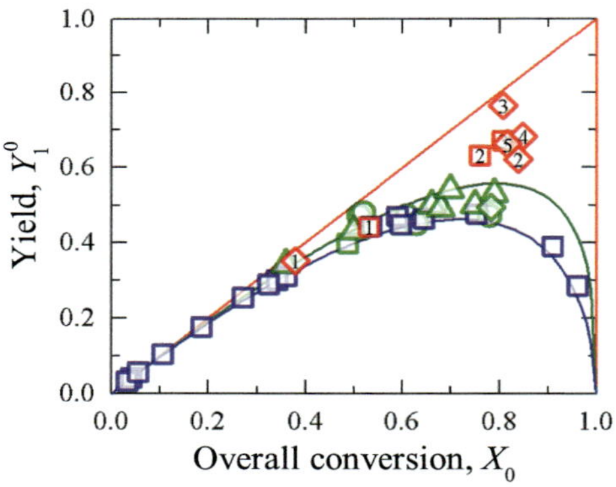

Figure 6.20 Conversion of lysozyme and yield of the mono-PEGylated protein for the lysozyme PEGylation reaction with 5 kDa methoxypolyethylene glycol. Adapted from [345], with permission from John Wiley & Sons. Experimental data include various concentrations, pH and temperatures obtained in batch mode. The data corresponding to the nonselective chemistry are presented with blue symbols, the selective N-terminal chemistry with green symbols and the integrated processes with recycling with red symbols. The numbers in the symbols for the integrated process with recycling represent the cycle number. The data were taken from: □ [345], ○ [342], □ [362], △ [363], ◇ [361], □ [364] and ◇ [345].

- *Nonselective PEGylation at the lysine residues.* Blue squares in Figure 6.20 show the data corresponding to the PEGylation of lysozyme with mPEG-SPA, each point representing different PEG/protein ratios ξ or different reaction times t^{BSTR} [345]. This approach undeniably suffers from its lack of selectivity for a given residue, leading to the formation of different positional isomers and various degrees of PEGylation. As already shown in Section 6.2.3, there is a clear trade-off between conversion and yield because high protein conversion is associated with the formation of multi-PEGylated proteins and therefore a low yield to the target product. The contour of this zone was computed with Equation 6.28 using $k_0 = 83.7$ mL/mol/min and $k_1 = 56.9$ mL/mol/min. It is represented as a blue line in Figure 6.20.
- *Selective PEGylation at the N-terminal.* Certain reactants, such as aldehyde functionalized PEG, are more selective to the N-terminal amine than to the lysine side chains under acidic conditions. Indeed, the difference in pKa between the α and ε amines can be exploited to increase the selectivity of the conjugation reactions. As compared to the nonselective chemistry, the ratio k_1/k_0 is decreased, thus leading to a higher yield for a given conversion. The corresponding data obtained under various operating conditions (temperature, pH, and ξ) are represented by green symbols in Figure 6.20 [342, 361–363]. It can be seen that only a slight improvement of the reaction performance is obtained as compared to the non-selective case considered earlier. A better yield could be obtained for a given conversion, but the improvement would never exceed 10%. Moreover, it should be noted that when using a different chemistry a different distribution of isoforms

may be produced. The upper limit of the yield/conversion values obtained with the selective chemistry was computed with Equation 6.28 by adjusting the ratio k_1/k_0 to 0.4 (instead of 0.7 in the previous case) to envelop the experimental points. It is represented as a green line in Figure 6.20.

- *Integration of the reaction and separation units by recycling of the unreacted protein.* In this case, the unreacted protein is recycled back to the reactor where it is further conjugated. This enables overtaking the trade-off usually observed for conjugation reactions. For the recovery of the unreacted protein, two approaches have been reported so far: ultrafiltration (red squares [364]) and ion exchange chromatography (red diamonds [345]). The numbers indicated in the symbols correspond to the cycle number. It is observed that with both approaches, a higher conversion and higher yield as compared to the selective chemistry are obtained after the second cycle. In principle, thanks to the recycling of the unreacted protein, the entire region below the red diagonal in Figure 6.20 is accessible.

A comparison among the different data shows that, although intrinsically limited in conversion by the multiplicity of the reactive sites, the performance of the PEGylation process can be strongly improved by combining the conjugation reactor with a separation unit and recycling the unreacted protein. However, there may be some concerns regarding the impact of recycling on the denaturation and aggregation of therapeutic proteins. Therefore, the protein structure and activity should be monitored in order to control that the protein being recycled is identical to the one freshly loaded into the reactor. This is currently challenging and the inline monitoring of protein quality attributes is a hot topic in the field of bioprocess engineering [365].

6.4.3 Process Performances: Yield and Productivity

In order to better understand the behavior of the integrated conjugation process presented in Figure 6.19, we report in the following a more detailed model-based analysis of the impact of the protein conversion inside the reactor on the yield and productivity of the process. We discuss first the results obtained in the batch case and then compare batch and continuous processes in the next section.

The BSTR, the single-column chromatographic process and the filtration unit were described in Sections 6.2.3, 6.3.2, and 6.3.3, respectively. The following process variables are considered for the three unit operations:

Reaction As in the case study of Section 6.2.3, the volume of the reactor is set here to $V^{BSTR} = 1.5$ mL, the initial concentration of protein to $C_0^F = 9$ mg/mL, and the reaction time to $t^{BSTR} = 50$ min. The conversion of the protein inside the reactor is modulated by varying the concentration of functionalized mPEG-SPA. For each value of ξ, the final concentrations of PEG, lysozyme, mono-PEGylated protein, and multi-PEGylated proteins at the reactor outlet are simulated. The reaction mixture is then injected as such into the chromatography column.

Purification The dimensions of the single-column and the elution protocol are kept unchanged independent of the composition of the feed mixture and set to those chosen in Section 6.3.2 (i.e., 5 × 16 mm column, with modifier gradient from 0 to 1 M NaCl over 25 CV at a flow rate of 1.4 mL/min). However, the collection windows are varied in order to obtain the highest possible recovery ensuring a purity of the mono-PEGylated product above 95%. In addition, it is crucial to set the purity of the unreacted lysozyme to be recycled as high as possible to avoid the presence of mono-PEGylated lysozyme in the recycle stream, which would further react into the reactor and thus lower the performance of the process.

Filtration Finally, the ultrafiltration/diafiltration step can be performed with centrifugational filters. Three diafiltration volumes are needed to reduce the concentration of modifier below 5% of its initial value (see Section 6.3.2). According to the supplier's information 5 min are necessary per-pass, so that the total time of the ultrafiltration/diafiltration step is 15 min. In order to investigate the influence of the recycling on the process performance, the sieving coefficient of the protein ($\mathcal{S}_0$) is varied between 0 (complete recycling) and 1 (no recycling). On the other hand, the sieving coefficient of the salt is set to one.

The simulation results are presented in Figure 6.21, where the overall yield for the mono-PEGylated lysozyme (continuous lines) and the productivity (dotted lines) are shown as a function of the protein conversion in the reactor. For the sake of simplicity, the productivity of the process is defined in milligrams of mono-PEGylated lysozyme per unit time and unit volume of the system (BSTR+single-column). Let us first consider the case of an ideal membrane where the protein sieving coefficient would be $\mathcal{S}_0 = 0$, which means that no protein is lost during the ultrafiltration/diafiltration operation so that all the unreacted protein is recycled. It can be seen that the highest yield is obtained at very low protein conversion inside the reactor. This implies that, in terms of yield

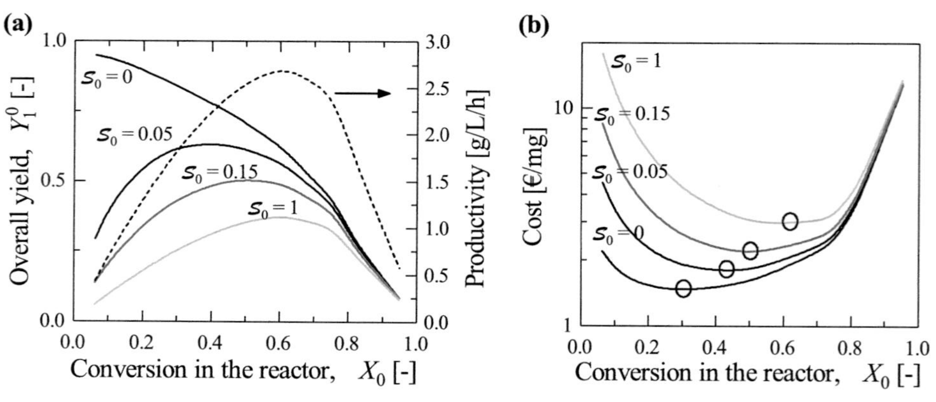

Figure 6.21 (a) Overall yield (continuous lines) and productivity (dotted line) as a function of the protein conversion inside the reactor. The productivity is defined with respect to the volume of the system ($V^{BSTR} + V_{col}$) and assuming a total batch time of 85 min (50 min for the reaction, 20 min for the chromatographic separation and 3 × 5 min for the filtration). (b) Cost associated with the production of monoconjugated protein as a function of the protein conversion inside the reactor considering $a = 1$ €/mg and $b = 7 \times 10^{-3}$ €/mL/h in Equation 6.41. Open circles represent the minimum cost for the different sieving coefficients considered.

of the mono-PEGylated protein, it is best to operate the reactor at very low conversion so that the synthesized product is almost exclusively the desired one. In this case, the recycling of the unreacted protein ensures that a high overall conversion is obtained, even though the reactor is operated at low conversion. The counterpart is that almost no conjugated protein is produced so that the productivity is quite low. When the sieving coefficient increases from 0 to 0.05, the protein recovery decreases to 85% after three diafiltration volumes (see Section 6.3.2). This has a detrimental impact on the overall yield, especially at low conversion inside the reactor. The decrease of the yield is getting even more pronounced when increasing the protein sieving coefficient until the limiting case where $\mathcal{S}_0$ equals 1, which means that the protein is not recycled at all. In this case, a similar curve as the one presented in Figure 6.20 in the case of nonselective PEGylation reaction is obtained. In the absence of recycling, the maximum yield is obtained for a conversion of nearly 60%.

In addition, it should be noted that the productivity is independent of the sieving coefficient $\mathcal{S}_0$ and only depends on the conversion of the protein inside the reactor. It can be seen that the productivity is maximal for a conversion of nearly 60%. There is therefore a trade-off here between the yield, which would require working at low protein conversion in the reactor, and the productivity, which would be maximal at about 60% conversion.

In order to determine which is the optimal conversion of the protein inside the reactor, it is necessary to evaluate the production cost as a function of conversion. Here, a simplistic cost function is considered: it is assumed that the total production cost depends only on the yield Y_1^0 and productivity φ_1 according to:

$$\text{cost} = \frac{a}{Y_1^0} + \frac{b}{\varphi_1} \tag{6.41}$$

The values of the parameters a and b depend on the cost of the raw material (in €/mg) and on the operating cost of the process (in €/mL/h), respectively. The latter may include manpower or the energy required to maintain a chromatographic unit at a high pressure, for example. According to Equation 6.41, a low cost can be obtained either at high yield or at high productivity, which are two antagonistic objectives (see Figure 6.21(a)). With expensive proteins it may be more interesting to work at high yield not to waste the reactants in the formation of undesired products rather than operating the process at high productivity.

For illustrative purposes, the coefficients a and b are arbitrarily set to 1 €/mg and 7×10^{-3} €/mL/h, respectively. With these values, the term a/Y_1^0 in Equation 6.41 roughly contributes to 75% of the total cost at the maximal productivity, which corresponds to a conversion per-pass of 60% in Figure 6.21(a). The selected values of a and b provide an order of magnitude of the cost breakdown in agreement with a recent study, which showed that, on average, the antibody contributes for about 75% to the overall cost of ADCs production, the rest being attributed to the operating costs [354].

The cost calculated with Equation 6.41 is shown as a function of the protein conversion inside the reactor in Figure 6.21(b) for different values of the sieving coefficient of the protein. It can be seen that, when the unreacted protein is entirely recycled ($\mathcal{S}_0 = 0$), the conversion per-pass leading to the lowest cost is about 30%. On the other hand, as the sieving coefficient increases, i.e., when more and more unreacted protein is lost during the ultrafiltration/diafiltration step, the optimal protein conversion shifts to higher values and reaches about 60% in the absence of recycling ($\mathcal{S}_0 = 1$). It is worth mentioning that the absolute value of the minimum cost depends upon the values of the coefficients a and b, which are of course specific to the considered system. For a given value of the parameter b (related to the operating cost), the minimum of the cost function moves to lower values of the protein conversion when increasing the value of a (related to the material cost). In other words, if the reactants are expensive, it is advised to operate the reactor at low conversion so as to limit the loss of valuable reactant. But if the operating costs dominate, it becomes less critical to waste some reactant. The concepts illustrated here are sufficiently general so that a similar analysis can be performed with any conjugation process. This approach allows estimating the optimal conversion inside the reactor and properly designing the reaction, purification, and filtration units.

6.4.4 Batch or Continuous Operation?

We have seen throughout this chapter that there is no apparent difficulty to realize the protein conjugation process in the continuous mode. The PFR (designed in Section 6.2.3) and the multicolumn 4C-MCSGP process (designed in Section 6.3.3) are examples of technologies that can be used for this purpose. In this section, we propose to investigate the potential benefits of continuous technologies over the traditional batch ones in terms of productivity for the integrated process with recycling presented in Figure 6.19.

Just like for the batch case presented in Section 6.4.3, the productivity of the integrated process is studied here at various protein conversions. The flow rates and the volume of the continuous reactor are kept constant, and only the feed concentration of the mPEG-SPA solution, C_{PEG}^F, is varied to obtain different protein conversion values. Similarly, the 4C-MCSGP process is kept as designed in Section 6.3.3, and only the composition of the feed mixture is varied depending on the conversion of the conjugation reaction at the reactor outlet. The simulated productivity is shown in Figure 6.22 with a black solid line as a function of the protein conversion in the reactor. As a comparison, the productivity of the batch process shown in Figure 6.21(a) is also reported in Figure 6.22 with a black dotted line. Note that the total volume of the system is $V^{BSTR} + V_{col}$ for the batch process, while it is $V^{PFR} + 4V_{col}$ for the continuous process because four columns are used in the 4C-MCSGP.

It is seen in Figure 6.22 that the continuous process is more productive than the batch process. This is in part because the batch process requires to wait for the filtration step to be completed so as to recycle the unreacted protein in the reactor. To investigate the impact of keeping the reactor idle while performing the filtration, the productivity was computed assuming that the ultrafiltration/diafiltration step could be performed instantaneously. The results are shown with the gray dotted line in Figure 6.22. It is seen that, in the case under consideration, comparable process performances can be

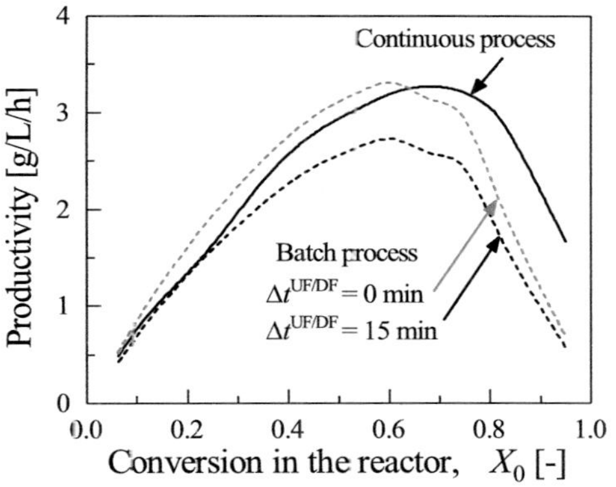

Figure 6.22 Productivity of the integrated continuous (solid line) and batch (dashed lines) processes as a function of the protein conversion inside the conjugation reactor. Two diafiltration/ultrafiltration times, namely 0 and 15 min, are considered for the batch process.

obtained with batch and continuous processes when neglecting the filtration time. This is understandable because the BSTR and PFR have the same residence time, and thus the same productivity. The superiority of the continuous chromatographic process over the single-column process appears at protein conversions above 60%, when the separation between the PEGylated product becomes more difficult due to the presence of a larger amount of di-PEGylated proteins. In that case, a better recovery and thus a higher productivity can be obtained with the 4C-MCSGP process as compared to the single-column chromatographic process.

Overall, in this example, the advantage of a transition from batch to continuous lies in two main reasons: *(i)* the absence of idle times for the reactor and *(ii)* the better performance of the countercurrent chromatographic process.

6.5 Conclusion

Protein conjugation offers a wide range of possibilities to improve the performance of a specific protein in various fields, going from *in vivo* imaging to oncology. Despite the large number of applications, common principles can be found to rationalize the synthesis of bioconjugates. Conjugation necessitates the proper selection of a protein, an add-on molecule to be grafted and the linker chemistry. The choice of the first two is driven by the desired application, while the latter depends upon the available functional groups on the protein surface. The presence of multiple functional groups in the protein amino-acid sequence typically leads to the synthesis of heterogeneous mixtures of conjugated compounds, going from the unreacted protein to highly conjugated products and spanning various positions of the add-on molecule on the protein. Depending on the level of details one wants to reach, several kinetic models are available, going from the description of the products only in terms of degree of conjugation, to the complete description of the distribution of positional isomers. These kinetic models are then used to design conjugation reactors in terms of reaction time and residence time for batch and continuous units, respectively.

Besides the reaction kinetics, another important aspect of bioconjugation processes is the purification of the target product. While filtration is often used to get rid of unreacted chemicals and small reaction by-products, chromatography is the technique of choice when it comes to isolate the target product, defined by proper specification in terms of number, and possibly position, of the grafted add-on molecules. The type of chromatographic medium is selected to best exploit the new properties brought to the protein by the add-on molecule: SEC/IEC for PEGylated proteins, HIC for ADC ...

Process integration and recycling of the unreacted protein enables overtaking the trade-off between yield and conversion exhibited by conjugation reactions, where the product of interest may further react to form overconjugated species. The use of integrated reaction and separation units with recycling allows in fact achieving simultaneously high yield and high conversion.

7 Protein Aggregation in Biopharmaceutical Processes

7.1 Introduction

Therapeutic proteins are only marginally stable in solution and are subject to both chemical instabilities (e.g., deamidation, oxidation, hydrolysis) and physical instabilities (e.g., denaturation, noncovalent aggregation). In biopharmaceutical processes, these instabilities are promoted by the exposure of proteins to a variety of stress conditions, such as elevated temperatures, low pH, or contact with hydrophobic surfaces [59, 366, 367]. In addition, it is also difficult to guarantee long-term stability of the final product because protein-based drugs are generally formulated at very high protein concentrations, what favors the formation of aggregates [368].

Because the presence of protein aggregates may impair drug potency and even trigger an adverse immune response of the organism [58, 59], the level of aggregates in the final drug product is regarded as a critical quality attribute that must be strictly controlled. However, removing aggregates to meet purity specifications may be very challenging because aggregation often occurs during downstream processing In this case, the competition between aggregate removal and aggregate formation may compromise the recovery of a monomeric protein solution. In addition, protein aggregation may have a detrimental effect on the course of the process operations, for example, due to membrane fouling and filter clogging [59, 369]. Overall, protein aggregation must be avoided in order to guarantee product quality and maintain process performances. Therefore, a solid understanding of the mechanisms underlying the aggregation process and of the impact of the various operating conditions on protein stability is needed to develop biopharmaceutical processes that minimize aggregate formation.

Depending on the protein and on the operating conditions, a broad spectrum of protein aggregates can be produced. Aggregates may differ in terms of their size (from dimers to visible precipitates), morphology (from filaments to compact amorphous aggregates), types of bonding (physical or chemical), reversibility (reversible or not), and protein conformation (native or denaturated) [370]. These aggregate characteristics may affect product therapeutic efficacy, product safety, and process performances in different ways [59]. For example, both product clearance from the organism and filter clogging during manufacturing are impacted by the aggregate size. Furthermore, there are evidences that the level of immunogenicity does not depend only on the amount of aggregates, but also on their physical properties, such as the size, degree

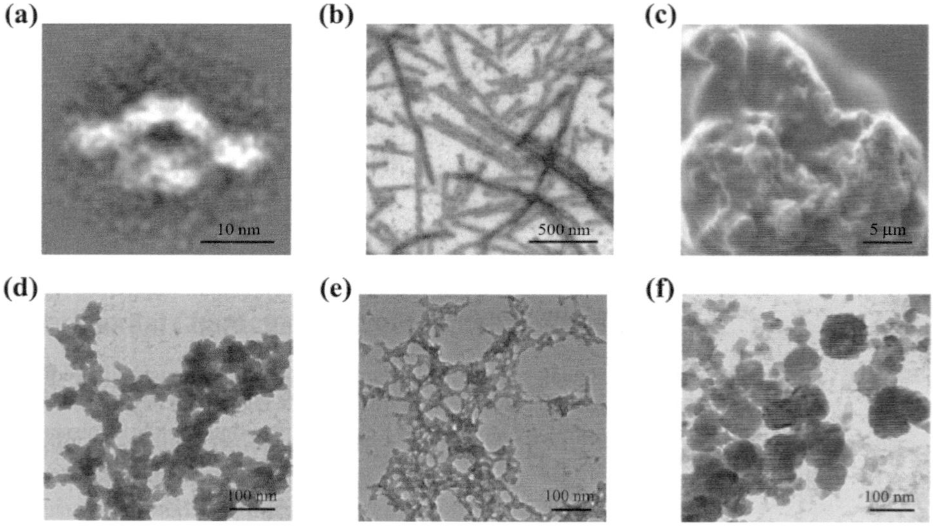

Figure 7.2 Electron micrographs of mAb aggregates differing in terms of size and morphology. (a) mAb dimer. Reprinted from [379]. With permission from Springer. (b) mAb filaments. Reprinted from [380]. With permission from Springer. (c) mAb particulates. Reprinted from [381], copyright 2012, with permission from Elsevier. (d–f) mAb amorphous aggregates obtained under heat stress, after freeze-thaw cycles, and after shaking, respectively. Reprinted from [382], copyright 2014, with permission from Elsevier.

electron microscopy, are very helpful because they allow a direct visualization of the spatial organization of the aggregates. Figure 7.2 shows some electron micrographs of mAb aggregates exhibiting different sizes and morphologies. Dimers, filaments, and large particulates are shown in Figure 7.2(a–c) [379–381], while Figure 7.2(d–f) shows aggregates of different compactness, from very open to very compact structures, which were obtained under various types of stress conditions (e.g., heating, freeze-thawing, shaking) [382].

A very useful parameter to describe the morphology of protein aggregates is the fractal dimension d_f. The concept of fractal scaling allows unifying in a simple way various aggregate morphologies by relating the aggregate mass with the aggregate radius according to the following equation [383, 384]:

$$\frac{MW_i}{MW_p} = k_f \times \left(\frac{R_i}{R_p}\right)^{d_f} \tag{7.1}$$

where MW_p is the molecular weight of the monomeric protein and R_p its radius. Similarly, MW_i is defined as the molecular weight of an aggregate containing i monomeric units (so that $MW_i/MW_p = i$) and R_i is its radius. There are different ways to define the radius of proteins and aggregates (e.g., gyration, hydrodynamic, smallest encompassing sphere) and Equation 7.1 can be used with any of these provided that R_p and R_i are defined in a consistent manner. In this chapter, we use exclusively the hydrodynamic radius (measured by DLS). The scaling prefactor k_f is usually in the order of unity.

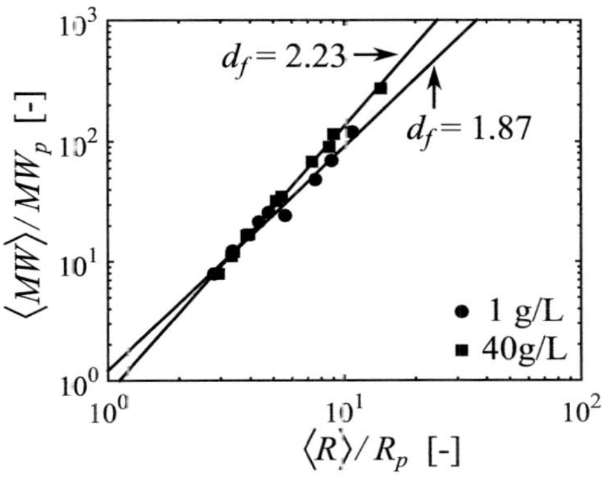

Figure 7.3 Determination of the fractal dimension of mAb aggregates at two different protein concentrations from the correlation between the molecular weight (measured by SLS) and the hydrodynamic radius (measured by DLS). Aggregation was induced by thermal stress and the different points correspond to different incubation times. Straight lines correspond to the fit with Equation 7.1. Adapted from [385], published by The Royal Society of Chemistry.

The value of d_f quantifies the aggregate compactness, with the extreme cases of $d_f = 1$ and $d_f = 3$ corresponding to fibrillar and spherical aggregates, respectively. Several techniques are available to estimate d_f experimentally, including light scattering, microscopy, and rheology [384]. The most direct method consists in measuring the molecular weight and the radius of aggregates of different sizes and fit d_f using Equation 7.1. In practice, it is hardly possible to isolate one single size of aggregates (i.e., containing exactly i monomeric units). Therefore, MW_i and R_i are replaced by $\langle MW \rangle$ and $\langle R \rangle$, which correspond to the average molecular weight and average radius of the aggregate population, respectively.

Figure 7.3 shows experimental data of the average aggregate molecular weight as a function of the average hydrodynamic radius obtained for mAb aggregates formed after various aggregation times starting from two different protein concentrations, namely 1 g/L and 40 g/L [385]. The straight lines correspond to fits with Equation 7.1, which provide the following values: $d_f = 1.87$ and $k_f = 1.21$ at 1 g/L, $d_f = 2.23$ and $k_f = 0.78$ at 40 g/L. Interestingly, a higher value of the fractal dimension is obtained at 40 g/L than at 1 g/L. This means that denser aggregates are formed under concentrated conditions than under dilute conditions. Besides protein concentration, the fractal dimension of the aggregates has been found to depend on a number of experimental conditions, including the solution pH and the presence of electrolytes [386].

7.2.2 Protein–Protein Interactions

In Section 7.2.1, we have seen how to characterize protein aggregates in solution, and in particular how to quantify the aggregate content, size, and morphology. In this section, we focus on the characterization of monomeric protein solutions, and more precisely on the evaluation of the type (i.e., net attractive or repulsive) and magnitude of protein–protein interactions, which play a key role in the protein aggregation propensity [387]. While strongly attractive protein–protein interactions promote aggregate formation, strongly repulsive protein–protein interactions tend to stabilize protein solutions against aggregation.

Proteins interact with each other via a variety of forces, which include electrostatic, van der Waals, hydration, and hydrophobic forces. The strength of electrostatic interactions can be assessed by evaluating the protein net charge from electrophoretic mobility measurements [388–390]. This technique is based on the principle that, in the presence of an electric field, molecules move with a velocity that depends on their zeta potential. The latter is the surface electric potential of a protein molecule surrounded by the layer of fluid accompanying it during its motion and is related to the protein net charge. Figure 7.4(a) illustrates how the net charge of an antibody decreases as the solution pH increases toward the protein isoelectric point (pI), which is defined as the pH value at which the protein net charge is null. These data indicate that the electrostatic repulsion between protein molecules is stronger at pH values far away from the protein pI than close to the pI. In the same figure, surface potential calculations at two pH values, namely pH 5 and pH 9 [391], are also shown. Such detailed simulations of the protein structure are very helpful to visualize how the charges are distributed on the surface of the protein. Here, it is observed that the protein surface is relatively uniform and positively charged at pH 5, while both positively charged and negatively charged regions are observed at pH 9, leading to a significantly reduced positive net charge. These simulations allows one to appreciate the strong heterogeneity of the charge distribution on the protein surface, especially in the vicinity of the protein pI.

Besides charge heterogeneities, the protein surface is also nonuniform in terms of hydrophobicity, exhibiting both aggregation-prone hydrophobic patches and more

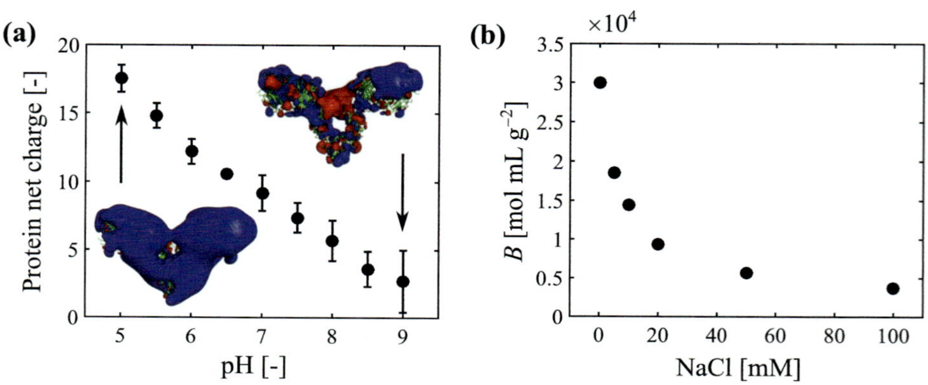

Figure 7.4 (a) Net charge of a mAb estimated from electrophoretic measurements as a function of the solution pH. Inserts show the charge distribution on the mAb surface, with positive and negative charges represented in blue and red, respectively. Adapted with permission from [391]. Copyright 2014 American Chemical Society. (b) Second virial coefficient of a mAb as a function of the concentration of sodium chloride. Adapted from [392], copyright 2010, with permission from Elsevier.

hydrophilic regions [393–395]. Overall, the simultaneous presence of charged, polar, and apolar residues makes the detailed characterization of protein–protein interactions extremely difficult. Therefore, the sum of all the protein–protein interactions is commonly quantified in a coarse-grained manner by the so-called second virial coefficient B. The latter is an effective thermodynamic parameter that reflects the nonideality of protein solutions. It is defined as the second coefficient in the virial expansion of the osmotic pressure Π as a function of the protein concentration [396, 397]:

$$\frac{\Pi}{k_B N_a T} = \frac{C_1}{MW_p}(1 + BC_1) \tag{7.2}$$

where k_B is the Boltzmann constant, T the temperature, N_a the Avogadro number. In addition, C_1 is the mass protein concentration (in g/L), so that B is expressed in L/g in the earlier expression.

Importantly, the second virial coefficient can be linked to the total protein–protein interaction potential U_T by [396]:

$$B = -\frac{N_a}{16\pi^2 MW_p} \int_{\Sigma} \int_{0}^{\infty} \left(\exp\left(-\frac{U_T}{k_B T} \right) - 1 \right) r^2 dr d\Omega \tag{7.3}$$

where U_T accounts for all kinds of intermolecular interactions, including electrostatics, van der Waals, hydration, and hydrophobic forces. The intermolecular potential is not only a function of the intermolecular center-to-center distance r, but also of the orientation of the interacting proteins due to the heterogeneity of the protein surface properties. The angular variables are collectively represented by Ω, which is integrated over the protein surface Σ in Equation 7.3. To summarize, the second virial coefficient B allows quantifying the two-body protein–protein interactions in a coarse-grained manner, lumping together all types of interactions and protein surface heterogeneities.

Interestingly, the sign of B reflects the overall nature of the two-body protein–protein interactions: a negative value of B indicates net attractive protein–protein interactions, while a positive B value corresponds to net repulsive protein–protein interactions. The absolute value of B instead quantifies the magnitude of these interactions, with large absolute B values indicating strong intermolecular interactions (either attractive or repulsive). It is important to note that B is a thermodynamic parameter, which provides information on the tendency of a protein solution to form aggregates or to remain in the monomeric state. However, B does not provide any information on the kinetics of aggregation, i.e., the rate at which aggregates are formed.

Static light scattering is the most common technique used to measure the second virial coefficient [388–390]. Indeed, the scattering of light does not depend only on the size of the molecules in solution, as we have seen in the previous section, but also on the intermolecular interactions. For illustrative purposes, Figure 7.4(b) shows how the second virial coefficient of a mAb decreases as a function of the concentration of sodium chloride in solution [392]. Such a behavior can be attributed to the screening of protein charges, which in turn lowers electrostatic intermolecular repulsion. Some more complex effects that may occur in the presence of salt will be discussed in Section 7.4.2.

Protein–protein interactions can also be investigated through the measure of the diffusion interaction coefficient k_D defined by:

$$D_1 = D_1^0(1 + k_D C_1) \tag{7.4}$$

where D_1 is the diffusion coefficient of the protein in the buffer solution measured by DLS at the protein concentration C_1, and D_1^0 is the value of the same coefficient at infinite dilution. As in Equation 7.2, C_1 is in g/L, so that k_D is expressed in L/g.

The parameter k_D is related to B according to [389, 392]:

$$k_D = 2B - k_S - v_{sp} \tag{7.5}$$

where k_S is the sedimentation coefficient of the protein and v_{sp} its partial specific volume, which is usually in the order of 0.7 mL/g [398]. Thanks to the development of DLS plate readers, k_D can be assessed in a rather high throughput manner, as compared to B measurements that are relatively time and material consuming. By using Equation 7.5, it is possible to estimate B from the independent determination of k_D by DLS and k_S by AUC, thus providing a complementary method to SLS [392]. Other alternative methods to determine B are emerging, such as self-interaction chromatography [399]. Although the absolute values of B measured with different techniques must be taken with some caution, comparative studies carried out with the same technique provide robust information about the changes in protein intermolecular interactions in different solution compositions [388–391].

7.2.3 Protein Structure

Besides protein–protein interactions, the protein spatial structure also significantly influences the protein aggregation propensity [387, 400]. The spatial arrangement of atoms in a protein is referred to as the protein secondary structure to describe the local conformation of particular portions of the protein, or to as the protein tertiary structure to refer to the entire protein structure. In contrast, the protein primary structure refers to the amino acid sequence, without any consideration in terms of spatial organization. The two main types of secondary structures are the α-helix and β-sheets, which are both mainly maintained by hydrogen bonding. The tertiary structure corresponds instead to the folding of the different regions of the protein (which may exhibit different secondary structures) into a spatially organized structure, mainly stabilized by disulfide bonding and hydrophobic interactions. Under physiological conditions, the hydrophobic and nonpolar parts of the protein tend to be buried in the core of the protein molecule in order to reduce their interaction with water molecules.

Information about the protein secondary and tertiary structures can be accessed by a wide variety of experimental techniques including circular dichroism (CD) [401], intrinsic tryptophan [402], and extrinsic dye fluorescence [403], as well as infrared [404] and Raman [405] spectroscopy. As an example, the insert of Figure 7.5(a) shows the CD spectrum of Apolipoprotein A-I, which is a protein essential in the transport and metabolism of cholesterol [406]. The signal measured by a CD instrument is termed

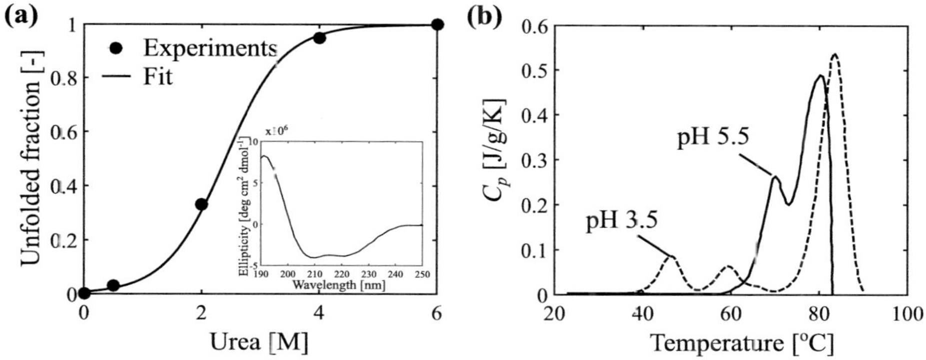

Figure 7.5 (a) Fraction of unfolded Apolipoprotein A-I as a function of urea concentration estimated from the CD ellipticity at 220 nm. The inset corresponds to the CD spectra of the protein in the absence of urea. Adapted from [406], copyright 2014, with permission from Elsevier. (b) DSC scans of a mAb at two pH values. Adapted from [407], copyright 2012, with permission from Elsevier.

CD ellipticity and is related to the difference in absorbance between right and left circularized polarized light by chromophores that are either chiral or in a chiral environment. The analysis of a CD spectrum in the far and near UV regions provide information about the secondary and tertiary structure of a protein, respectively [401]. For instance, the far UV spectrum presented in Figure 7.5(a) shows a maximum at 190 nm and two minima at 208 and 222 nm, which are characteristic features of α-helical structures. By monitoring the CD ellipticity at a properly selected wavelength, it is possible to estimate the fraction of unfolded protein upon a change in environmental conditions. This is shown in Figure 7.5(a) for Apolipoprotein A-I, where the CD intensity was monitored at 220 nm at different urea concentrations. It is seen that the fraction of unfolded protein increases with an increase in the concentration of urea [406], which is a common protein denaturant.

Besides chemical denaturants, protein unfolding is also promoted by high temperatures. Protein thermal unfolding can be followed by the aforementioned techniques (CD, fluorescence, and vibrational spectroscopy) as well as by differential scanning calorimetry (DSC) [408]. DSC is a thermoanalytical technique in which the amount of heat required to bring a sample at a given temperature is measured as a function of temperature. The comparison of these data with a reference material provides information about the endothermic and/or exothermic transitions that the sample undergoes. As an example, Figure 7.5(b) shows DSC scans of a mAb at two different pH values [407], namely 3.5. and 5.5. The presence of multiple peaks in the apparent heat capacity indicates that the various domains of the mAb unfold at different temperatures [409]. It is also interesting to note that such unfolding transitions occur at lower temperatures at pH 3.5 than at pH 5.5. A likely explanation is that protein unfolding is promoted at pH values far below the protein pI due to intramolecular electrostatic repulsion.

7.2.4 Solution Viscosity

The viscosity of protein solutions may increase dramatically with the protein concentration, which can heavily complicate product handling during downstream processing and drug administration by subcutaneous injection [368]. Regarding chromatographic operations, we have seen in Section 2.4.3 that the pressure drop in a column increases linearly with the fluid viscosity. This implies that the purification of viscous protein solutions may require selecting a low feed flow rate in order to maintain the pressure drop below the maximal recommended value. Viscous protein solutions are also troublesome during diafiltration/ultrafiltration steps because high viscosities reduce the permeate flux and may thus limit the ability of filtration systems to concentrate them up to the desired values [368].

Therefore, the viscosity of concentrated protein solutions is an important property that needs to be evaluated experimentally. This can be done with traditional rheological measurements, or from the determination of the diffusion coefficient of tracer polymeric nanoparticles with DLS [385, 410]. The latter technique allows to measure viscosities at the microliter scale and does not require shearing the sample, which simplifies data interpretation. Importantly, the size of the tracer particles should be carefully selected to be sufficiently large with respect to the protein molecules (otherwise the viscosity felt by the tracers will be lower than the macroscopic viscosity) and sufficiently small to diffuse by Brownian motion. In practice, particle sizes in the order of 150–200 nm have been shown to provide satisfactory results both for monomeric and aggregated protein solutions [385, 410]. Figure 7.6 shows the impact of mAb concentration on the solution viscosity, evaluated with both rheological and tracer measurements. It is seen that the two techniques give very similar results.

It is observed in Figure 7.6 that viscosity values almost ten times larger than the viscosity of water are obtained at the mAb concentration of 160 g/L. It is worth mentioning that the viscosity of protein solutions is highly system specific because it

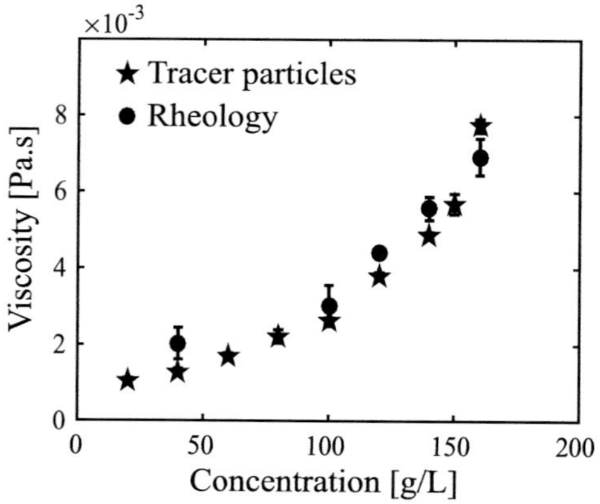

Figure 7.6 Solution viscosity as a function of mAb concentration estimated with a rheometer using a cone and plate geometry and with DLS using polymeric tracer nanoparticles. Adapted from [385], published by The Royal Society of Chemistry.

Table 7.1. Summary of the experimental techniques most commonly used to characterize protein solutions.

Property	Experimental techniques
Monomer content	Size exclusion chromatography Ultracentrifugation Field-flow fractionation
Aggregate size	Static and dynamic light scattering Microscopy Light obscuration Coulter counter
Aggregate compactness	Light scattering Microscopy
Protein–protein interactions	Electrophoresis Static and dynamic light scattering
Protein structure	Differential scanning calorimetry Circular dichroism Intrinsic tryptophan and extrinsic dye fluorescence Infrared and Raman spectroscopy
Solution viscosity	Rheology Dynamic light scattering (with tracer particles)

is affected by protein–protein interactions [411], which depend both on the protein under examination and on the solution conditions. For example, the viscosity of concentrated mAb solutions is generally found to be pH dependent [412, 413]. In addition, the presence of sugars, which are commonly added in protein solutions to increase long-term stability, has been shown to significantly increase the solution viscosity [414], which on the other hand may be reduced by increasing the ionic strength [414].

7.2.5 Summary

Table 7.1 provides a summary of the most common experimental techniques used to characterize protein solutions, as discussed in the previous sections.

7.3 Protein Aggregation Mechanisms

7.3.1 Colloidal Stability and Conformational Stability

Proteins exhibit very complex structures that are instrumental to their biological activity. In particular, proteins are characterized by very heterogeneous regions, which may be hydrophobic or hydrophilic, neutral, positively or negatively charged. In addition, the polypeptide chains adopt a well-defined three-dimensional arrangement that is stabilized by intramolecular interactions. All this eventually controls the propensity of a protein to aggregate, unfold, or adsorb on some surfaces. Single-point mutations have been shown to dramatically impact aggregation rates [415], thus confirming that even a slight change in the protein amino acid sequence can drastically affect the protein aggregation propensity. In this respect, molecular dynamic simulations are highly valuable to describe complex protein structures. For example, we have seen in Figure 7.4(a) that surface potential calculations are helpful to study the charge distribution at the protein surface. Molecular dynamics simulations have also proven useful to obtain information on the location of aggregation-prone regions [393–395]. However, the high resolution of the protein structure in atomistic simulations is achieved at the expense of a considerable computational effort, making in silico approaches not suitable to study protein aggregation kinetics over time scales relevant for drug manufacturing and storage, at least with the computing power currently available.

In an attempt to decrease system complexity, protein molecules can be characterized in a coarse-grained manner at the molecular level by averaging surface heterogeneities. In this frame, concepts and techniques from colloid science provide powerful tools to gain understanding on the thermodynamics and kinetics of protein aggregation [383]. Colloidal dispersions are two-component systems consisting of a discontinuous phase finely dispersed in a continuous medium, in which the elements of the dispersed phase are sufficiently small (roughly below 1 μm) to be affected by Brownian motion more than by gravity [416, 417]. Familiar examples of colloidal systems are clay slurries (solid particles in a liquid), tobacco smoke (solid particles in a gas), and aerosols (liquid droplets in a gas). Because the effect of gravity is limited with respect to that of Brownian motion, the elements of the dispersed phase do not sediment over the observable time scale. However, they typically aggregate in order to reduce the surface contact with the surrounding continuous phase. In other words, the aggregated state has most often a lower energy than the dispersed state and is thus the thermodynamically stable state. Nevertheless, it is possible to considerably delay the aggregation process by building a sufficiently high energy barrier with respect to the kinetic energy (i.e., the thermal energy) between the elements of the dispersed phase, for instance by increasing their charge and therefore their electrostatic repulsion. Figure 7.7 shows a scheme of the energy profile related to the transition from the dispersed colloidal state to the aggregated state that summarizes the concepts introduced earlier. One can identify the thermodynamic driving force for aggregation and the energy barrier delaying the kinetics of aggregation. If the aggregates are smaller than roughly 1 μm, they are still considered as colloidal and remain suspended in solution. Otherwise, the aggregates eventually sediment, and the system is not colloidal any longer.

Strictly speaking, protein solutions are not colloidal systems because they are monophasic. However, proteins are considerably larger than solvent molecules, so that they can be regarded as dispersed objects diffusing in a continuous solution. This allows borrowing some concepts from colloid science to investigate and characterize the stability of protein solutions, as described throughout this chapter.

We have seen in Section 7.2.2 that proteins in solution interact via numerous forces. In the frame of the classical DLVO (Derjaguin, Landau,Verwey, Overbeek) theory

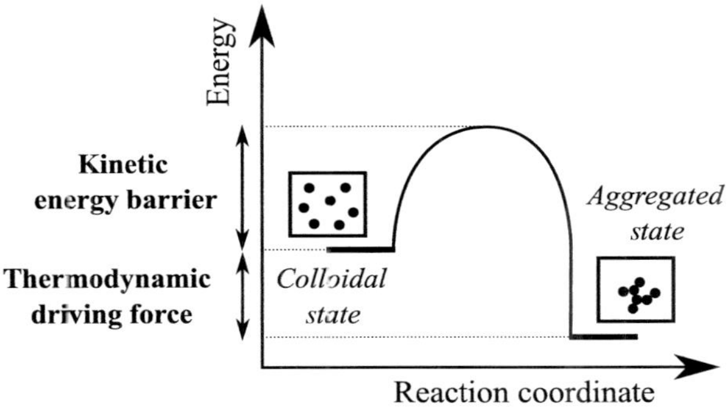

Figure 7.7 Qualitative energy profile, illustrating the thermodynamic and kinetic aspects of the transition from colloidal dispersions to aggregated systems. Adapted from [383], with permission from John Wiley & Sons.

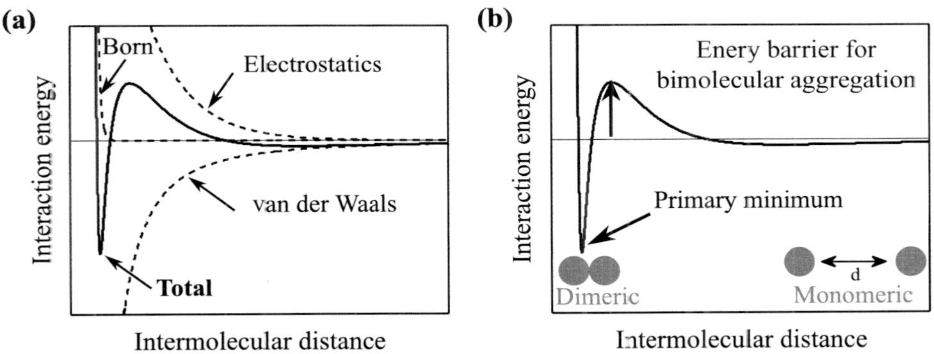

Figure 7.8 (a) Intermolecular interaction potentials in the frame of the DLVO theory. (b) Energy barrier and energy well of the total intermolecular potential. Adapted from [383], with permission from John Wiley & Sons.

[416, 417], the stability of colloidal systems results from the balance between long-range electrostatic repulsion, short-range van der Waals attraction, and very short-range hard-core (Born) repulsion. The DLVO theory neglects any other force (as, for example, hydrophobic forces) and assumes that the colloidal particles are spherical and uniformly charged. It is therefore often too simplistic to describe quantitatively the aggregation behavior of proteins, but it provides a good starting point to analyze qualitatively some relevant physical phenomena. Figure 7.8(a) shows a typical scheme of the interaction potential between two molecules as a function of their center-to-center distance in the frame of the DLVO theory. The dotted lines correspond to the three interaction potentials taken separately, while the thick solid line corresponds to the total interaction potential, given by the sum of electrostatic, van der Waals, and Born potentials. The two main features of the total interaction potential are highlighted in Figure 7.8(b): a deep energy well at short intermolecular distances (often referred to as the primary minimum) and an energy barrier at longer intermolecular distances.

interconnected, and must therefore be considered simultaneously when assessing the protein aggregation propensity. Because the solution physicochemical properties regulate the type and strength of both inter- and intramolecular interactions, the two types of stabilities turn out to be very sensitive to the solution composition, as described in Section 7.4.

7.3.2 Aggregation Pathway

In order to account for the contributions of both protein colloidal and conformational stabilities to the overall protein aggregation mechanism, both unimolecular conformational changes and bimolecular aggregation events must be considered in the reaction pathway [418, 419].

Accordingly, the scheme of a typical protein aggregation mechanism can be represented as shown in Figure 7.10. It starts with monomeric conformational changes, which result in the formation of a nonnative aggregation-prone state of the protein (reaction I). This intermediate promotes the nucleation of oligomeric species (reactions II and III), which subsequently grow to form larger aggregates, either by monomer addition (reaction IV) or by aggregate–aggregate assembly (reaction V). For the sake of clarity, aggregation events are represented only with a nonnative protein conformation. However, native aggregation may also occur, either as a dominant mechanism or in competition with nonnative protein aggregation.

This scheme, which is referred to as the modified Lumry–Eyring Nucleated Polymerization (LENP) model, is sufficiently general to describe a wide variety of aggregation mechanisms [418, 419]. The relative importance of the individual steps depends on the system under investigation [420], in terms of both the protein amino acid sequence [376] and the environmental conditions [387]. In particular cases, some steps of this general aggregation scheme can be omitted. For example, reaction II (reversible

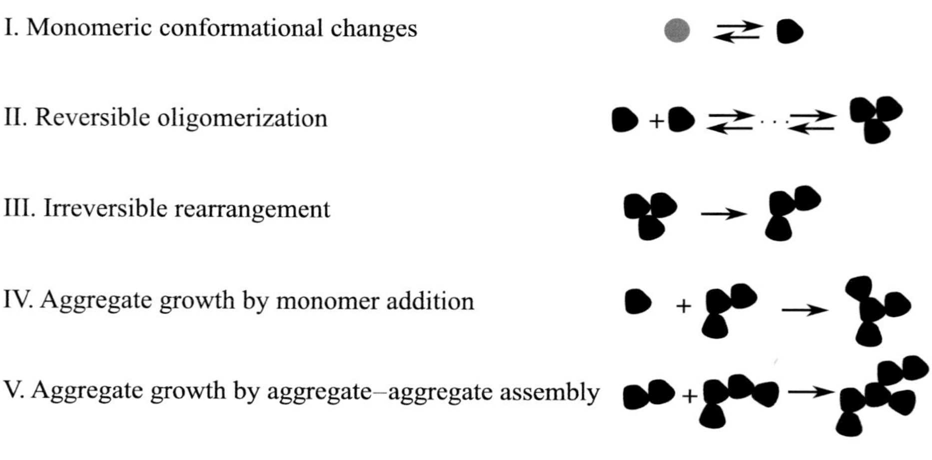

Figure 7.10 Schematic representation of the events involved in protein aggregation. Adapted from [383], with permission from John Wiley & Sons.

oligomerization) is not present if oligomers are irreversible (i.e., they do not dissociate during the time scale of interest). In contrast, if the aggregation process is mediated by specific effects, such as interactions with surfaces [421], protein hydrolysis [422], or aggregate breakage [423], additional reactions should be included.

In order to determine which step in the kinetic scheme is limiting the rate of the overall aggregation process, it is convenient to describe the kinetics of monomer depletion in terms of an apparent reaction order n_{app} defined as:

$$\mathcal{R}_1 = -k_{app} C_1^{n_{app}} \tag{7.6}$$

where $\mathcal{R}_1$ is the rate of monomer consumption, C_1 is the concentration of monomeric protein, and k_{app} is the apparent rate constant for monomer depletion. In the particular case of a well-mixed protein solution contained in a closed vessel, which is analogous to a batch stirred tank reactor (BSTR) in the case of a chemical reaction, $\mathcal{R}_1$ is equal to $\mathrm{d}C_1/\mathrm{d}t$. Note that the quantity accessible experimentally (e.g., by SEC) is usually the total monomer concentration, i.e., the sum of the concentrations of folded and unfolded protein conformations. This means that, although the model shown in Figure 7.10 differentiates between folded/unfolded proteins, model simulations can most often only be compared with experimental data in terms of total monomer concentration.

Values of n_{app} lying between 1 and 2 are typically observed. In particular, $n_{app} = 1$ and $n_{app} = 2$ suggest that unimolecular protein unfolding and bimolecular dimer formation are limiting the rate of monomer depletion, respectively [424]. Let us have a closer look at these two limiting cases. The integration of Equation 7.6 with $n_{app} = 1$ in a BSTR gives $\ln(C_1/C_1^F) = -k_{app}t$, where C_1^F is the initial ("feed") concentration of monomeric protein. This implies that, when the aggregation kinetics follows a first order process, monomer depletion kinetics obtained at various protein concentrations C_1^F should all collapse on one single straight line when $\ln(C_1/C_1^F)$ is plotted as a function of time. On the other hand, the integration of Equation 7.6 with $n_{app} = 2$ in a BSTR gives $1/C_1 - 1/C_1^F = k_{app}t$. This implies that, when the aggregation kinetics follows a second order process, monomer depletion kinetics obtained at various protein concentrations should all collapse on one single straight line when $1/C_1 - 1/C_1^F$ is plotted as a function of time. Overall, the proper linearization of monomer depletion kinetics obtained at several initial protein concentrations allows one to identify the reaction order, and thus the rate-limiting step for monomer depletion. Note that it is essential to perform experiments at several protein concentrations because obtaining a linearization of C_1 as a function of time at a single protein concentration is not sufficient to conclude on the reaction order. In addition, a sufficiently large monomer conversion should be reached in order to properly determine n_{app}. When none of the two linearizations shown earlier describes the set of experimental data, this may suggest that unimolecular unfolding and bimolecular aggregation may occur at a similar rate. This might also indicate that nonideal effects lead to concentration dependent reaction rate constants, as discussed for concentrated protein solutions in Section 7.4.4.

The apparent reaction order is strongly affected both by the protein structure and by the environmental conditions. For example, the monomer depletion kinetics of two different monoclonal antibodies were shown to be rate-limited by protein unfolding and

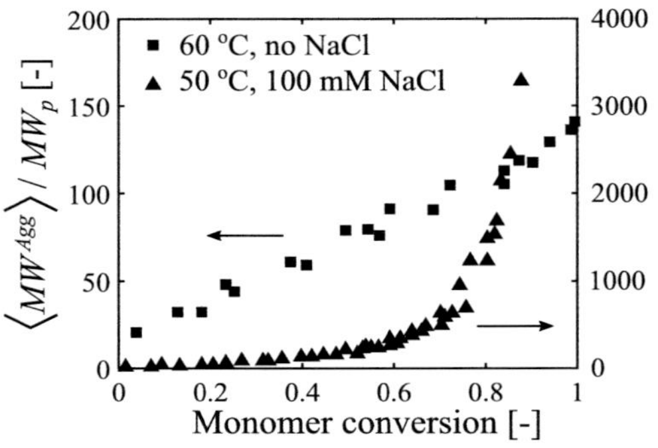

Figure 7.11 Aggregate molecular weight as a function of monomer conversion for α-chymotrypsinogen under two different experimental conditions, as indicated in the figure. Reprinted from [375], copyright 2009, with permission from Elsevier.

dimerization, respectively, under the same experimental conditions [376]. On the other hand, the aggregation of a colony-stimulated factor was shown to be rate-limited either by protein unfolding or by dimerization depending on the solution composition [400].

In addition to determining the reaction order, it is valuable to examine the mechanism of aggregate growth. Using experimental data obtained by SEC-MALS, it is possible to investigate the average molecular weight of the aggregates as a function of the monomer conversion, defined as $1 - C_1/C_1^F$. A linear increase of the aggregate molecular weight with conversion is characteristic of a monomer addition mechanism (i.e., reaction IV prevails over reaction V in Figure 7.10), whereas an upward profile indicates that aggregate–aggregate assembly is appreciable (i.e., reaction V prevails over reaction IV) [374, 375]. This is illustrated in Figure 7.11 with data obtained with α-chymotrypsinogen under two different experimental conditions [375]. At 60 °C in the absence of NaCl, it is observed that the aggregate molecular weight increases linearly with the monomer conversion, which suggests that aggregate growth occurs via monomer addition. In contrast, at 50 °C and in the presence of 100 mM NaCl, the aggregate molecular weight first increases linearly with monomer conversion but then shows a sharp upturn. This indicates that aggregate growth occurs via monomer addition at low conversion, and then via aggregate–aggregate assembly when most of the monomer is depleted.

7.3.3 Aggregation Rate Constants

The description of the aggregation mechanism shown in Figure 7.10 requires the definition of appropriate reaction rate expressions. While protein conformational changes can be described by a first-order unfolding/refolding reaction, aggregation events are second order, and the corresponding rate constants depend on the size of the two colliding aggregates. If only oligomers are formed, only few aggregation rate constants are necessary to describe the aggregation kinetics. However, if large aggregates are formed, the number of reaction rate constants to be defined may be considerably high. The aggregate size is in fact expected to have an impact on the aggregation rate constant, so that each combination of aggregation events will be characterized by its own reaction rate constant. The following expression has been derived in the frame of colloid science for the aggregation rate constant of two aggregates [383]:

$$k_{i,j} = \frac{8}{3}\frac{k_B T}{\eta} \times \frac{1}{4}\left(i^{-1/d_f} + j^{-1/d_f}\right)\left(i^{1/d_f} + j^{1/d_f}\right) \times (ij)^{1-1/d_f} \times \frac{1}{W} \tag{7.7}$$

where i and j are the number of monomeric units in the colliding aggregates, η is the solvent viscosity, and d_f is the fractal dimension of the aggregates, which is assumed to be independent of the aggregate size (see Equation 7.1). Finally, W is the Fuchs stability ratio whose physical meaning is discussed in the following.

It is remarkable that according to Equation 7.7, only two parameters are required to quantify the aggregation rate constant between two aggregates of any size under given experimental conditions (i.e., T and η fixed): the fractal dimension d_f and the Fuchs stability ratio W. As a comparison, the number of reaction rate constants necessary to describe the formation of aggregates constituted of n primary particles increases with $n^2/4$. For example, one would need to estimate 2500 model parameters to describe the formation of aggregates constituted of 100 monomers, which roughly corresponds to mAb aggregates of 50 nm in radius considering $d_f = 2$. By using Equation 7.7, the number of model parameters is reduced to 2.

The aggregation rate constant shown in Equation 7.7 results from several contributions: the term $2k_BT/(3\eta) \times \left(i^{-1/d_f} + j^{-1/d_f}\right)$ describes the aggregate diffusivity, while the term $\left(i^{1/d_f} + j^{1/d_f}\right)$ reflects the aggregate collision cross section, and the term $(ij)^{1-1/d_f}$ accounts for the probability of finding one monomeric protein on the external surface of the aggregate. Under given experimental conditions, these three terms depend only on the fractal dimension d_f, which quantifies the aggregate compactness. For a given aggregate mass i, the larger d_f, the smaller the aggregate radius ($R_i \propto i^{1/d_f}$). Therefore, the diffusion coefficient increases with increasing d_f ($D_i \propto 1/R_i$), while the aggregate cross section decreases with increasing d_f.

The Fuchs stability ratio W is defined as follows:

$$W = 2R_p \int_{2R_p}^{\infty} \exp\left(\frac{U_T(r)}{k_B T}\right)\frac{dr}{r^2} \tag{7.8}$$

It represents the ratio between the aggregation rate constant between two molecules in the presence of an energy barrier and its corresponding value in the absence of a repulsive barrier. In other words, W represents the reciprocal of the probability that the collision between two interacting molecules would lead to the formation of an aggregate, and is thus equal to or larger than one.

Equation 7.7 was originally derived for spherical and uniform colloidal particles aggregating under dilute conditions. Nevertheless, it was shown to successfully describe aggregation kinetics of several protein systems [383]. In addition, it has been modified to cover not only the dilute range but also moderate protein concentrations [386].

It is worth pointing out that the definition of W given by Equation 7.8 is relatively similar to the one used for the second virial coefficient B given by Equation 7.3 in the sense that both involve the protein–protein interaction potential U_T. However, they are not sensitive to the same characteristics of the interaction potential. In particular, B is a thermodynamic parameter that strongly depends on the primary energy well, i.e., it characterizes the difference in energy between the monomeric state and the aggregated state, which is the most stable state. On the other hand, W is a kinetic parameter that quantifies the energy barrier that colliding proteins need to overcome in order to reach the aggregated state. Both the values of B and W tend to be very sensitive to small uncertainties in the interaction potential, making practically unfeasible their computation from Equations 7.3 and 7.8. The experimental determination of B has been addressed in Section 7.2.2, while W is usually evaluated by comparing model simulations with experimental data, as discussed in the next section.

7.3.4 Population Balance Equations

In order to simulate the kinetics of protein aggregation, the reaction scheme of Figure 7.10 can be translated into mass balance equations, where each elementary reaction is characterized by a proper reaction rate. The collection of material balances describing the time evolution of each aggregate size is commonly referred to as a set of population balance equations (PBE). The PBE for an aggregating system are given below:

$$\mathcal{R}_i = \frac{1}{2}\sum_{j=1}^{i-1} k_{j,i-j} C_j C_{i-j} - C_i \sum_{j=1}^{\infty} k_{i,j} C_j \tag{7.9}$$

where C_i is the concentration of aggregates containing i monomeric units (in number per m^3) and $\mathcal{R}_i$ is the effective rate at which aggregates of mass i are formed (if $\mathcal{R}_i > 0$) or consumed (if $\mathcal{R}_i < 0$). In the particular case of a well-mixed protein solution aggregating in a closed vessel, $\mathcal{R}_i$ is equal to $\mathrm{d}C_i/\mathrm{d}t$. The positive term on the right-hand side accounts for the formation of aggregates of mass i due to the aggregation of two aggregates of mass j and $i - j$. The factor $1/2$ is necessary not to count twice the aggregation events appearing in the formation term because $k_{i,j} = k_{j,i}$. The negative term instead accounts for consumption of aggregates of mass i due to their aggregation with aggregates of any mass j.

The PBE given in Equation 7.9 can be modified to account for nucleation events occurring in the early stages of protein aggregation, such as protein unfolding, reversible oligomerization, oligomer structural rearrangement [376], or breakage [422, 423].

As an example, let us now consider the case of a monoclonal antibody whose aggregation behavior at neutral pH under heat stress has been analyzed in details elsewhere [376]. Kinetic experiments were performed by incubating a mAb solution in closed vials at elevated temperature. Figure 7.12(a) shows the time evolution of the monomer, dimer, and trimer concentrations followed by SEC at a mAb concentration of 2 g/L. The species eluting from the SEC column were analyzed by an inline MALS detector and Figure 7.12(b) shows the increase in the weight-average molecular weight

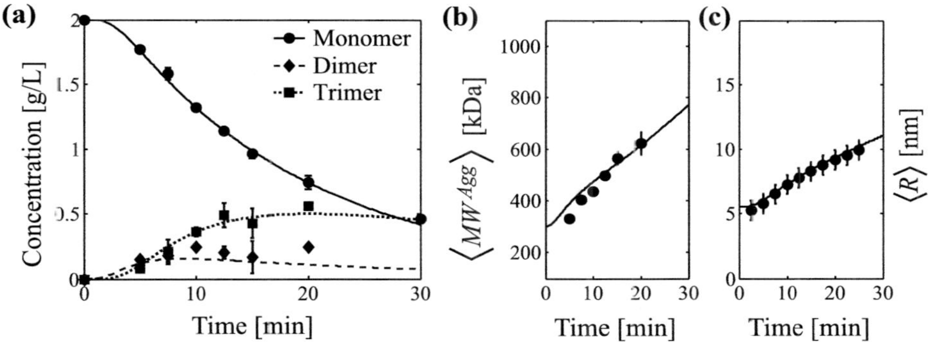

Figure 7.12 Comparison between experimental data (symbols) and model simulations (lines) for a mAb aggregating in a closed vial under thermal stress (70 °C). Adapted with permission from [376]. Copyright 2014 American Chemical Society. (a) Monomer, dimer and trimer concentrations measured by SEC. (b) Weight-average molecular weight of the aggregates measured by inline SLS after elution from the SEC column. (c) Hydrodynamic radius of the overall (nonfractionated) population measured by offline DLS. The model shown in Figure 7.13 was used with the aggregation rate constant given in Equation 7.7 and the following parameter values: $k_U = 1 \times 10^{-3}\ \mathrm{s}^{-1}$, $d_f = 1.85$, $W_{11} = 8.5 \times 10^6$, $W_{1j} = 4 \times 10^7$, $W_{ij} = 8 \times 10^8$.

Protein unfolding	$F \xrightarrow{k_U} U$	
Dimer formation	$U + U \xrightarrow{k_{1,1}} U_2$	
Monomer addition	$U + U_i \xrightarrow{k_{1,i}} U_{i+1}$	$i \geqslant 2$
Cluster–cluster aggregation	$U_i + U_j \xrightarrow{k_{i,j}} U_{i+j}$	$i,j \geqslant 2$

Figure 7.13 Aggregation pathway of a mAb exposed to thermal stress, corresponding to the experimental data shown in Figure 7.12. Adapted with permission from [376]. Copyright 2014 American Chemical Society.

of the aggregates. Finally, Figure 7.12(c) shows the average hydrodynamic radius of the nonfractionated sample measured by DLS.

By performing experiments at several initial protein concentrations, it was found that the kinetics of monomer depletion follows a first-order process. In addition, it was found that the aggregates do not dissociate upon dilution, which suggests that the aggregation events are irreversible. Based on these observations, the mechanism shown in Figure 7.13 was proposed, where F and U denote the folded (i.e., native) and an unfolded conformation of the monomer, respectively.

Then, this mechanism was translated into population balance equations, and the expression of the aggregation rate constant shown in Equation 7.7 was used to quantify the rate of the aggregation steps. The fractal dimension d_f was estimated experimentally with light scattering experiments, while the unfolding rate constant and the Fuchs stability ratio were estimated by fitting the PBE simulations to the experimental data [376]. It is worth recalling that Equation 7.7 was derived for spherical particles characterized by a uniform reactivity. However, the situation is more complex with proteins, where reactivity strongly depends on the protein spatial conformation and on the exposition of aggregation-prone patches. Consequently, it is expected that the protein reactivity decreases with the aggregate size due to the coverage of hydrophobic patches. This effect can be accounted for by introducing several W values that reflect the differences in reactivity of the various subpopulations of species, which however comes at the expense of a higher number of model parameters to be estimated [376]. In the mechanism considered here, the unfolded monomer U can be regarded as an unstable intermediate that has hydrophobic patches exposed to the aqueous solvent and therefore a higher reactivity compared to the aggregates. Therefore, three types of aggregation events characterized by species of different reactivities can be identified: monomer–monomer, monomer–aggregate, and aggregate–aggregate. Accordingly, three values of the Fuchs stability ratio were considered: W_{11}, W_{1j} and W_{ij}, to describe dimer formation, growth by monomer addition and growth by cluster–cluster aggregation, respectively. The following values were obtained by fitting model simulations to experimental data: $W_{11} = 8.5 \times 10^6$, $W_{1j} = 4 \times 10^7$, $W_{ij} = 8 \times 10^8$. Interestingly, $W_{11} < W_{1j} < W_{ij}$,

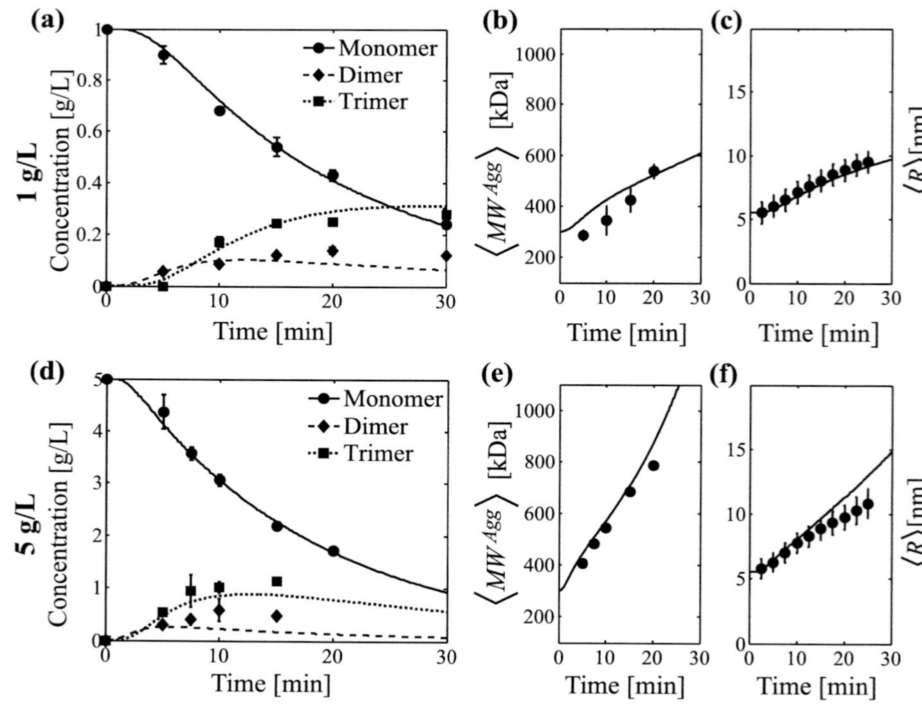

Figure 7.14 Comparison between experimental data (symbols) and model predictions (lines) for a mAb aggregating in a closed vial under thermal stress. The model parameters were determined at a protein concentration of 2 g/L (see Figure 7.12) and used to predict the aggregation kinetics at 1 g/L (a–c) and 5 g/L (d–f). Adapted with permission from [376]. Copyright 2014 American Chemical Society.

which reflects the high reactivity of the unfolded monomer compared to the aggregates, in agreement with the reduction of the number of aggregation-prone patches discussed earlier.

Figure 7.12 shows the comparison between model simulations and experimental results. It is seen that the proposed model describes well the set of experimental data. This shows that, by using the expression of the aggregation rate constant derived in the frame of colloid science, only five model parameters were sufficient to characterize the aggregation behavior of this mAb under thermal stress.

Moreover, once the model parameters have been determined at a given protein concentration (here 2 g/L), the kinetic model could be used to successfully predict the mAb aggregation kinetics at other protein concentrations, namely 1 and 5 g/L [376]. The comparison between model predictions and experimental data can be appreciated in Figure 7.14. This illustrates that model-based approaches are very powerful to predict the impact of protein concentration on the kinetics of aggregation.

7.4 Impact of Operating Conditions on Protein Aggregation

In order to identify which parts of biopharmaceutical processes are critical for aggregate formation, it is essential to understand how the operating conditions affect the protein aggregation propensity. In this section, we discuss the effect of solution pH, salt type and concentration, temperature, and protein concentration. The reader interested in the role of pharmaceutically relevant surfaces in protein adsorption and protein stability is referred to the following review [425]. Here, we focus on aggregation studies performed under stagnant conditions, while protein aggregation in flows (e.g., chromatography, filtration) will be examined in Section 7.5.

7.4.1 Solution pH

For illustrative purposes, let us consider the mAb whose aggregation mechanism was presented in Figure 7.13. This mAb was found to aggregate according to a two-step mechanism, which is constituted by a monomolecular unfolding step followed by bimolecular aggregation events. Figure 7.15(a) shows the monomer depletion kinetics of this mAb at three different pH values below the isoelectric point (pI 8.5) [386]. It is observed that the rate of monomer depletion is lower at pH 6.5 than at pH 5.5 and 7.4. In order to rationalize this intriguing observation, the values of the unfolding rate constant (which is related to the conformational stability) and of the Fuchs stability ratios (which are related to the colloidal stability) were determined by fitting model simulations to experimental data at each pH values. The comparison between model simulations and experimental data is shown in Figure 7.15(a) for the monomer depletion kinetics, while results about dimer concentration, trimer concentration and aggregate molecular weight are reported in the original study [386]. The obtained parameter values of k_U and W are shown in Figure 7.15(b) and (c), respectively, where it is seen that both the unfolding rate constant and the Fuchs stability ratios exhibit the largest value at pH 5.5.

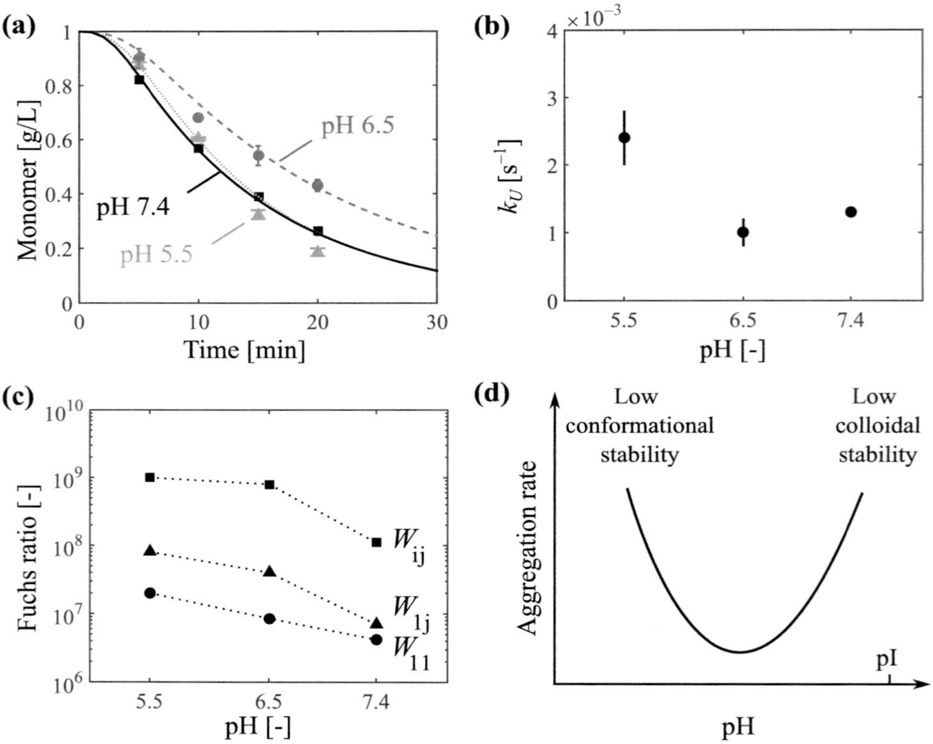

Figure 7.15 Effect of solution pH on the aggregation kinetics of a monoclonal antibody at pH values below the mAb pI (equal to 8.5). Adapted with permission from [386]. Copyright 2016 American Chemical Society. (a) Monomer depletion kinetics followed by SEC. The mAb solution was incubated in closed vials at 70 °C (b) Unfolding rate constant as a function of pH. (c) Fuchs stability ratios as a function of pH. (d) Qualitative behavior of the rate of protein aggregation (defined in terms of monomer depletion) as a function of the pH of the solution.

These results can be explained by considering the qualitative energy diagram shown in Figure 7.9(a). At pH values far below the protein pI (such as pH 5.5 here), the protein carries a high positive net charge. This results in a strong electrostatic repulsion between protein molecules (i.e., high colloidal stability and low aggregation rate), but also in a strong electrostatic repulsion within the protein molecule (i.e., low conformational stability and high unfolding rate) [387]. When the solution pH is close to the protein pI, the opposite situation is encountered: The protein net charge is close to zero, which results in a high conformational stability, but a low colloidal stability [387]. These considerations are summarized on the drawing of Figure 7.15(d). Note that a similar reasoning could be applied at pH values above the protein pI, leading to an increase of the unfolding rate constant and Fuchs stability ratios with the solution pH.

Overall, a high aggregation propensity is usually observed both at pH values far away from the protein pI and at pH values close to the protein pI, leading to a narrow pH range where protein stability is maximized [387]. However, this qualitative analysis is valid only in the case of nonspecific electrostatic effects, i.e., when an increase in the protein charge destabilizes the folded state. In some cases, specific electrostatic effects such as ion pairing or salt bridges might occur, potentially triggering a stabilization of the folded state with increasing protein charge, and thus resulting in a more complex effect of the solution pH on the aggregation propensity.

7.4.2 Salt Type and Concentration

Besides solution pH, the type and concentration of salt also affect protein stability. In this section, we discuss the influence of electrolytes first on the protein colloidal stability and then on the protein conformational stability.

At low to moderate ionic strength, salt has predominantly a charge screening effect, which reduces the magnitude of electrostatic intermolecular interactions. This is generally evidenced by a decrease in the protein electrophoretic mobility or a decrease in the second virial coefficient with salt concentration, as discussed in the context of Figure 7.4(b). This screening effect reduces the distance at which molecules start to interact with each other, which is referred to as the Debye length. In the case where net intermolecular electrostatic interactions are repulsive, this leads to a reduction of the repulsive energy barrier, which in turn results in an increase in the protein aggregation propensity.

In the case of the mAb whose aggregation mechanism is presented in Figure 7.13, it was shown by comparing model simulations with experimental data that the presence of a low concentration of NaCl (10 mM) leads to a decrease of the mAb colloidal stability without impacting the mAb conformational stability [426]. These conclusions were drawn from the observation that the presence of salt leads to a decrease of the Fuchs stability ratio without affecting the unfolding rate constant. Nevertheless, the protein conformational stability is likely to be affected at higher salt concentrations.

The impact of a moderate salt concentration on the structure of a mAb can be appreciated in Figure 7.16(a), which shows CD spectra acquired in various solution conditions [427]. At neutral pH, the spectrum exhibits a minimum at around 216 nm that is characteristic of the ordered β-sheet structures of immunoglobulins. When the solution pH is lowered to 3, the shape of the CD spectrum remains similar but the intensity is significantly reduced, which indicates a loss in the content of ordered β-sheet structures due to intramolecular repulsion forces. Interestingly, the addition of 150 mM of sodium chloride or sodium sulphate at pH 3 leads to CD spectra intermediate to those obtained at pH 3 and pH 7 in the absence of salt. This indicates that the presence of salt is able to regenerate partially the original β-sheet structure, probably due to charge screening or anion binding that reduce electrostatic intramolecular repulsion.

Besides interacting with charged residues, salts may also interact with polar and apolar regions of the protein. Following the approach of Timasheff and Arakawa, the net interaction of salts (and cosolutes in general) with the protein surface is commonly described in terms of a preferential interaction parameter [428]. This parameter is a measure of the excess of salt in the immediate neighborhood of the protein as compared to its concentration in the bulk solvent. It reflects the change in free energy of the

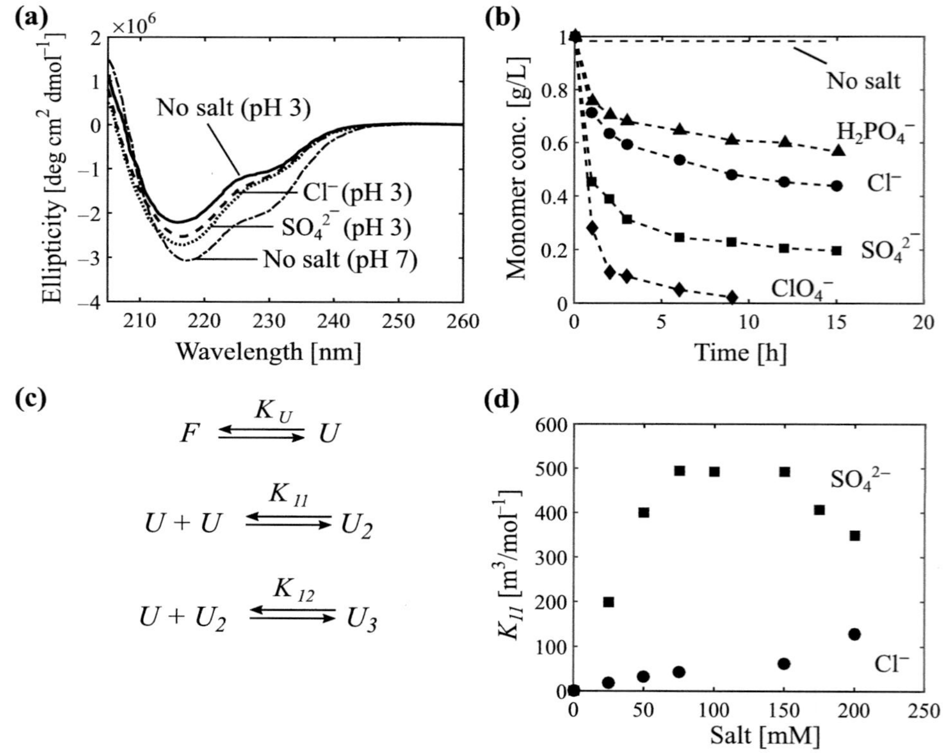

Figure 7.16 Role of salts on the aggregation kinetics of a monoclonal antibody. Adapted from [427], copyright 2012, with permission from Elsevier. (a) CD spectra in the solution buffer without salt (at pH 3 and 7) and in the presence of 150 mM sodium chloride and sodium sulfate (at pH 3). (b) Kinetics of monomer depletion in the absence of salt, and in the presence of 150 mM of several sodium salts at pH 3. Aggregation experiments were performed in closed vials at room temperature. (c) Mechanism describing oligomer formation at low pH. K_U, $K_{1,1}$, and $K_{1,2}$ denote the unfolding, dimerization, and trimerization equilibrium constants, respectively. (d) Dimerization equilibrium constant as a function of salt concentration at pH 3.

protein surface due to salt addition. When interactions between the salt and protein surface are nonfavorable, the salt is preferentially excluded from the protein surface, which leads to a stabilization of the protein structure. Indeed, nonfavorable interactions between the salt and the protein surface favor native compact protein conformations over more open unfolded structures, which would increase the contact surface with the solvent. On the other hand, preferential binding of the salt to the protein surface tends to destabilize the protein structure. It is worth stressing that the preferential interaction parameter is a global effective parameter, which reflects the net interaction of salts with the protein surface. Because the protein surface is notoriously heterogeneous, a given ion may preferentially bind or be excluded in different regions of the protein [429]. When the ion is overall preferentially excluded, the protein surface energy increases upon salt addition. On the one hand, this stabilizes the protein structure because the protein tends to reduce its contact with the solvent. On the other hand, it also promotes aggregation because the surface exposed to the solvent of a dimer is smaller than that of two monomers. These considerations help to understand the nonintuitive observation that the ions that stabilize well the protein structure are also often those that promote bimolecular aggregation [428].

Overall, the interplay between intermolecular and intramolecular charge screening, specific ion binding and preferential interactions leads to complex aggregation behaviors. As an example, Figure 7.16(b) shows the monomer depletion kinetics of a mAb followed by SEC at pH 3 (which is below the protein pI) with 150 mM of different sodium salts [427]. It is observed that the counter ions Cl^-, $H_2PO_4^-$, and ClO_4^- have very different effects on the protein stability although they have the same valence.

Under these experimental conditions, it was shown that the mAb under investigation forms dimers and trimers that are reversible upon dilution [427]. Accordingly, the aggregation mechanism shown in Figure 7.16(c) was proposed. Note that the constants K introduced in this figure correspond to equilibrium constants, which are defined as the ratio between the forward and backward reaction rate constants. The experimental data were simulated in the frame of this aggregation scheme, and the estimated values of the dimerization equilibrium constant ($K_{1,1}$) are plotted as a function of salt concentration for sodium sulphate and sodium chloride in Figure 7.16(d). In the case of NaCl, it is observed that the dimerization equilibrium constant increases monotonically with the salt concentration in the investigated range of concentrations. On the other hand, a maximum is observed in the case of Na_2SO_4, which indicates a restabilization effect of the protein structure at high salt concentration [427].

To summarize, an increase in the aggregation propensity is usually observed at low salt concentration due to intermolecular charge screening. At higher salt concentrations, a stabilization or destabilization of the protein structure may be observed depending on the relative contribution of various effects including intramolecular charge screening, ion binding and interactions with noncharged residues. If the salt has an overall destabilizing effect on the protein structure, the aggregation propensity is expected to increase monotonically with the salt concentration. On the other hand, if the salt stabilizes the protein structure, a decrease of the aggregation propensity with the salt concentration may be observed at high salt concentrations. This qualitative overall picture summarizes the main aspects to be considered when assessing protein stability, but does not describe a universal behavior of the impact of salt on the protein aggregation propensity. The net result of the addition of salt on protein stability is indeed dictated by the delicate balance between the protein colloidal and conformational stabilities, which is strongly system specific. Moreover, salt effects are pH-dependent because pH regulates the protein charge distribution [427].

Another aspect to consider when dealing with salts is their effect on protein solubility [428]. In general, protein solubility shows a maximum as a function of salt concentration [21]. At low salt concentration, an increase in salt concentration usually increases protein solubility, which is referred to as a "salting in" effect. On the other hand, at high salt concentration, an increase in salt concentration often reduces protein solubility, which is referred to as a "salting out" effect. Once more, the magnitude of these effects depends on the salt type and on the solution pH [21]. In particular, lower solubilities are commonly observed at pH values close to the protein pI.

7.4.3 Temperature

High temperatures may be encountered in biopharmaceutical processes, as, for example, during certain viral inactivation steps, or during accelerated studies aiming at evaluating the stability of various drug formulations in a short time. A solid understanding of the impact of temperature on the kinetics of protein aggregation is therefore paramount to reduce the protein aggregation propensity in biopharmaceutical processes and to extrapolate results from accelerated stability studies to relevant temperature conditions. Although it is generally recognized that a temperature increase promotes protein denaturation and aggregation, it is almost impossible to quantitatively predict its impact. As a matter of fact, protein aggregation is a complex phenomenon involving several elementary steps (such as monomolecular unfolding and bimolecular collisions), which are likely to be affected by temperature in different ways. While Arrhenius behaviors have been observed over narrow ranges of temperatures [430, 431], a plot of $\ln(k_{app})$ as a function of the reciprocal of the temperature is usually nonlinear when covering the whole range of temperatures from accelerated (let's say above 50 °C) to long-term storage conditions (at around 5–8 °C) [432–434]. Figure 7.17 illustrates the complex temperature dependence of the apparent rate constant for monomer depletion, k_{app}, for two mAbs. A possible reason for such a behavior is rooted in a change of the aggregation mechanism with temperature [432, 433]. It is, for example, expected that native protein aggregation plays a more significant role in the aggregation mechanism at low temperature than under thermal stress, where protein unfolding is promoted [435]. Besides, additional complex phenomena may occur at low temperature, such as cold denaturation [367, 436].

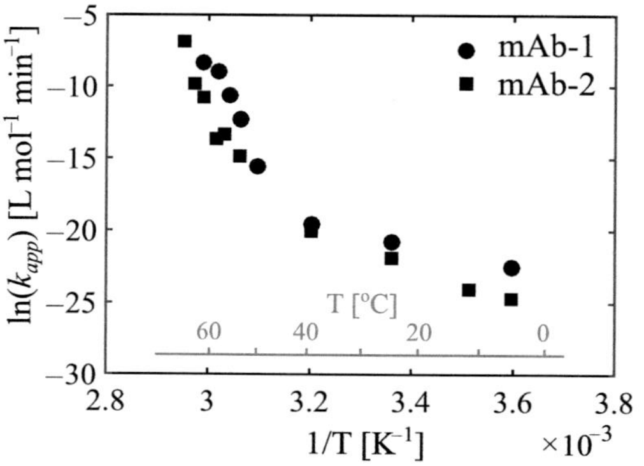

Figure 7.17 Arrhenius plot of the apparent reaction rate constant for monomer depletion observed for two monoclonal antibodies, where two distinct slopes can be identified in the investigated temperature range. Adapted from [433], copyright 2011, with permission from Elsevier.

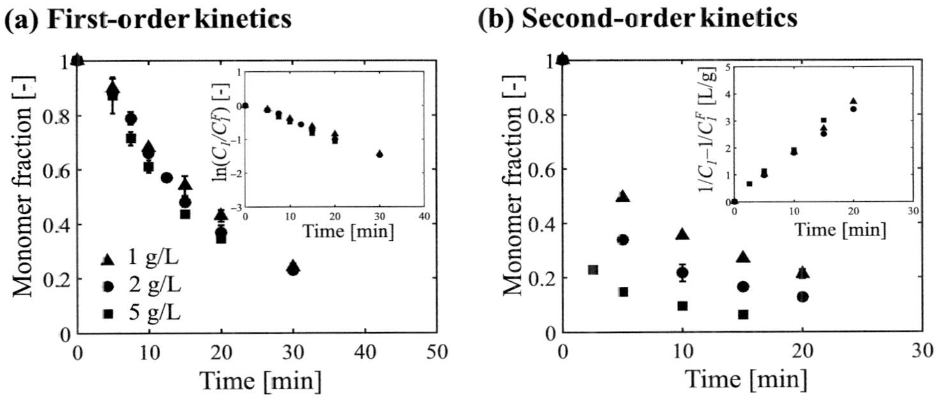

Figure 7.18 Normalized monomer depletion kinetics of two mAbs followed by SEC. Experiments were performed by incubating protein solutions at 1, 2, and 5 g/L in closed vials at 70 °C. The data linearization in the inserts reveal a first-order and second-order process in case (a) and (b), respectively. Adapted with permission from [376]. Copyright 2014 American Chemical Society.

7.4.4 Protein Concentration

Protein concentration may significantly vary during the course of the production process, from the cell culture step to drug formulation. It is thus essential to have a good knowledge of the impact of protein concentration on the rate of aggregation. Figure 7.18 shows experimental data of the monomer depletion kinetics of two monoclonal antibodies followed by SEC [376]. The monomer content has been normalized by the total protein concentration in order to better assess the effect of concentration. Let us first consider the situation presented in Figure 7.18(a). The linearization in the inset shows that the monomer depletion kinetics follows a first-order process, as discussed in Section 7.3.2. This observation suggests that, for the system under investigation, the monomer depletion kinetics is rate-limited by monomolecular protein unfolding. On the other hand, in the case presented in Figure 7.18(b), the linearization in the inset shows that the monomer depletion kinetics follows a second-order process, which suggests that, for this second system, the monomer depletion kinetics is rate-limited by bimolecular dimer formation.

It is interesting to note that the impact of protein concentration on the normalized monomer depletion kinetics significantly differs depending on the reaction order (Figure 7.18). In the case of a first-order process, monomer depletion kinetics obtained at various protein concentrations collapse on one single curve when normalized by the initial protein concentration C_1^F (Figure 7.18(a)). In other words, at a given time t, the relative content of monomer (C_1/C_1^F) is independent of the initial protein concentration. In the case of a second-order process instead, the relative concentration of monomer at a given time t decreases with an increase in initial protein concentration (Figure 7.18(b)). To summarize, an increase in the protein concentration accelerates (respectively does not affect) the normalized kinetics of monomer depletion in the case of a second-(respectively first) order kinetics.

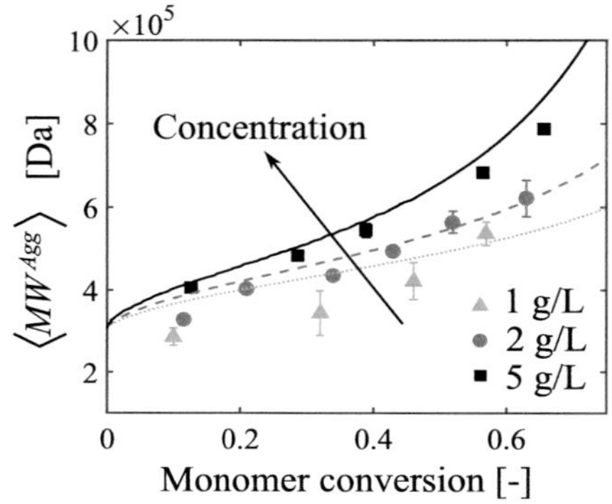

Figure 7.19 SEC-MALS data showing the average molecular weight of mAb aggregates as a function of monomer conversion (defined as $1 - C_1/C_1^F$) at three initial mAb concentrations (symbols) in the case of monomer depletion limited by protein unfolding. Lines correspond to model simulations performed with the model presented in Figure 7.13 and the parameters given in the caption of Figure 7.12. Adapted with permission from [376]. Copyright 2014 American Chemical Society.

However, it must be realized that, even in the case where the monomer depletion follows a first-order kinetics, aggregation events per se are accelerated by an increase in the initial protein concentration because they are bimolecular collision events. Therefore, at a given time, or equivalently at a given monomer conversion, an increase in protein concentration leads to fewer but larger aggregates. To illustrate this point, the molecular weight of the aggregates formed in the case of Figure 7.18(a), which was measured by SEC-MALS, is plotted in Figure 7.19 as a function of the monomer conversion (defined as $1-C_1/C_1^F$). It is seen that at a given conversion, larger aggregates are formed when the protein concentration increases. This is because an increase in protein concentration leads to a decrease in the characteristic time of aggregation (defined as $1/(k_{1,1}C_1^F)$) without affecting the characteristic time of unfolding (defined as $1/k_U$). In other words, when the protein concentration increases, the unfolded monomers are produced over a similar time scale, but they aggregate faster. This qualitative analysis is confirmed by model simulations, which are represented by lines in Figure 7.19.

It must be underlined that the analysis of the concentration impact carried earlier was performed under dilute conditions, i.e., in the range of few g/L. At higher protein concentrations, some "nonideal" effects appear, which complicate the picture and also pose analytical challenges [368, 437]. By combining experimental and theoretical investigations, it was found that the bimolecular aggregation rate constant of a mAb decreases with the protein concentration due to thermodynamic and hydrodynamic nonidealities. It has been demonstrated that this effect can be accounted for by multiplying the aggregation rate constant evaluated under dilute conditions by a correction factor that accounts for the variation of the diffusion coefficient with the protein concentration under concentrated conditions [386]. This correction factor can be measured with light scattering and depends on the solution pH and composition. An additional correction must be included when aggregation is accompanied by an increase in solution viscosity.

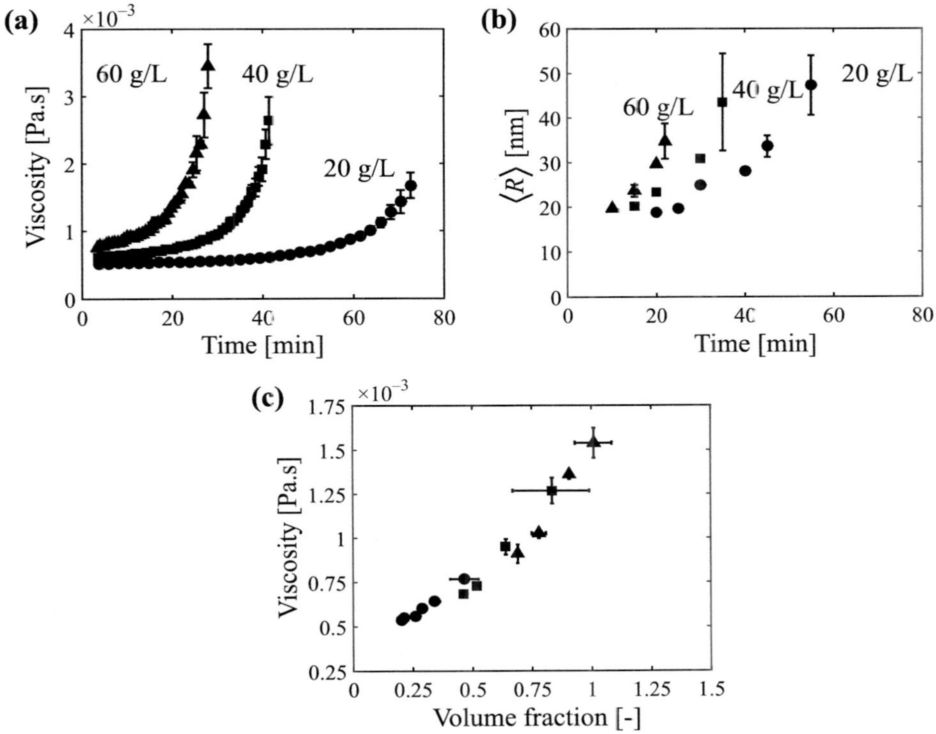

Figure 7.20 Viscosity effects during mAb aggregation occurring in closed vials incubated at 70 °C. Adapted from [385], published by The Royal Society of Chemistry. (a) Viscosity increase as a function of time at three mAb concentrations measured from the diffusion coefficient of polymeric tracer nanoparticles. (b) Increase in the hydrodynamic radius measured with DLS in the absence of polymeric tracer particles at the three considered mAb concentrations. (c) Viscosity as a function of the volume fraction defined in Equation 7.10. Symbols are identical to those used in panels (a) and (b).

Both reversible self-association and irreversible aggregation of proteins in concentrated solutions can indeed lead to a viscosity increase [377, 385, 438]. As an example, Figure 7.20(a) shows the increase in viscosity of a mAb solution irreversibly aggregating under thermal stress at three different mAb concentrations: 20, 40, and 60 g/L [385]. These experimental data were obtained by measuring the diffusion coefficient of polymeric tracer particles with DLS, as explained in Section 7.2.4. It is observed that the solution viscosity increases for the three concentrations under investigation and that the viscosity increase is more pronounced at higher protein concentrations. To rationalize these data, it is useful to introduce the volume fraction occupied by the aggregates Φ. The occupied volume fraction can be roughly evaluated by approximating the aggregate population to a monodisperse population of average radius $\langle R \rangle$ and average number concentration $\langle C \rangle$. The latter can be estimated by dividing the initial number concentration of monomeric protein C_1^F by the average number of monomers per aggregate, which is given by $\langle MW\rangle/MW_p$. By relating $\langle MW\rangle$ to $\langle R\rangle$ through the fractal scaling given by Equation 7.1, it follows that [385]:

$$\Phi \approx \frac{4\pi}{3}\frac{C_1^F}{k_f}R_p^{d_f}\langle R\rangle^{3-d_f} \tag{7.10}$$

It is seen that, at a given protein concentration, the volume fraction increases with the aggregate size if the fractal dimension is lower than 3. In this case, the aggregates are not compact structures but entrap some void, and they consequently occupy more space than their single constituting molecules taken separately. In the present example, the aggregate fractal dimension was estimated to 2.10, 2.23, and 2.36 at 20, 40, and 60 g/L [385], respectively. The increase in the aggregate radius with time measured with DLS is shown for the three protein concentrations in Figure 7.20(b). These results can be used to compute the occupied volume fraction at each time point according to Equation 7.10. The viscosity data of Figure 7.20(a) are replotted as a function of the so-computed volume fractions in Figure 7.20(c). Remarkably, it is observed that the data obtained at different concentrations all collapse on a single curve.

It is rather intuitive that an increase of the solution viscosity will slow down the aggregation process by reducing the diffusivity of the aggregates. However, it is quite challenging to predict the quantitative impact of the viscosity increase on the aggregation rate because the local viscosity experienced by an aggregate depends on its size. Indeed, large aggregates experience the macroscopic viscosity of the protein solution (e.g., the viscosity shown in Figure 7.20(c)), while smaller aggregates feel a lower viscosity, which can go down till the solvent viscosity for sufficiently small aggregates. Such a viscosity scaling could be identified by means of Brownian dynamics simulations and introduced in the size dependent aggregation rate constant given by Equation 7.7 to model protein aggregation kinetics [386, 439].

7.5 Critical Steps for Aggregation in Biopharmaceutical Processes

After having elucidated the impact of some operating conditions on the protein aggregation propensity, we identify which parts of the protein production process are critical for the formation of aggregates. As discussed in Chapter 1, there is a large variety of biopharmaceutical processes, which differ both in their upstream and downstream parts. However, in the case of monoclonal antibodies, a kind of general production platform can be identified. Therefore, in the following, we will analyze the potential causes of protein aggregation following the sequence of operations typically encountered in mAb production processes.

7.5.1 Upstream

The protein production process starts with a cell culture step, which is generally performed at mild temperature (above 30 °C) and neutral pH. These conditions may promote aggregation, especially if the pI of the expressed protein is close to the pH at which the bioreactor is operated [59]. Moreover, cell culture media contain a variety of dissolved species that may potentially impact protein folding and protein aggregation [366]. Finally, the renewal of the air–water interface has been shown to induce protein denaturation and aggregation [440]. Therefore, air sparging, which is needed for oxygen supply and carbon dioxide removal, is also a possible cause of protein instability in bioreactors. The preceding considerations deal with extracellular protein stability, but intracellular aggregation may also occur, especially when the level of expression is very high as in bacterial fermentation [365].

7.5.2 Capture Step

For monoclonal antibodies, the capture step is usually performed by affinity chromatography after clarification of the supernatant. This step requires the use of low pH eluents at a sufficiently high ionic strength to desorb the protein from the affinity ligands. As described in the previous section, acidic pH environments and the presence of salt are known to promote aggregate formation. In addition, experimental studies suggest that protein interactions with the affinity ligands and the resin backbone may lead to protein unfolding, which then contributes to the formation of aggregates [12, 441, 442]. Overall, protein capture by affinity chromatography represents a critical unit operation during which aggregation is likely to occur [59, 443]. Depending on the protein under consideration, either oligomers, soluble high molecular weight aggregates, or insoluble precipitates can be formed during the mAb recovery step [444].

7.5.3 Polishing Steps

Although mAb capture by affinity chromatography probably represents the most troublesome chromatographic step in terms of aggregate formation, protein aggregation may also occur during the subsequent chromatographic polishing steps. For instance, ion exchange and hydrophobic interaction chromatography may require the use of high ionic strength eluents, which promote aggregation (see Section 7.4.2). Moreover, proteins are often stable over a narrow range of pH only (Section 7.4.1), so that eluents at pH values either too close or too far from the protein pI may result in aggregate formation.

The interplay between aggregate removal and aggregate formation in chromatographic columns can be described by combining chromatographic models presented in Chapter 2 and aggregation models presented earlier in this chapter. As an example, let us consider the simple case of the binary separation between protein monomers and dimers in a chromatographic column and examine the situation where elution conditions promote dimer formation in the liquid phase. The PD model is selected to describe hydrodynamics in the chromatographic column (see Equation 2.43), while the kinetics of internal mass transfer is accounted for by using the LDF approximation (see Equation 2.58). For the sake of simplicity, we consider linear adsorption isotherms. Finally, we add a dimerization reaction in the liquid phase, which is characterized by

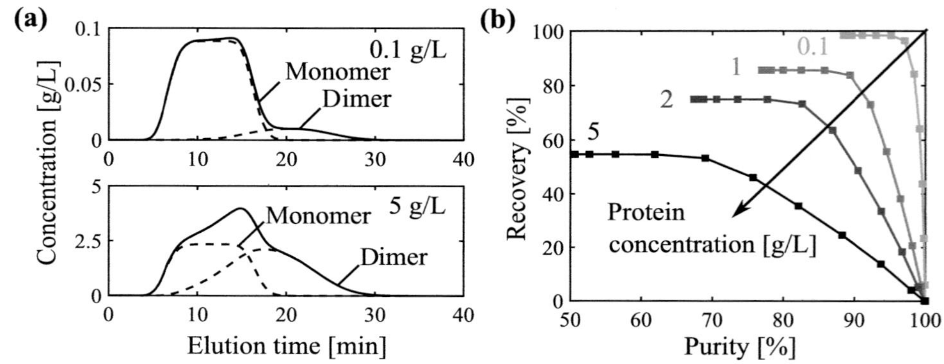

Figure 7.21 Impact of aggregation on the separation performance. (a) Simulated chromatograms obtained under isocratic conditions using a PD model (Equation 2.43), the LDF approximation (Equation 2.58) and linear adsorption isotherms with the following parameter values: $\varepsilon_e = 0.4$, $\varepsilon_i = 0.7$, $\overline{\overline{K}}_M = 10$, $\overline{\overline{K}}_D = 30$, $Pe_c = 5600$, $t_{i,M} = 5$ s, $t_{i,D} = 10$ s, where M denotes the monomer and D the dimer, respectively. A column with $L_{col} = 74$ cm and $D_{col} = 15.3$ cm was selected and operated at a flow rate $Q = 317$ L/h. The injection time was 9.6 min. The rate constant of dimer formation in the liquid phase is $k_{1,1} = 0.2$ L/g/min. Two feed protein concentrations were considered, namely 0.1 g/L (top panel) and 5 g/L (bottom panel). (b) Recovery–purity plots at several feed protein concentrations. The reported protein concentration values correspond to the total mass concentrations of protein, including both monomers and dimers.

the rate $k_{1,1}C_1C_1$. The selected values of the model parameters are summarized in the caption of Figure 7.21. To perform the simulations, we investigate several protein concentrations, ranging from 0.1 to 5 g/L, with 10% dimers in the feed mixture. The simulated chromatograms obtained at protein concentrations of 0.1 and 5 g/L are shown in Figure 7.21(a), in the top panel and bottom panel, respectively. The solid lines correspond to the total protein concentration, while the dashed lines show the peaks of the monomeric and dimeric protein. It is seen that a good separation is achieved between the monomer and dimer at 0.1 g/L. However, when the protein concentration is increased to 5 g/L, aggregation is accelerated and significant overlap is observed between the peaks, thus compromising the final product purity. The decrease of the purity with increasing protein concentration can be exclusively attributed to the kinetics of aggregation because we considered linear adsorption isotherms.

Let us now quantify the impact of aggregation on the performance of this chromatographic process in terms of purity and recovery. To do so, for each protein concentration, we consider a collecting window centered on the monomer peak, and we compute the purity and the recovery for various collection durations. The results are presented in Figure 7.21(b), where it is seen that, at a given feed concentration, the protein recovery decreases when the purity increases. In the case under investigation, this trade-off does not represent a real problem at low protein concentration. However, when the protein concentration increases, a high purity can only be obtained at the expense of a very low recovery. To rationalize these results, it is interesting to compare the characteristic time of aggregation (defined as $1/(k_{1,1}C_1)$) with respect to the mean retention time of the front (roughly 6 min here). With the considered value of the aggregation rate constant, namely $k_{1,1} = 0.2$ L/g/min, it follows that the characteristic time of aggregation is reduced from 50 min to 2 min when the protein feed concentration is increased from 0.1 to 5 g/L. Therefore, at high concentration, the time the protein spends in the column is longer than the time it takes to aggregate, which explains the poor separation. Overall, these simulation results show that the ratio between the characteristic time of aggregation and that of retention is crucial to design chromatographic processes where protein aggregation occurs. In particular, this implies that increasing the column length or reducing the fluid flow rate with a view to increasing the number of plates may actually worsen the quality of the separation.

In the earlier example, we assumed that aggregation occurs in the liquid phase only. Nevertheless, it has been reported that interactions with the column matrix may promote protein conformational changes, potentially triggering aggregate formation [445]. This phenomena has been observed with ion exchange chromatographic media [446–450] and may be especially pronounced when protein purification relies on hydrophobicity, such as hydrophobic interaction or reversed-phase chromatography [451–457].

To illustrate the impact of surfaces on protein unfolding and aggregation, experimental results obtained with cation exchange media are presented in Figure 7.22 [447]. In this study, a protein solution was loaded in columns packed with different chromatographic media: Source 30 S (Figure 7.22(a)), Eschmuno S (b), POROS 50 HS (c), and Nuvia S (d). In a first set of experiments, the protein was recovered directly after loading by applying a linear gradient of NaCl in the eluent (dashed lines). It is observed in Figure 7.22 that for all the chromatographic media under investigation, the protein elutes as a single peak. In a second set of experiments, the eluent flow was stopped after loading the protein solution, and the column was held idle for 1000 min before recovering the protein with the same elution gradient (solid lines). In this case, it is seen in Figure 7.22 that the protein elutes as two peaks, the first one being at the same position than the peak obtained without hold prior to protein recovery. It is interesting to note that the relative areas of these peaks depend on the chromatographic medium: the percentage of protein eluting in the second peak was 14%, 44%, 51%, and 73% for Source, Eshmuno, POROS, and Nuvia media, respectively. In a previous study, it was shown by SEC analyses that the first peak contains only protein monomers, while the second peak contains both protein monomers and aggregates [450]. In addition, amide hydrogen–deuterium exchange mass spectrometry (HX–MS) experiments were performed in order to examine the protein structure [447, 450]. As a matter of fact, the conventional experimental techniques described in Section 7.2.3 (e.g., circular dichroism, intrinsic fluorescence) cannot be applied easily to examine protein structures on the adsorbed phase due to the opacity of the chromatographic media [453]. HX–MS experiments instead exploit the dependence of the hydrogen–deuterium exchange rate of amides on their degree of solvent exposure, which in turn provides a measure of protein unfolding. Such investigations revealed that the aggregates contain proteins in a nonnative conformation [447].

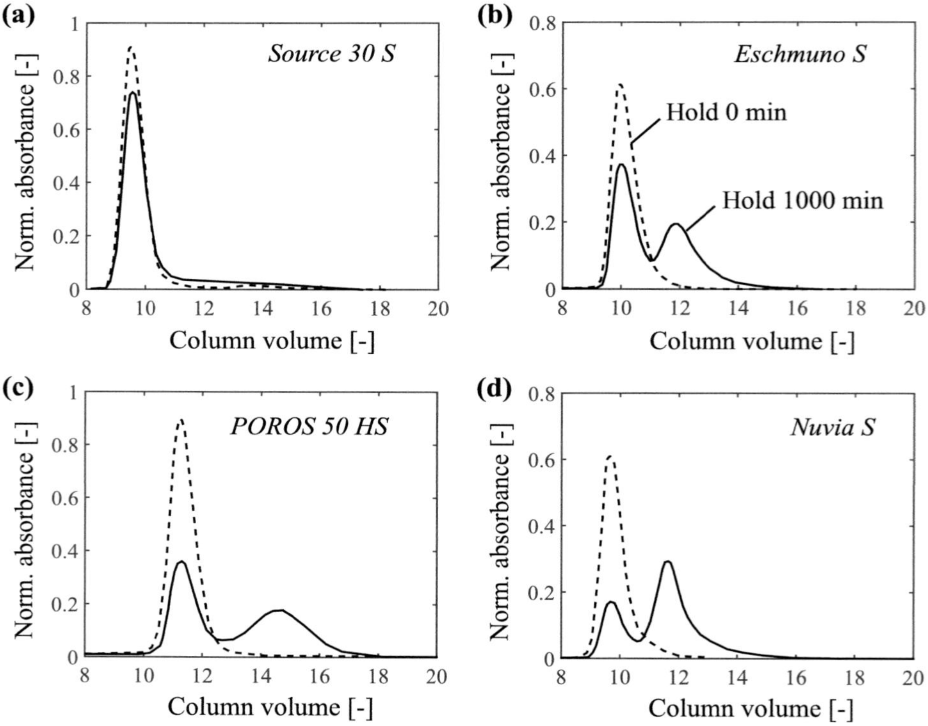

Figure 7.22 Chromatographic behavior of a mAb on columns packed with (a) Source 30 S, (b) Eschmuno S, (c) POROS 50 HS, and (d) Nuvia S cation exchange media. The load buffer was 20 mM $NaCH_3COOH$ at pH 4.8 in all cases. After loading, either the protein was recovered directly (dashed line) or the flow was stopped and the column was held idle for 1000 min before recovering the protein (solid lines). In both cases, protein recovery was triggered by applying a linear gradient of NaCl (from 0 to 1 M in 20 column volumes). Adapted from [447], copyright 2015, with permission from Elsevier.

A wide range of experimental conditions were explored, including various chromatographic media, protein concentrations, load flow rate, solution compositions, and monoclonal antibodies [447]. In particular, it was found that the two-peak elution behavior is more pronounced for chromatographic media that are composed of porous particles grafted with polymer chains, such as Eschmuno (Figure 7.22(b)) or Nuvia (Figure 7.22(d)). In these materials, protein binding is believed to occur throughout the grafted polymer layer, i.e., it is not restricted to monolayer coverage as in the Source chromatographic medium, for example. A strong two-peak elution behavior was also observed for the POROS medium (Figure 7.22(d)). This material has no grafted polymers but is characterized by a bimodal pore size distribution whose micropores are smaller than the pores of the other media considered in this study. Overall, these experimental data suggest that protein unfolding on chromatographic media occurs more significantly on media grafted with polymer chains or containing very small pores. Although very long hold times were selected for this study (1000 min), a two-peak elution behavior was also observed for shorter times that may be relevant for preparative chromatography [446, 447].

7.5.4 Viral Inactivation

The downstream processing of monoclonal antibodies usually requires two viral inactivation procedures (see Section 1.2.5). A first viral inactivation step is often performed by keeping the protein solution after the capture step under acidic conditions for a certain period of time [22, 40, 41]. As discussed in the previous section, low pH solutions are known to promote protein unfolding and aggregation. As an example, mAb aggregation experiments performed in closed vials under acidic stress conditions are presented in Figure 7.23 [458]. In particular, Figure 7.23(a) and Figure 7.23(b) show the

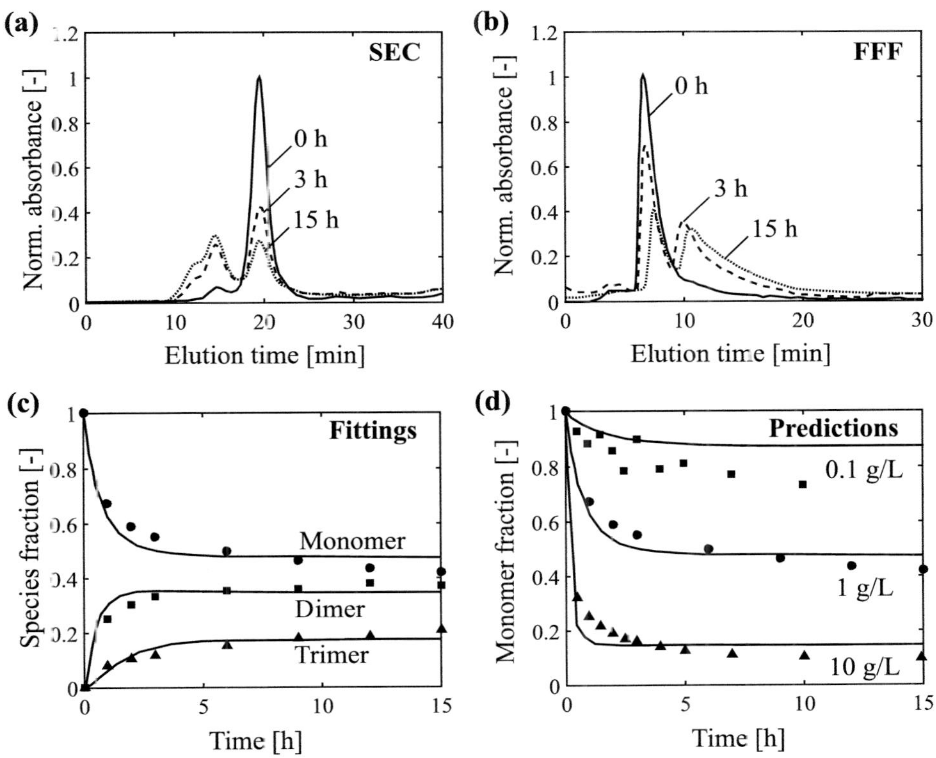

Figure 7.23 Case study of a mAb aggregating due to acidic stress. Adapted from [458]. With permission from Springer. Experiments were performed in closed vials at room temperature with a mAb concentration of 1 g/L, in a 25 mM citric buffer at pH 3.0 containing 150 mM NaCl. (a) SEC chromatograms at different times. (b) FFF analysis of the same samples. (c) Comparison between model simulations (lines) and experiments (symbols) in terms of monomer, dimer and trimer content at mAb concentration of 1 g/L. (d) Comparison between model predictions (lines) and experiments (symbols) at 0.1, 1, and 10 g/L using the model parameters determined at 1 g/L.

SEC chromatogram and FFF fractogram, respectively, recorded at various time points. Note that aggregates elute before the monomer in the case of SEC, but the situation is opposite for FFF. It is observed both from SEC and FFF results that the protein is initially mainly monomeric and then forms dimers and trimers, but no larger aggregates. The underlying aggregation mechanism is the one presented in Figure 7.16(c). The equilibrium constants K_U, $K_{1,1}$, and $K_{1,2}$ were determined by fitting model simulations to experimental data obtained at the mAb concentration of 1g/L, and the quality of the fit can be appreciated in Figure 7.23(c). Then, the same equilibrium constants were used to predict monomer depletion kinetics at two other mAb concentrations (namely 0.1 and 10 g/L) without any additional parameter adjustment. The comparison between model predictions and experiments is shown in Figure 7.23(d). Remarkably, the model is able to capture the aggregation behavior of this mAb over two orders of magnitude in protein concentration, thus confirming the potential of modeling for the prediction of the impact of operating conditions.

As mentioned previously, the experiments presented in Figure 7.23 were performed in closed vials, which is representative of the situation encountered in a batch process. In a continuous process, one could envisage two main types of units ("reactors") to implement the viral inactivation step: a continuous stirred tank reactor (CSTR) or a continuous plug flow reactor (PFR), which have been introduced in Section 1.4. In both cases, it is essential to select properly the flow rate and the reactor size in order to ensure that the protein solution is kept for a sufficient time at low pH to obtain the desired log reduction. This is easily achieved in a PFR because all viruses spend exactly the same time at low pH. In addition, a continuous plug flow reactor is analogous to a batch stirred tank reactor (BSTR), where the distance along the PFR would correspond to the time in a BSTR. This implies that a similar aggregate content is expected at the outlet of a PFR and at the end of the viral inactivation step performed in a BSTR. On the other hand, a CSTR is characterized by a broad residence time distribution, as illustrated in Figure 1.7(c). It results that a fraction of viruses spends a very short time in the reactor no matter how large is the average residence time. Therefore, continuous stirred tank reactors seem poorly suited for viral inactivation.

Besides low-pH treatments, other strategies for viral clearance include heat inactivation (e.g., Enbrel®) and filtration [40, 41]. We have seen in the previous section that the exposure to elevated temperatures accelerates protein aggregation, especially if the temperature is close to the protein denaturation temperature (which can, for example, be determined by DSC). As an example, the aggregation behavior of a mAb induced by heat stress was presented in Figure 7.13. Protein aggregation during filtration is instead discussed in the next section.

7.5.5 Filtration

Filtration is routinely employed in biopharmaceutical processes, for instance to clarify the cell culture harvest (microfiltration), remove viruses (nanofiltration), exchange the solution buffer (diafiltration), or concentrate the final product (ultrafiltration). In the first two examples, the protein solution is collected in the permeate flow of the filtration

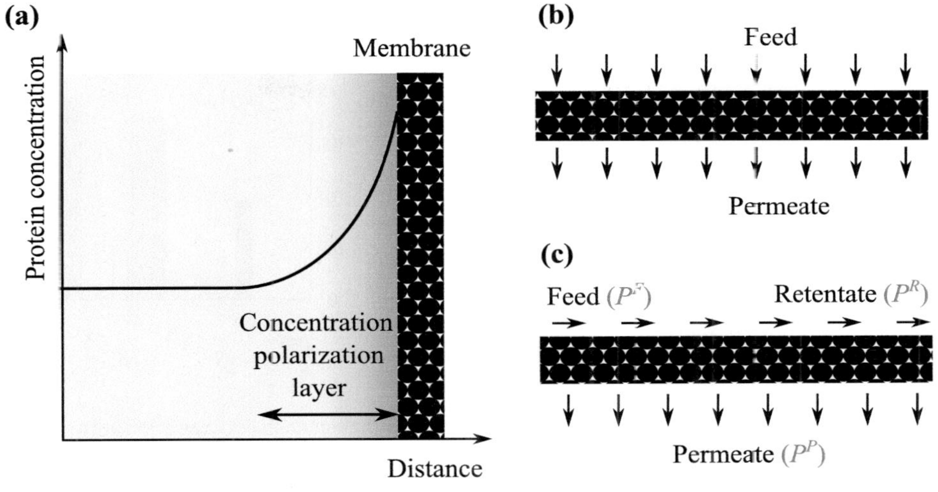

Figure 7.24 (a) Illustration of the concentration polarization layer in a filtration system. (b) Schematic representation of a frontal filtration system. (c) Schematic representation of a tangential filtration system.

system, while it is collected in the retentate in the other two. Filtration has been reported to promote aggregate formation due to the contact with membrane surfaces, to the presence of small air bubbles, and to shear stress [59]. Moreover, aggregation during ultrafiltration may be significantly accelerated due to the high protein concentrations that are commonly encountered toward the end of the production process.

A typical profile of the protein concentration observed in the retentate of a filtration system is shown in Figure 7.24(a). It is seen that the protein concentration increases in the region close to the surface of the membrane, which is referred to as the concentration polarization layer [459]. Such local increase in protein concentration may result in filter fouling and promote protein aggregation. To reduce protein accumulation at the membrane surface, frontal filtration systems where the feed solution flows perpendicular to the membrane may be replaced by tangential filtration systems, where the feed solution flows parallel to the membrane. Tangential filtration systems are typically used for the ultrafiltration/diafiltration steps, while clarification of the supernatant and sterile filtration are more commonly performed in frontal mode. The conventional and tangential filtration systems are schematically represented in Figure 7.24(b) and (c), respectively. For the tangential filtration system, we denote P^F, P^R, and P^P the pressures in the feed, retentate, and permeate, respectively. We can then define the pressure drop between the feed and the retentate as $P^F - P^R$ and the transmembrane pressure as $(P^F + P^R)/2 - P^P$.

For illustrative purposes, let us consider the data shown in Figure 7.25, which were obtained during the concentration of a mAb with tangential ultrafiltration [460]. The initial protein concentration was 5 g/L, and the goal was to reach 140 g/L. To do so, experiments were performed at two different pressure drops between the feed and the retentate, namely 1.2 and 3.0 bars. These pressure drops correspond to average fluxes

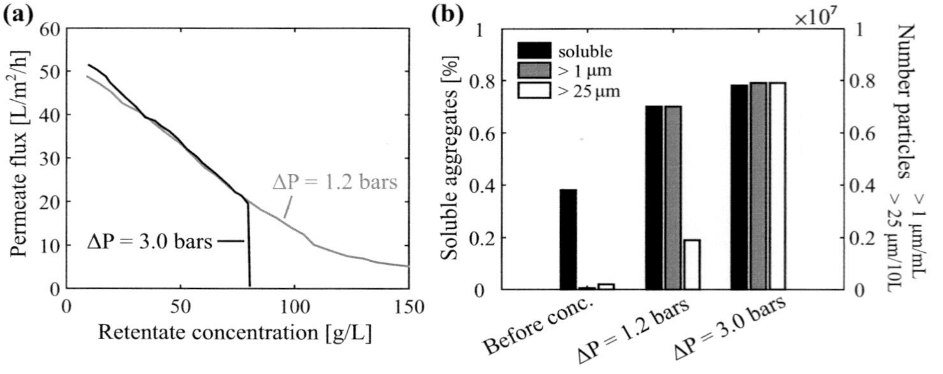

Figure 7.25 (a) Permeate flux as a function of the retentate concentration at two different pressure drops between the feed and the retentate. The transmembrane pressure was kept constant at 0.6 bar. (b) Aggregate formation before concentration and when reaching 80 g/L using two pressure drop values. The mass percentage of soluble aggregates was measured by SEC. The number of particles was measured with light obscuration. Adapted from [460], copyright 2009, with permission from Elsevier.

in the retentate in the order of 310 and 845 L/m^2/h, respectively. The transmembrane pressure, was kept equal to 0.6 bar in these two experiments. Figure 7.25(a) shows the permeate flux as a function of the protein concentration for the experiments performed at the two different pressure drops. It is observed that the permeate flux decreases when the protein concentration in the retentate increases. Up to 80 g/L, the permeate flux is similar in the two experiments because the pressure across the membrane is the same (0.6 bar). However, it is observed that the permeate flux suddenly drops to 0 at 80 g/L for the experiments performed at 3.0 bars. To understand this observation, let us examine the results shown in Figure 7.25(b), where the amount of aggregates is reported before concentration (at 5 g/L) and when reaching 80 g/L with the two different pressure drops. The content of soluble aggregates was measured by SEC, while the number of particles larger than 1 μm and larger than 25 μm was evaluated by light obscuration. For comparison, an antibody is about 10 nm in diameter, which is more than 100 times smaller than the reported particle sizes. It is seen in Figure 7.25(b) that the percentage of soluble aggregates only slightly increased during filtration. However, the number of particles dramatically increased in both experiments. Interestingly, the formation of particles and in particular of very large particles (>25 μm), is more pronounced at the highest pressure drop. This seems to indicate that the sudden decrease of the permeate flux observed in Figure 7.25(a) was due to filter clogging. In both experiments, the protein residence time in the filtration system was similar, namely around 2 h. However, the shear stress and the number of passes were higher in the experiment performed at higher pressure drop due to the larger flow rate. This suggests that the formation of aggregates in a filtration system does not only depend on the protein residence time but also on the shear stress and/or number of passes.

Table 7.2. Summary of critical steps for monoclonal antibody aggregation in biopharmaceutical processes and the potential corresponding causes of aggregate formation.

Process step	Potential causes for aggregate formation
Cell culture	Temperature above 30 °C Cell culture media components Air bubbles
Chromatography	pH far from pI (e.g., affinity) or close to pI (e.g., IEC) High ionic strength Interaction with the surface of the chromatographic medium
Viral inactivation	Low pH (acidic treatment) High temperature (heat treatment)
Filtration	High protein concentration Interaction with the surface of the filter Air bubbles Shear stress

Filter clogging due to aggregate formation is an important issue in biopharmaceutical processes [59, 369]. In particular, aggregate formation during sterile filtration is critical for product quality because this is the last step of the purification process.

7.5.6 Summary

In biopharmaceutical processes, therapeutic proteins experience a wide range of pH values, ionic strengths, protein concentrations, and temperatures and may be exposed to various types of interfaces. In this section, the main steps that are critical for aggregate formation have been identified for a general mAb production process, and the conclusions are summarized in Table 7.2. The organization of this table reflects the manufacturing of monoclonal antibodies, but the factors identified as potential causes of aggregate formation are relatively general and apply to other therapeutic proteins. Moreover, additional steps that are not listed in Table 7.2 but may also cause protein aggregation include protein refolding, freezing-thawing cycles, drying and final filling [366, 461]. After production, protein aggregation is also likely to occur during long-term storage, especially for concentrated protein solutions [368, 462].

7.6 Methods to Reduce the Aggregate Content

7.6.1 Limiting Aggregate Formation

As described in the preceding sections, protein stability depends on a variety of factors, which can be classified as intrinsic (i.e., related to the protein properties) and extrinsic (i.e., related to the environmental conditions) [461]. Therefore, several strategies can be envisaged to limit aggregate formation in biopharmaceutical processes: improve the stability of the protein itself, improve the design of the unit operations, or improve the operating conditions. Each of these approaches is discussed in the following sections.

Aggregation-Resistant Proteins

A large effort is devoted to the development of atomistic simulation tools to identify aggregation-prone regions of therapeutic proteins [393–395]. Such techniques essentially localize hydrophobic patches that are exposed at the protein surface and assess their size and shape. This information allows performing target mutations on these aggregation "hot spots" to engineer aggregation-resistant proteins [393, 463]. This strategy was successfully applied to enhance the stability of antibodies and therefore represents a promising approach to support the rational design of protein drug candidates. It is nevertheless crucial to test the biological activity of the mutants because performing mutations close to the protein active sites may result in a loss of protein specific activity.

Another possibility to enhance protein stability is to graft polymer molecules to the protein [46], such as PEG, which has been discussed in Chapter 6. For illustrative purposes, Figure 7.26 shows the impact of PEGylation on the stability of the granulocyte colony stimulating factor (GCSF) [464], which is a therapeutic protein in the order of 19 kDa. It is seen that PEGylation with 5 kDa PEG significantly reduced aggregate formation as compared to the non-PEGylated protein, most likely due to the coverage of aggregation-prone regions. It is also observed that the use of 20 kDa PEG only slightly improved protein stability as compared to the 5 kDa PEG. This seems to indicate that a 5 kDa PEG molecule is already sufficiently large to shield an important portion of the protein surface. It was shown in the original work that PEGylation did not only reduce the aggregate content but also the aggregate size. Indeed, large precipitates were formed for the non-PEGylated protein while only small soluble aggregates were formed in the case of the PEGylated proteins (both with 5 and 20 kDa PEG).

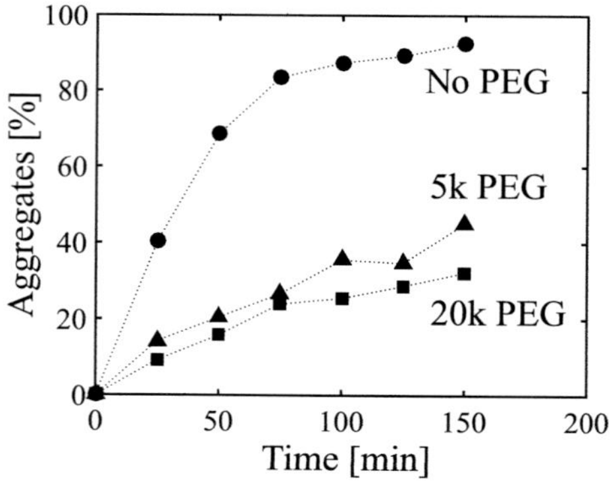

Figure 7.26 Aggregation kinetics of the granulocyte colony stimulating factor in closed vials incubated at 37 °C and neutral pH, without PEGylation, with 5 kDa and 20 kDa PEG. The content of aggregates was determined by SEC. Adapted from [464] by permission of John Wiley & Sons, Inc.

Process Design

Besides improving the intrinsic stability of the protein, another strategy consists of designing processes that reduce aggregate formation. A good example is represented by the development of perfusion bioreactors [24], which were described in Section 1.4. Although protein aggregation during cell culture is rather limited in the case of mAbs, this might become a critical issue for structurally unstable proteins, such as blood coagulation factors or fusion proteins. In these cases, the use of perfusion bioreactors allows limiting the formation of aggregates by reducing the protein residence time in the bioreactor from approximately one week for a batch or fed-batch process to less than one day.

Matrix-assisted protein refolding provides another good example of a process that can be used to reduce aggregate formation. As discussed in Section 1.2, protein refolding is typically required when the protein is expressed by bacteria as intracellular insoluble deposits, called inclusion bodies, that contain the protein in a misfolded aggregated state [15]. The challenge here is to dissolve these inclusion bodies and properly refold the inactive protein *in vitro*. This step is particularly delicate because protein aggregation competes with protein folding. Because folding is a first-order process, while aggregation is a second-order process, a high dilution of the protein solution in the refolding buffer favors folding with respect to aggregation. Although this strategy is widely used in practice, the ensuing large volumes and low protein concentrations complicate downstream processing and increase operating costs. Refolding in chromatographic columns, and in particular in SEC columns, provides an efficient alternative to reduce the extent of dilution [53]. The underlying mechanisms of on-column refolding are still unclear, but it has been hypothesized that the gradual buffer exchange and the steric hindrance of the aggregates within the porous resin allow a larger portion of the protein to refold properly. Whether or not the chromatographic surface influences protein refolding is instead still a matter of debate [53].

When aggregation takes place in chromatographic columns, it may be advantageous to consider continuous countercurrent processes. As a matter of fact, we have seen throughout this book that countercurrent chromatographic processes allow achieving better separations than their single-column equivalent. This implies that a smaller number of plates is needed in a countercurrent system to obtain a certain productivity for given target purity and recovery values. Therefore, smaller columns and/or higher flow rates may be used, which results in shorter residence times. Such decrease of the protein residence time in the chromatographic column with respect to the characteristic time of aggregation is expected to reduce the formation of aggregates. To illustrate this point, we compare the process performance of single-column and simulated moving bed processes in the case where aggregation occurs.

For this, we continue the example of Section 7.5.3 dealing with the separation of protein monomers and dimers. The design of the single-column and SMB processes is performed according to the methodology given in Box 2.1 and Box 3.2, respectively. A target purity of 99%, a target recovery of 90%, a maximum pressure drop of 10 bar, and an average feed flow rate of 100 L/h (i.e., a production requirement of 90 g/h at a feed concentration of 1 g/L) were considered. To ease the determination of the process

Table 7.3. Values of the process variables for the single-column and SMB processes for a monomer/dimer separation. The processes were designed in the absence of aggregation according to the methodology given in Box 2.1 and Box 3.2, respectively, considering $\chi_M = 99\%$, $\vartheta_M = 90\%$, $\dot{q}_M = 90$ g/h, $\Delta P = 10$ bar, $C_M^F = 1$ g/L, $\delta_K = 0.8$ bar.h/m^2, and $c_{VD} = 3.7$ s, where M stands for the monomer.

Single-column			SMB		
n^T		1	n_T		8
L_{col}	[cm]	74.0	L_{col}	[cm]	10.2
D_{col}	[cm]	15.3	D_{col}	[cm]	9.4
Q	[L/h]	317	Q^F	[L/h]	100
$t_{c,M}^{beg}$	[min]	3.7	Q^D	[L/h]	89
$t_{c,M}^{end}$	[min]	15.0	Q^R	[L/h]	90
Δt^F	[min]	9.6	Q^{IV}	[L/h]	68
Δt	[min]	28.0	Δt	[min]	1.5
φ_M	[g/L/day]	180	φ_M	[g/L/day]	382

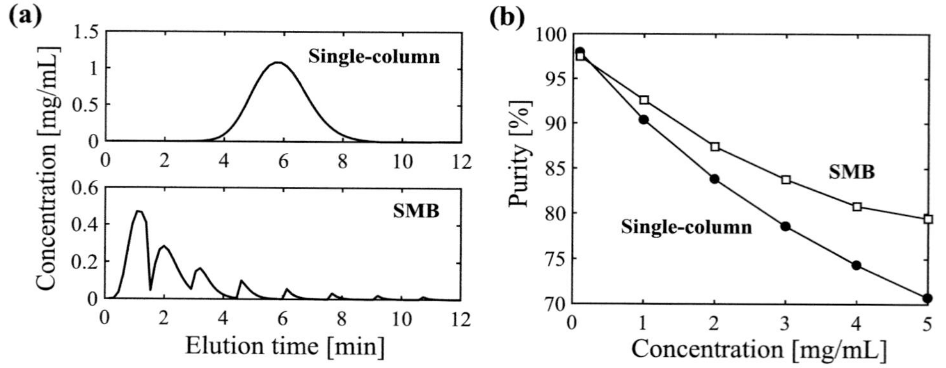

Figure 7.27 (a) Residence time distribution obtained with a pulse injection of monomeric protein in the single-column (top) and SMB (bottom) processes. The monomeric protein elutes in the raffinate stream of the SMB. The simulations were performed in the absence of aggregation. (b) Purity at the outlet of the single-column and in the raffinate of the SMB process at various feed concentrations considering that the dimer formation takes place in the liquid phase with a rate constant $k_{1,1} = 0.2$ L/g/min.

variables, aggregation was not accounted for in the shortcut method design. The results are summarized in Table 7.3, where it is seen that the productivity can be doubled in the SMB process as compared to the single-column process. This increase in productivity comes from the reduction in the total column volume required to perform the separation. Such reduction of the total column volume is also expected to lead to shorter residence times. This can be verified in Figure 7.27(a), where the concentrations at the outlet of the single-column and in the raffinate of the SMB for a pulse injection of monomer are presented. It is observed that, on average, the monomeric protein exits from the SMB process about two times faster than from the single-column process.

In a second step, the dimerization reaction was added to the chromatographic model. As in Section 7.5.3, it is considered that the dimer formation takes place in the liquid phase with a rate constant $k_{1,1} = 0.2$ L/g/min. The single-column chromatograms obtained with the process variables reported in Table 7.3 are shown in Figure 7.21(a) for two different concentrations, namely 0.1 and 5 g/L. The purity of the fraction collected between the cut-times given in Table 7.3 is plotted in Figure 7.27(b) as a function of the feed concentration with filled symbols. It is observed that the purity strongly decreases with the protein feed concentration due to the formation of aggregates. As a comparison, the purity in the raffinate stream of the SMB process is shown with open symbols. Interestingly, the SMB process better copes with aggregate formation and allows maintaining a higher purity than that obtained in the single-column process. This can be attributed to the shorter times that the protein spends in the unit. This example illustrates how multicolumn countercurrent processes can be beneficial to separation involving unstable proteins.

Operating Conditions

A third strategy to reduce aggregate formation in biopharmaceutical processes is to properly select the operating conditions, and in particular the solution pH and buffer composition [45, 387, 461, 465]. For instance, most of the methods aiming at reducing aggregation during mAb capture focus on moderating the solution pH during the recovery step by using additives such as sodium chloride, urea, or amino acids [466]. A case in point is arginine, which has been shown to allow higher elution pH values than citrate during desorption from Protein A chromatographic columns [467]. Suitable additives may also be used to reduce the aggregate content during ion exchange or hydrophobic interaction chromatography. For example, urea has been found to induce oligomer dissociation of apolipoprotein A-I during anion exchange chromatography [406, 469]. Operating the column at low temperature is an alternative strategy that may effectively reduce the aggregate content. It has, for example, been reported for mAb capture [466] and for the purification of Factor VIII with hydrophobic interaction chromatography [468]. However, reducing temperature might reduce protein solubility, potentially leading to the formation of insoluble particulates.

Nonionic surfactants such as polysorbate (20 and 80) are instead commonly used to prevent agitation-induced and freezing-induced aggregation [59, 461, 465]. The stabilization effect of nonionic surfactants is primarily attributed to their capacity to outcompete proteins at interfaces with air, hydrophobic solid surfaces, or ice, thereby reducing surface-induced protein denaturation and aggregation [465]. It has also been suggested that the protective effect of surfactants may come from their direct adsorption on aggregation-prone hydrophobic regions of the proteins [465].

Finally, polyols (such as sorbitol or trehalose) are effective stabilizers against heat stress and drying [461, 465], and are thus widely used as formulation excipients both in the liquid and dried states. During drying, polyol molecules serve as water substitutes to replace the hydrogen bonding requirements of polar groups on the protein surface. In liquid formulations instead, the stabilization mechanism results from the exclusion of polyol molecules at the protein surface [428, 470]. As discussed in Section 7.4.2 in the

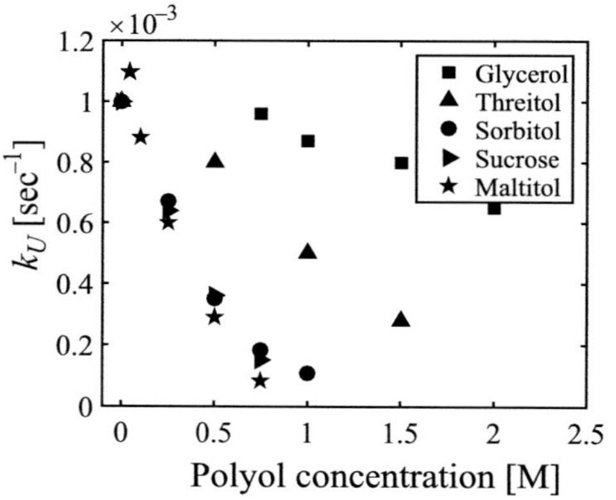

Figure 7.28 Impact of polyols on the unfolding rate constant of a monoclonal antibody aggregating in closed vials incubated at 70 °C. Adapted from [430], copyright 2015, with permission from Elsevier.

context of salt effects, nonfavorable interactions between the polyol molecules and the protein surface favor native compact protein conformations over more open unfolded structures. Figure 7.28 shows how the unfolding rate constant of a monoclonal antibody aggregating under thermal stress decreases with polyol concentration for various polyol molecules [430]. It is seen that the stabilization effect increases with the polyol size from glycerol (3 carbon atoms), to threitol (4 carbon atoms) and sorbitol (6 carbon atoms). Sucrose and maltitol, each containing 12 carbon atoms, instead show a stabilization effect similar to that observed in the case of sorbitol. These results show that the stabilization effect does not depend only on the volume fraction filled by the polyol molecules, but also on their chemical nature.

7.6.2 Removing Aggregates

Techniques

The aforementioned strategies aim at reducing aggregate formation in biopharmaceutical processes. However, it is almost unavoidable that aggregates will form to some extent, thus requiring the implementation of appropriate methods for aggregate removal. Large insoluble aggregates are commonly removed by filtration, while oligomers and high-molecular-weight soluble aggregates typically necessitate the use of chromatography [59]. Ion exchange chromatography (IEC) and hydrophobic interaction chromatography (HIC) are very often selected at the production scale, either in elution or frontal mode [22].

Figure 7.29(a) shows chromatograms obtained in frontal mode with IEC and HIC [471]. Three phases can be identified: a load phase, during which the feed solution is injected in the column, a wash phase during which the unbound species elute, and a cleaning phase during which the species bound to the chromatographic medium are recovered by changing the solution conditions. This is done by increasing the salt content for IEC and decreasing the salt content for HIC. Figure 7.29(b) shows the

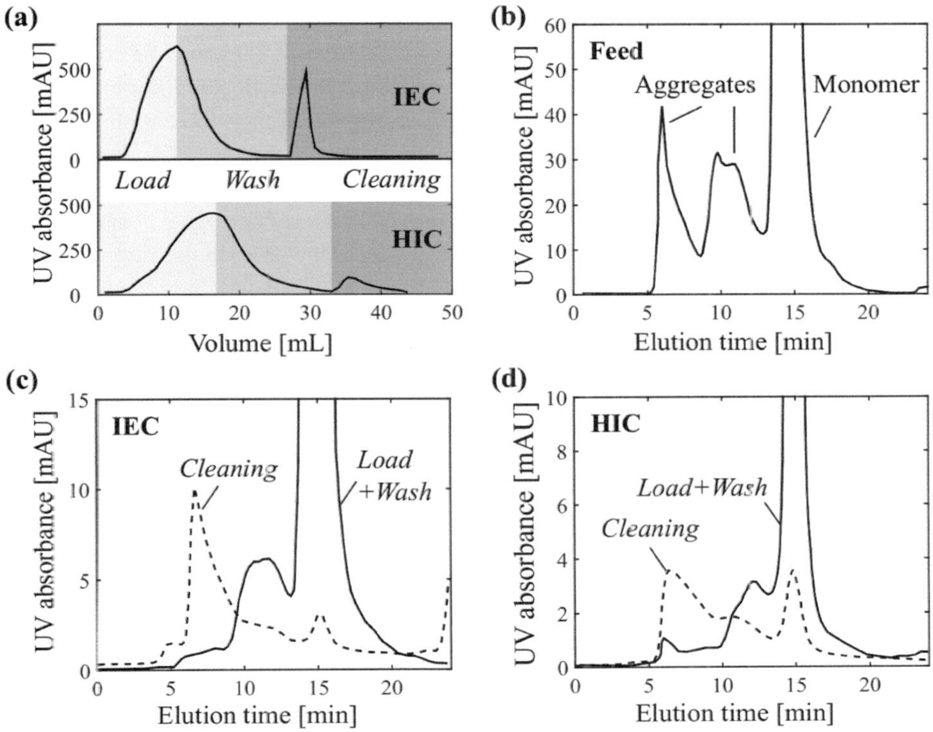

Figure 7.29 Aggregate removal with IEC and HIC in frontal mode. Adapted from [471], with permission from Elsevier. (a) Chromatogram recorded at the outlet of the preparative columns. (b) Analytical SEC chromatogram of the feed mixture. (c) Analytical SEC chromatograms of the fractions collected during the load phase and the wash phase in the IEC run. (d) Analytical SEC chromatograms of the fractions collected during the load phase and the wash phase of the HIC run.

analytical SEC chromatogram of the feed mixture, which contains the monomeric protein and aggregates of various sizes, while the analytical SEC chromatograms of the purified solution obtained by IEC and HIC are shown in Figure 7.29(c) and Figure 7.29(d), respectively. The solid lines correspond to the solution flowing through the column (i.e., collected during the load and wash phases), while the dashed lines correspond to the solution recovered during cleaning.

It is observed that both in the cases of IEC and HIC, the monomeric protein is mainly recovered in the flow through, while the aggregates are retained in the column and recovered upon cleaning. It is often observed that aggregates are more retained by IEC and HIC chromatographic media than the monomeric protein. This seems to indicate that the aggregates have a higher charge and a larger hydrophobic surface than the monomer due to their larger size. However, because aggregates are thought to form due to interactions between hydrophobic patches of the monomeric proteins, it is likely that the hydrophobic surface of one single protein is lower in the aggregated state than in the monomeric state. It is also worth mentioning that aggregates do not necessarily all share the same physicochemical characteristics, because they may be constituted of different variants of the monomeric protein, which may in particular differ in terms of net charges. Such heterogeneities in the aggregate population may potentially result in a class of aggregates less retained than the monomer, and another class more retained [472]. It was shown that the use of multimodal chromatography, combining IEC and HIC, could help to remove heterogeneously charged antibody aggregates [472]. In addition, if the size of the aggregates is comparable to that of the pores of the chromatographic medium, a steric exclusion effect may arise, potentially reducing the retention of the aggregates in the column.

IEC and HIC offer the possibility to remove simultaneously various impurities, such as charge variants, DNA, and host cell proteins, in addition to protein aggregates. In contrast, the resolution of SEC is rather low and does not allow separating impurities with a size comparable to the protein of interest. Nevertheless, protein denaturation and aggregation may occur during purification by IEC and HIC, as discussed in Section 7.5. In this case, a final aggregate removal step may be performed by SEC, which has the advantage to be operated under relatively mild conditions. This is, for example, the case for the following mAbs: CEA - Scan® (Arcitumomab), Zenapax® (Daclizumab), and Campath - 1H® (Alemtuzumab) [97]. In the following, we discuss how to perform a SEC process using a countercurrent multicolumn unit and compare its performance with that of a single-column process.

SMB for Continuous Aggregate Removal

We have already shown in Chapters 3 to 6 that continuous countercurrent chromatography allows improving separation performances in a wide variety of cases. In this section, we apply the concepts that were introduced in the previous chapters to the particular case of aggregate removal. As an example, we consider the SEC chromatogram shown with the black solid line in Figure 7.30(a), which was obtained during mAb purification with a single-column process at the pilot scale [473]. Because we are interested in performing a binary separation under isocratic conditions, the classical simulated moving bed (SMB) process can be selected here to move from single-column to continuous countercurrent chromatography.

We perform model simulations to compare the SMB and single-column processes for the particular separation considered in Figure 7.30. To do so, it is first necessary to determine the thermodynamic, hydrodynamic, and kinetic model parameters for the monomer and for the aggregates. Because the UV signal was saturated in the experiment performed at the pilot scale (see Figure 7.30), the estimation of model parameters is done based on analytical chromatograms obtained with the same chromatographic medium (Superdex 200). For this, we select the PD model and use the LDF approximation to describe internal mass transfer. We obtain $\overline{K}_M = 0.33$, $\overline{K}_A = 0.15$, $\varepsilon_e = 0.33$, $\varepsilon_i = 0.88$, $t_{i,M} = 5$ s, $t_{i,A} = 7$ s, $Pe_{p,M} = Pe_{p,A} = 0.05$, where M and A denote the monomer and the aggregates, respectively. Moreover, the area of the monomer peak is evaluated so as to obtain a total aggregate content of 3.5%, as determined by an experiment performed at the laboratory scale at lower protein concentration [473].

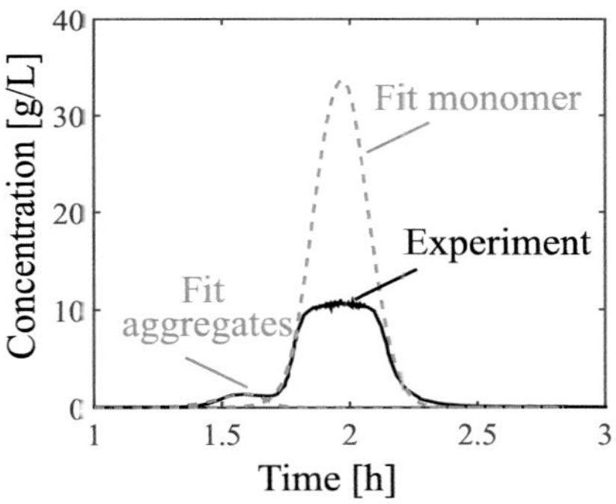

Figure 7.30 SEC chromatogram obtained at the pilot scale with a Superdex 200 column of 10 cm diameter and 85 cm length (solid lines) compared to simulations with the PD model using the LDF approximation. The following parameters were used: $\overline{K}_M = 0.33$, $\overline{K}_A = 0.15$, $\varepsilon_e = 0.33$, $\varepsilon_i = 0.88$, $d_p = 40$ µm, $t_{i,M} = 5$ s, $t_{i,A} = 7$ s, $Pe_{p,M} = Pe_{pA} = 0.05$, where M and A denote the monomer and the aggregates, respectively. The UV signal is not usable above a concentration of approximately 10 g/L because it saturates. Experimental data reprinted from [473], copyright 2007, with permission from Elsevier.

Finally, a total feed concentration of approximately 50 g/L and an injection volume of 375 mL are considered. The corresponding model simulations (dashed lines) are compared the experimental results (continuous lines) in Figure 7.30.

We have seen in Section 2.4.2 that fluid–solid equilibria can be assumed linear in size exclusion chromatography. Then, we have seen in Section 3.2 that for linear equilibria and in the frame of the equilibrium theory, the zone of complete separation is defined by $\overline{K}_A \le m^{II} < m^{III} \le \overline{K}_M$, which is a triangle in the m^{II}–m^{III} plane. In this example, we select the operating point located at the vertex of the triangle (i.e., $m^{II} = \overline{K}_A$, $m^{III} = \overline{K}_M$), which corresponds to the maximum normalized feed flow rate allowing complete separation. In addition, we have seen that the zone of complete regeneration is defined by $m^I \ge \overline{K}_M$ and $m^{IV} \le \overline{K}_A$. Here, we choose to operate the system at the minimum normalized eluent flow rate allowing complete regeneration of the liquid and solid phases in the frame of the equilibrium theory (i.e., $m^I = \overline{K}_M$, $m^{IV} = \overline{K}_A$).

In the single-column process, the column length was 85 cm, and the maximal flow rate that could be used to maintain an acceptable pressure drop was 32.7 mL/min (25 cm/h) [473]. We design the SMB process following the method summarized in Box 3.1. Accordingly, we impose a similar pressure drop and average fluid velocity in both processes. For illustrative purposes, we select a SMB process with two columns per zone with the same column diameter and total column length than the single-column process (i.e., $L_{col} = 10.6$ cm in the SMB). From the definition of the m parameters, it can be shown that the average flow rate $\langle Q \rangle$ is related to the average value of the m parameters $\langle m \rangle$ according to $\langle Q \rangle \Delta t = V_{col}(\varepsilon_e + (1 - \varepsilon_e)\langle m \rangle)$. Considering the values of the m^I to m^{IV} parameters given earlier lead to $\langle m \rangle = 0.24$, which results in $\Delta t = 13.9$ min. Then, the internal flow rates as well as the inlet and outlet flow rates can be computed from mass balance equations (see Table 3.2 and Equations 3.10–3.13),

Table 7.4. Comparison between a single-column process and a SMB process for the removal of aggregates by SEC. The SMB process was designed based on an experimental chromatogram obtained at the pilot scale.

	Single-column	SMB
Purity [%]	99.6	99.6
Recovery [%]	81.7	90.6
Productivity [g/L/day]	34	68.3

which provides 7.25 mL/min for the feed, eluent, extract, and raffinate. Knowing these parameters is sufficient to simulate the SMB process, and thus obtain the values of the purity, recovery and productivity at steady state.

The performance parameters of the single-column and SMB processes are summarized in Table 7.4. For the single-column process, a purity of 99.6% and a recovery of 81.7% were obtained experimentally [473]. Considering that the single-column process can be operated in 1.5 h (see Figure 7.30), a productivity of 34 g/L/day is estimated by neglecting the turnaround times. As a comparison, the SMB process allows reaching a recovery of 90.6% and a productivity of 68.3 g/L/day for the same target purity. This increase of productivity by a factor 2 can be attributed to the countercurrent movement between the fluid and solid phases in the SMB process.

Overall, this example illustrates the high potential of countercurrent multicolumn separation units for aggregate removal by SEC. The SMB process has also been shown to be very attractive for SEC assisted protein refolding [474, 475]. In particular, the use of a tandem SMB for the processing of recombinant proteins (one SMB for refolding followed by one SMB for purification) demonstrated that a significantly higher productivity and lower buffer consumption could be obtained in the integrated continuous operation as compared to the single-column process [474].

7.7 Conclusion

The formation of protein aggregates in biopharmaceuticals is a critical issue occurring during upstream processing, downstream processing, and long-term storage. It is crucial to have a deep understanding of the phenomena underlying the aggregation process in order to develop strategies aiming at improving protein stability. In particular, it is essential to remember that the aggregation propensity is dictated by a delicate balance between the protein colloidal stability (related to protein–protein intermolecular interactions) and the protein conformational stability (related to protein intramolecular interactions and protein–solvent interactions), which are both highly sensitive to environmental conditions.

The aggregation of therapeutic proteins involves phenomena that span several orders of magnitude in terms of length scales, and a multiscale representation of the aggregation process is proposed in Figure 7.31.

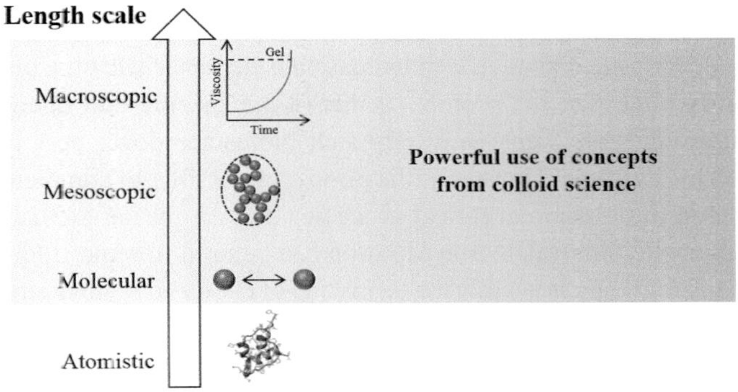

Figure 7.31 Multiscale view of therapeutic protein aggregation.

At the atomistic scale, molecular dynamics simulations can provide detailed information on the location of hydrophobic patches, on the surface electrostatic potential as well as on the protein spatial conformation. However, the time scales accessible by such simulations are extremely short, making the technique nonsuitable to follow protein aggregation during conditions relevant for manufacturing and storage. Concepts from colloid science instead represent powerful tools to investigate the protein aggregation behavior at the molecular scale and above. In particular, the Fuchs stability ratio (which is a kinetic parameter) and the second virial coefficient (which is a thermodynamic parameter) allow to quantify the protein–protein interactions in a coarse-grained manner. At the mesoscopic scale, the aggregate morphology can be quantified in the frame of the fractal scaling, and population balance equations allow one to describe monomer depletion and aggregate growth kinetics. Importantly, the comparison between PBE simulations and experimental data provide insights into the aggregation mechanisms and on the impact of operating conditions. By applying this approach, one can rationalize the impact of several operating conditions (such as the solution pH, the presence of cosolutes, the protein concentration, and the temperature) on the kinetics of aggregation. Finally, protein aggregation may impact the macroscopic properties of protein solution, for example due to the formation of visible particulates or to the viscosity increase.

After having introduced these key notions, we have analyzed the risk of aggregate formation in the different steps of biopharmaceutical processes with particular emphasis on mAb production. Basically, aggregation is susceptible to occur at all steps of the production process due to various stress conditions, including the exposure to low pH solutions, the presence of salts or hydrophobic surfaces. The case where aggregation occurs in chromatographic columns is particularly troublesome because the competition between aggregate removal and aggregate formation may prevent the recovery of a pure monomeric solution.

Several strategies can be envisaged to reduce the formation of protein aggregates in biopharmaceutical processes: design aggregation-resistant proteins (e.g., perform target mutations or conjugate the protein with polymer molecules), select an appropriate process (e.g., perfusion bioreactors or multicolumn chromatographic unit), or tune the environmental conditions (e.g., adding additives to the protein solutions). In any of these approaches, the underlying objective is to increase the characteristic time of aggregation with respect to the time that the protein spends in the specific unit operation, which may vary from around one week in a fed-batch bioreactor down to 1 h for a viral inactivation step for example. However, it is almost impossible to completely suppress protein aggregation in biopharmaceutical processes, thus requiring the use of efficient strategies for aggregate removal. While filtration can be used to remove large insoluble aggregates, chromatography is usually the technique of choice to remove smaller soluble aggregates (mainly IEC, HIC, SEC). We have shown in this chapter that aggregate removal by size exclusion chromatography could greatly benefit from countercurrent chromatographic processes.

8 Conclusion

Economical pressure, wish to improve performance, and successes from other industrial sectors are as many drivers that push the biopharmaceutical industry to give serious considerations to continuous processes. Throughout this book, we have shown how a rational approach based on chemical engineering principles can help to evaluate the potential benefits of moving from a batch to a continuous process.

We focused on three main aspects: chromatography, bioconjugation, and protein stability. For each of these topics, we started by analyzing the key underlying physical mechanisms. For example, in the case of chromatography, the separation of solutes results from the coupling between the thermodynamics of adsorption, the hydrodynamics and the kinetics of mass transfer. For what concerns bioconjugation, an additional aspect comes into play, namely the kinetics of chemical reaction. Finally, the formation of protein aggregates is dictated by the thermodynamics and kinetics of various elementary processes including monomolecular unfolding and bimolecular aggregation. Once the fundamental phenomena have been identified, suitable mass balance equations were derived to model the processes of interest. A deep understanding of the subject matters has been achieved by using model simulations to better explain experimental data. Modeling has also been proven highly valuable to investigate unexplored regions of conditions and process variables, thus reducing the experimental effort.

Building on the knowledge so-acquired, we critically compared continuous processes to their more traditional batch equivalent. In the case of chromatography, we have seen that the main advantage of continuous multicolumn processes is their ability to simulate a countercurrent movement between the solid and liquid phases. This allows reaching a higher productivity and lower eluent consumption for high target purity and recovery values. From the invention of the simulated moving bed in the petrochemical industry, many countercurrent multicolumn processes have been developed over the years to tackle the challenges faced by the biopharmaceutical industry, such as the necessity to cope with highly complex mixtures and to accommodate for changes in the eluent composition. A pallet of such processes has been presented, both in the context of protein capture and polishing applications. We have proposed two main types of methods to design multicolumn processes, namely an empirical one and a model-based one. The former relies on chromatograms obtained with single-column experiments and allows rapidly identifying process variables that deliver the product with acceptable specifications in terms of purity and recovery for fixed column dimensions. The so-obtained multicolumn process is generally not optimized with regard to productivity and eluent consumption, but provides a reasonable point to start with. On the other hand, the model-based approach allows a more complete design in terms of column dimensions and process variables with a view to optimizing process performance. In the frame of bioconjugation, we have seen that continuous technologies allow reducing idle times, which has a positive impact on productivity. In addition, we have shown that the choice of the reactor (batch, continuously stirred, or tubular) strongly influences the distribution of conjugated proteins. Finally, regarding protein aggregation, the main advantage of continuous technologies lies in the fact that the residence time in a unit can be reduced without affecting process performance, as it has been discussed for both perfusion bioreactors and countercurrent chromatographic units. If the residence time in a unit can be sufficiently reduced with respect to the characteristic time of aggregation, implementing continuous processes may be a way to reduce the extent of aggregation. The general considerations listed help understand how the biopharmaceutical industry could benefit from continuous technologies for what concerns chromatography, bioconjugation, and protein aggregation.

We have seen through several examples that a rational approach to design processes requires defining precise objectives. For instance, the process maximizing the productivity is most likely not the one maximizing product recovery nor the one minimizing eluent consumption. The choice of a process is thus a rather delicate task that necessitates accepting compromises. After reading this book, the reader should be equipped with some of the tools necessary to make a sound decision regarding process selection, although we are aware that we covered only few of the topics pertinent to the biopharmaceutical industry. Besides the mechanistic approach used throughout this monograph, further understanding biopharmaceutical systems will require bringing together different fields, such as biology, molecular dynamics, and complex fluid dynamics. We hope that this book has brought some answers to the fascinating challenges that remain to be solved.

Bibliography

[1] Carter PJ. Introduction to current and future protein therapeutics: A protein engineering perspective. Exp Cell Res. 2011; 317(9): 1261–1269.

[2] Leader B, Baca QJ, Golan DE. Protein therapeutics: A summary and pharmacological classification. Nat Rev Drug Discov. 2008; 7(1): 21–39.

[3] Samanen J. Similarities and differences in the discovery and use of biopharmaceuticals and small-molecule chemotherapeutics. In: Ganellin C, Jefferis R, Roberts S, editors. Introduction to biological and small molecule drug research and development. Elsevier; 2013. p. 161–203.

[4] Wurm FM. Production of recombinant protein therapeutics in cultivated mammalian cells. Nat Biotech. 2004; 22(11): 1393–1398.

[5] Demain AL, Vaishnav P. Production of recombinant proteins by microbes and higher organisms. Biotechnol Adv. 2009; 27(3): 297–306.

[6] Kaspar AA, Reichert JM. Future directions for peptide therapeutics development. Drug Discov Today. 2013; 18(17–18): 807–817.

[7] Pina AS, Lowe CR, Roque ACA. Challenges and opportunities in the purification of recombinant tagged proteins. Biotechnol Adv. 2014; 32(2): 366–381.

[8] Arnau J, Pedersen J. A universal platform for the purification of therapeutic proteins using affinity tags: The use of engineered aminopeptidases for His-tag removal. BioProcess J. 2006; 5(3): 59–65.

[9] Uversky V, Orengo C, Bateman A. Protein families: Relating protein sequence, structure, and function. John Wiley & Sons, Inc.; 2013.

[10] Sammut SJ, Finn RD, Bateman A. Pfam 10 years on: 10 000 families and still growing. Brief Bioinform. 2008; 9(3): 210–219.

[11] Finn RD, Coggill P, Eberhardt RY, Eddy SR, Mistry J, Mitchell AL, et al. The Pfam protein families database: Towards a more sustainable future. Nucleic Acids Res. 2016; 4(44): 279–285.

[12] Shukla AA, Hubbard B, Tressel T, Guhan S, Low D. Downstream processing of monoclonal antibodies – Application of platform approaches. J Chromatogr B. 2007; 848(1): 28–39.

[13] Ecker DM, Jones SD, Levine HL. The therapeutic monoclonal antibody market. MAbs. 2015; 7(1): 9–14.

[14] Huang CJ, Lin H, Yang X. Industrial production of recombinant therapeutics in *Escherichia coli* and its recent advancements. J Ind Microbiol Biotechnol. 2012; 39(3): 383–399.

[15] Clark EdB. Protein refolding for industrial processes. Curr Opin Biotech. 2001; 12(2): 202–207.

[16] Walsh G, Jefferis R. Post-translational modifications in the context of therapeutic proteins. Nat Biotech. 2006; 24(10): 1241–1252.

[17] Bailey JE, Ollis DF. Biochemical engineering fundamentals. Second edition. McGraw-Hill; 1986.

[18] Katoh S, Horiuchi J, Yoshida F. Biochemical engineering. Second edition. Wiley-VCH; 2015.

[19] Shuler ML, Kargi F, Delisa M. Bioprocess engineering. Basic concepts. Third edition. Prentice Hall; 2017.

[20] Liu HF, Ma J, Winter C, Bayer R. Recovery and purification process development for monoclonal antibody production. MAbs. 2010; 2(5): 480–499.

[21] Carta G, Jungbauer A. Protein chromatography: Process development and scale-up. John Wiley & Sons, Inc.; 2010.

[22] Curling J. The development of antibody purification technologies. In: Gottschalk U, editor. Process scale purification of antibodies. John Wiley & Sons, Inc.; 2008. p. 25–52.

[23] Hober S, Nord K, Linhult M. Protein A chromatography for antibody purification. J Chromatogr B. 2007; 848(1): 40–47.

[24] Boedeker BGD. Recombinant Factor VIII (Kogenate®) for the treatment of hemophilia A: The first and only world-wide licensed recombinant protein produced in high-throughput perfusion culture. In: Knablein J, editor. Modern biopharmaceuticals. Wiley-VCH Verlag GmbH & Co. KGaA; 2013. p. 429–443.

[25] Burnouf T, Radosevich M. Affinity chromatography in the industrial purification of plasma proteins for therapeutic use. J Biochem Biophys Meth. 2001; 49(13): 575–586.

[26] Allen RR, Felix F. Affinity chromatography of enzymes. In: Hage DS, editor. Handbook of affinity chromatography, Second edition. Chromatographic science series. CRC Press; 2005. p. 313–346.

[27] Arunakumari A, Wang J. Purification of human monoclonal antibodies: Non-protein A strategies. In: Gottschalk U, editor. Process scale purification of antibodies. John Wiley & Sons, Inc.; 2008. p. 103–123.

[28] Zhou JX, Tressel T. Basic concepts in Q membrane chromatography for large-scale antibody production. Biotechnol Prog. 2006; 22(2):341–349.

[29] Ghosh R. Protein separation using membrane chromatography: Opportunities and challenges. J Chromatogr A. 2002; 952(12): 13–27.

[30] Pabby AK, Rizvi SSH, Sastre AM. Handbook of membrane separations: Chemical, pharmaceutical, food and biotechnological applications. CRC Press; 2015.

[31] Saraswat M, Musante L, Ravid A, Shortt B, Byrne B, Holthofer H. Preparative purification of recombinant proteins: Current status and future trends. BioMed Res Int. 2013; 2013: 312709.

[32] Hustedt H, Kroner KH, Menge U, Kula MR. Protein recovery using 2-phase systems. Trends Biotechnol. 1985; 3(6): 139–144.

[33] Rosa PAJ, Ferreira IF, Azevedo AM, Aires-Barros MR. Aqueous two-phase systems: A viable platform in the manufacturing of biopharmaceuticals. J Chromatogr A. 2010; 1217(16): 2296–2305.

[34] Oelmeier SA, Dismer F, Hubbuch J. Application of an aqueous two-phase systems high-throughput screening method to evaluate mAb HCP separation. Biotechnol Bioeng. 2011; 108(1): 69–81.

[35] Asenjo JA, Andrews BA. Aqueous two-phase systems for protein separation: A perspective. J Chromatogr A. 2011; 1218(49): 8826–8835.

[36] Glynn J. Process-scale precipitation of impurities in mammalian cell culture broth. In: Gottschalk U, editor. Process scale purification of antibodies. John Wiley & Sons, Inc.; 2008. p. 309–324.
[37] Thömmes J, Gottschalk U. Alternatives to packed-bed chromatography for antibody extraction and purification. In: Gottschalk U, editor. Process scale purification of antibodies. John Wiley & Sons, Inc.; 2008. p. 293–308.
[38] Gagnon P. Technology trends in antibody purification. J Chromatogr A. 2012; 1221: 57–70.
[39] Mollerup I, Jensen SW, Larsen P, Schou O, Snel L. Insulin purification. In: Flickinger MC, Drew SW, editors. Encyclopedia of bioprocess technology. John Wiley & Sons, Inc.; 2010. p. 1–7.
[40] Mehta A, Lovato Tse M, Fogle J, Len A, Shrestha R, Fontes N, et al. Purifying therapeutic monoclonal antibodies. Society for Biological Engineering; 2008.
[41] Remingtonm KM. Fundamental strategies for viral clearance – Part 2: Technical approaches. BioProcess International; 2015.
[42] European Medicines Agency. Guideline on virus safety evaluation of biotechnological investigational medicinal products; 2006.
[43] Johnson SA, Brown MR, Lute SC, Brorson KA. Adapting viral safety assurance strategies to continuous processing of biological products. Biotechnol Bioeng. 2017; 114(1): 21–32.
[44] Daugherty AL, Mrsny RJ. Formulation and delivery issues for monoclonal antibody therapeutics. Adv Drug Deliver Rev. 2006; 58(56): 686–706.
[45] Ohtake S, Kita Y, Arakawa T. Interactions of formulation excipients with proteins in solution and in the dried state. Adv Drug Deliver Rev. 2011; 63(13): 1053–1073.
[46] Frokjaer S, Otzen DE. Protein drug stability: A formulation challenge. Nat Rev Drug Discov. 2005; 4(4): 298–306.
[47] Mach H, Meyer BK, Shameem M. Properties of protein formulations. In: Meyer BK, editor. Therapeutic protein drug products. Woodhead Publishing; 2012. p. 47–65.
[48] Houp RC. Ultrafiltration and diafiltration. J Validation Technol; 2009 15(4): 40–45.
[49] Zydney AL. Membrane technology for purification of therapeutic proteins. Biotechnol Bioeng. 2009; 103(2): 227–230.
[50] Meyer BK, Coless L. Compounding and filling: Drug substance to drug product. In: Meyer BK, editor. Therapeutic protein drug products. Woodhead Publishing; 2012. p. 83–95.
[51] Meyer BK, Shameem M. Commercial therapeutic protein drug products. In: Meyer BK, editor. Therapeutic protein drug products. Woodhead Publishing; 2012. p. 1–11.
[52] Carpenter JF, Chang BS, Garzon-Rodriguez W, Randolph TW. Rational design of stable lyophilized protein formulations: Theory and practice. In: Carpenter JF, Manning MC, editors. Rational design of stable protein formulations: Theory and practice. Springer US; 2002. p. 109–133.
[53] Jungbauer A, Kaar W, Schlegl R. Folding and refolding of proteins in chromatographic beds. Curr Opin Biotech. 2004; 15(5): 487–494.
[54] Stephanopoulos N, Francis MB. Choosing an effective protein bioconjugation strategy. Nat Chem Biol. 2011; 7(12): 876–884.
[55] Pfister D, Morbidelli M. Process for protein PEGylation. J Control Release. 2014; 180: 134–149.
[56] Teicher BA, Chari RV. Antibody conjugate therapeutics: Challenges and potential. Clin Cancer Res. 2011; 17(20): 6389–6397.
[57] ICH Harmonised tripartite guideline. Q6B. Specifications: Test procedures and acceptance criteria for biotechclogical/biological products. CPMP/ICH/365/96; 1999.

[58] Ratanji KD, Derrick JP, Dearman RJ, Kimber I. Immunogenicity of therapeutic proteins: Influence of aggregation. J Immunotoxicol. 2014; 11(2): 99–109.

[59] Vázquez-Rey M, Lang DA. Aggregates in monoclonal antibody manufacturing processes. Biotechnol Bioeng. 2011; 108(7): 1494–1508.

[60] Consortium SG. Protein production and purification. Nature Methods. 2008; 5(2): 135–146.

[61] Johnston A, Adcock W. The use of chromatography to manufacture purer and safer plasma products. Biotechnol Genet Eng Rev. 2000; 17: 37–70.

[62] Denizli A. Plasma fractionation: Conventional and chromatographic methods for albumin purification. Hacettepe J Biol Chem. 2011; 39(4): 315–341.

[63] Andersson L, Blomberg L, Flegel M, Lepsa L, Nilsson B, Verlander M. Large-scale synthesis of peptides. Biopolymers. 2000; 55(3): 227–250.

[64] Remingtonm KM. Fundamental strategies for viral clearance – Part 1: Exploring the regulatory implications. BioProcess International; 2015.

[65] ICH Harmonised tripartite guideline. Q5A(R1). Viral safety evaluation of biotechnology products derived from cell lines of human or animal origin. CPMP/ICH/295/95; 1999.

[66] Cordoba-Rodriguez RV. Aggregates in mAbs and recombinant therapeutic proteins: A regulatory perspective. BioProcess International; 2008.

[67] Fekete S, Beck A, Fekete J, Guillarme D. Method development for the separation of monoclonal antibody charge variants in cation exchange chromatography – Part II: pH gradient approach. J Pharm Biomed Anal. 2015; 102: 282–289.

[68] Epstein MS, Ehrenpreis ED, Kulkarni PM, the FDARMCotACoG. Biosimilars: The Need, the challenge, the Future: The FDA perspective. Am J Gastroenterol. 2014; 109(12): 1856–1859.

[69] Zhao YY, Wang N, Liu WH, Tao WJ, Liu LL, Shen ZD. Charge variants of an avastin biosimilar isolation: Characterization, *in vitro* properties and pharmacokinetics in rat. PLoS ONE. 2016; 11(3): e0151874.

[70] Voisard D, Meuwly F, Ruffieux PA, Baer G, Kadouri A. Potential of cell retention techniques for large-scale high-density perfusion culture of suspended mammalian cells. Biotechnol Bioeng. 2003; 82(7): 751–765.

[71] Kompala DS, Ozturk SS. Optimization of high cell density perfusion bioreactors. In: Ozturk SS, Hu WS, editors. Cell culture technology for pharmaceutical and cell-based therapies. CRC Press; 2005. p. 387–416.

[72] Reay D, Ramshaw C, Harvey A. Process intensification: Engineering for efficiency, sustainability and flexibility. Second edition. Butterworth-Heinemann; 2013.

[73] Anderson NG. Practical use of continuous processing in developing and scaling up laboratory processes. Org Process Res Dev. 2001; 5(6): 613–621.

[74] Vervaet C, Remon JP. Continuous granulation in the pharmaceutical industry. Chem Eng Sci. 2005; 60(14): 3949–3957.

[75] Poechlauer P, Manley J, Broxterman R, Gregertsen B, Ridemark M. Continuous processing in the manufacture of active pharmaceutical ingredients and finished dosage forms: An industry perspective. Org Process Res Dev. 2012; 16(10): 1586–1590.

[76] Byrn S, Futran M, Thomas H, Jayjock E, Maron N, Meyer RF, et al. Achieving continuous manufacturing for final dosage formation: Challenges and how to meet them. May 20–21, 2014 Continuous Manufacturing Symposium. J Pharm Sci. 2015; 104(3): 792–802.

[77] Adamo A, Beingessner RL, Behnam M, Chen J, Jamison TF, Jensen KF, et al. On-demand continuous-flow production of pharmaceuticals in a compact, reconfigurable system. Science. 2016; 352(6281): 61–67.

[78] Konstantinov KB, Cooney CL. White paper on continuous bioprocessing. May 20–21, 2014 Continuous Manufacturing Symposium. J Pharm Sci. 2015; 104(3): 813–820.

[79] Nicoud RM. The amazing ability of continuous chromatography to adapt to a moving environment. Ind Eng Chem Res. 2014; 53(10): 3755–3765.

[80] Langer ES, Rader RA. Continuous bioprocessing and perfusion: Wider adoption coming as bioprocessing matures. BioProcess J. 2014; 13(1): 43–49.

[81] Jungbauer A. Continuous downstream processing of biopharmaceuticals. Trends Biotechnol. 2013; 31(8): 479–492.

[82] Pollock J, Ho SV, Farid SS. Fed-batch and perfusion culture processes: Economic, environmental, and operational feasibility under uncertainty. Biotechnol Bioeng. 2013; 110(1): 206–219.

[83] Steinebach F, Müller-Späth T, Morbidelli M. Continuous counter-current chromatography for capture and polishing steps in biopharmaceutical production. Biotechnol J. 2016; p. 1126–1141.

[84] Pall Life Science. Is continuous bioprocessing the future? BioProcess International. Sponsored supplement; 2016.

[85] Walther J, Godawat R, Hwang C, Abe Y, Sinclair A, Konstantinov K. The business impact of an integrated continuous biomanufacturing platform for recombinant protein production. J Biotechnol. 2015; 213: 3–12.

[86] Xenopoulos A. A new, integrated, continuous purification process template for monoclonal antibodies: Process modeling and cost of goods studies. J Biotechnol. 2015; 213: 42–53.

[87] Klutz S, Holtmann L, Lobedann M, Schembecker G. Cost evaluation of antibody production processes in different operation modes. Chem Eng Sci. 2016; 141: 63–74.

[88] Bunnak P, Allmendinger R, Ramasamy SV, Lettieri P, Titchener-Hooker NJ. Life-cycle and cost of goods assessment of fed-batch and perfusion-based manufacturing processes for mAbs. Biotechnol Prog. 2016; 32(5): 1324–1335.

[89] Subramanian G, editor. Continuous processing in pharmaceutical manufacturing. Wiley; 2015.

[90] Croughan MS, Konstantinov KB, Cooney C. The future of industrial bioprocessing: Batch or continuous? Biotechnol Bioeng. 2015; 112(4): 648–651.

[91] Subramanian G. Continuous biomanufacturing: Innovative technologies and methods. John Wiley & Sons; 2017.

[92] Karst DJ, Steinebach F, Morbidelli M. Continuous integrated manufacturing of therapeutic proteins. Curr Opin Biotechnol. 2018; 53: 76–84.

[93] Sawyer D, Sanderson K, Lu R, Daszkowski T, Clark E, Mcduff P, et al. Biomanufacturing technology roadmap – Executive summary. BioPhorum Operations Group; 2017.

[94] Jayaraman K. India's cipla sets sights on Avastin, Herceptin and Enbrel. Nat Biotech. 2010; 28(9): 883–884.

[95] Kuhn TS. The structure of scientific revolutions. University of Chicago Press; 1962.

[96] Chatterjee S. FDA perspective on continuous manufacturing. IFPAC Annual Meeting, Baltimore; 2012.

[97] Allison G, Cain YT, Cooney C, Garcia T, Bizjak TG, Holte O, et al. Regulatory and quality considerations for continuous manufacturing. May 20–21, 2014 Continuous Manufacturing Symposium. J Pharm Sci. 2015; 104(3): 803–812.

[98] Levenspiel O. Chemical reaction engineering. Third edition. John Wiley & Sons, Inc.; 1999.

[99] Fogler HS. Elements of chemical reaction engineering. Fourth edition. Prentice Hall; 2005.

[100] Villiger TK, Roulet A, Perilleux A, Stettler M, Broly H, Morbidelli M, et al. Controlling the time evolution of mAb N-linked glycosylation, Part I: Microbioreactor experiments. Biotechnol Prog. 2016; 32(5): 1123–1134.

[101] Xu S, Chen H. High-density mammalian cell cultures in stirred-tank bioreactor without external pH control. J Biotechnol. 2016; 231: 149–159.

[102] Karst DJ, Steinhoff RF, Kopp MRG, Serra E, Soos M, Zenobi R, et al. Intracellular CHO cell metabolite profiling reveals steady-state dependent metabolic fingerprints in perfusion culture. Biotechnol Prog. 2017; 33(4): 879–890.

[103] Nicoud RM. From batch to continuous processes: A good answer, but what is the question? Part I – General considerations and the reaction unit perspective. Chemistry Today. 2016; 34(4): 38–43.

[104] Nicoud RM. From batch to continuous processes: A good answer, but what is the question? Part II – The purification unit operations, specific situations, the complete process perspective. 2016. 2016; 34(5): 33–38.

[105] Yang WC, Lu J, Kwiatkowski C, Yuan H, Kshirsagar R, Ryll T, et al. Perfusion seed cultures improve biopharmaceutical fed-batch production capacity and product quality. Biotechnol Prog. 2014; 30(3): 616–625.

[106] Hecht V, Duvar S, Ziehr H, Burg J, Jockwer A. Efficiency improvement of an antibody production process by increasing the inoculum density. Biotechnol Prog. 2014; 30(3): 607–615.

[107] Karst DJ, Serra E, Villiger TK, Soos M, Morbidelli M. Characterization and comparison of ATF and TFF in stirred bioreactors for continuous mammalian cell culture processes. Biochem Eng J. 2016; 110: 17–26.

[108] Hutter S, Villiger TK, Bruhlmann D, Stettler M, Broly H, Soos M, et al. Glycosylation flux analysis reveals dynamic changes of intracellular glycosylation flux distribution in Chinese hamster ovary fed-batch cultures. Metab Eng. 2017; 43(Pt A): 9–20.

[109] Karst DJ, Scibona E, Serra E, Bielser JM, Souquet J, Stettler M, et al. Modulation and modeling of monoclonal antibody N-linked glycosylation in mammalian cell perfusion reactors. Biotechnol Bioeng. 2017; 114(9): 1978–1990.

[110] Chrysanthopoulos PK, Goudar CT, Klapa MI. Metabolomics for high-resolution monitoring of the cellular physiological state in cell culture engineering. Metab Eng. 2010; 12(3): 212–222.

[111] Steinhoff RF, Karst DJ, Steinebach F, Kopp MR, Schmidt GW, Stettler A, et al. Microarray-based MALDI-TOF mass spectrometry enables monitoring of monoclonal antibody production in batch and perfusion cell cultures. Methods. 2016; 104: 33–40.

[112] Bertrand V. Intracellular profiling for biopharmaceutical cultivation processes. ETH Zurich; 2018.

[113] Vernardis SI, Goudar CT, Klapa MI. Metabolic profiling reveals that time related physiological changes in mammalian cell perfusion cultures are bioreactor scale independent. Metab Eng. 2013; 19: 1–9.

[114] Karst DJ, Steinhoff RF, Kopp MRG, Soos M, Zenobi R, Morbidelli M. Isotope labeling to determine the dynamics of metabolic response in CHO cell perfusion bioreactors using MALDI-TOF-MS. Biotechnol Prog. 2017; 33(6): 1630–1639.

[115] Konstantinov K, Goudar C, Ng M, Meneses R, Thrift J, Chuppa S, et al. The "push-to-low" approach for optimization of high-density perfusion cultures of animal cells. Adv Biochem Eng Biot. 2006; 101: 75–98.

[116] Vogel JH, Nguyen H, Pritschet M, Van Wegen R, Konstantinov K. Continuous annular chromatography: General characterization and application for the isolation of recombinant protein drugs. Biotechnol Bioeng. 2002; 80(5): 559–568.

[117] Hilbrig F, Freitag R. Continuous annular chromatography. J Chromatogr B. 2003; 790(12): 1–15.

[118] Uretschläger A, Jungbauer A. Preparative continuous annular chromatography (P-CAC), a review. Bioproc Biosyst Eng. 2002; 25(2): 129–140.

[119] Wolfgang J, Prior A. Continuous annular chromatography. Adv Biochem Eng Biotechnol. 2002; 76: 233–255.

[120] Bloomingburg GF, Carta G. Separation of protein mixtures by continuous annular chromatography with step elution. Chem Eng J Biochem Eng. 1994; 55(1): B19–B27.

[121] Brozio J, Bart HJ. A rigorous model for annular chromatography. Chem Eng Technol. 2004; 27(9): 962–970.

[122] Evans DH, Piermarini PM, Choe KP. The multifunctional fish gill: Dominant site of gas exchange, osmoregulation, acid-base regulation, and excretion of nitrogenous waste. Physiol Rev. 2004; 85(1): 97–177.

[123] Broughton DB, Gerhold CG, inventors; Continuous sorption process employing fixed bed of sorbent and moving inlets and outlets. US 2985589 A; 1961, filing date.

[124] Majozi T. Batch chemical process integration. Analysis, synthesis and optimization. Springer; 2010.

[125] El-Halwagi MM. Process integration. vol. 7. Elsevier; 2006.

[126] Smith R. Chemical process: Design and integration, Second edition. Wiley; 2016.

[127] Warikoo V, Godawat R, Brower K, Jain S, Cummings D, Simons E, et al. Integrated continuous production of recombinant therapeutic proteins. Biotechnol Bioeng. 2012; 109(12): 3018–3029.

[128] Godawat R, Konstantinov K, Rohani M, Warikoo V. End-to-end integrated fully continuous production of recombinant monoclonal antibodies. J Biotechnol. 2015; 213: 13–19.

[129] Steinebach F, Ulmer N, Wolf M, Decker L, Schneider V, Wälchli R, et al. Design and operation of a continuous integrated monoclonal antibody production process. Biotechnol Prog. 2017; 33(5): 1303–1313.

[130] Gomes J, Chopda VR, Rathore AS. Integrating systems analysis and control for implementing process analytical technology in bioprocess development. J Chem Technol Biotechnol. 2015; 90(4): 583–589.

[131] Jiang M, Severson KA, Love JC, Madden H, Swann P, Zang L, et al. Opportunities and challenges of real-time release testing in biopharmaceutical manufacturing. Biotechnol Bioeng. 2017; 114(11): 2445–2456.

[132] Rögner M. Size exclusion chromatography. In: Michael K, editor. Protein liquid chromatography. Volume 61 of Journal of chromatography library. Elsevier; 2000. p. 89–145.

[133] Carpenter JF, Randolph TW, Jiskoot W, Crommelin DJ, Middaugh CR, Winter G. Potential inaccurate quantitation and sizing of protein aggregates by size exclusion chromatography: Essential need to use orthogonal methods to assure the quality of therapeutic protein products. J Pharm Sci. 2010; 99(5): 2200–2208.

[134] Arakawa T, Ejima D, Li T, Philo JS. The critical role of mobile phase composition in size exclusion chromatography of protein pharmaceuticals. J Pharm Sci. 2010; 99(4): 1674–1692.

[135] Jacob LR. Hydrophobic interaction chromatography. In: Michael K, editor. Protein liquid chromatography. Volume 61 of Journal of chromatography library. Elsevier; 2000. p. 235–269.

[136] Schlüter H. Reversed-phase chromatography. In: Michael K, editor. Protein liquid chromatography. Volume 61 of Journal of chromatography library. Elsevier; 2000. p. 147–234.

[137] Helfferich FG. Ion exchange. Courier Corporation; 1962.

[138] Roos PH. Ion exchange chromatography. In: Michael K, editor. Protein liquid chromatography. Volume 61 of Journal of chromatography library. Elsevier; 2000. p. 3–88.

[139] Haddad PR, Jackson PE. Ion-exclusion chromatography. In: Haddad PR, Jackson PE, editors. Ion chromatography – Principles and applications. Volume 46 of Journal of chromatography library. Elsevier; 1990. p. 195–222.

[140] Fritz JS. Principles and applications of ion-exclusion chromatography. J Chromatogr A. 1991; 546: 111–118.

[141] Starkenstein E. Über Fermentwirkung und deren Beeinflussung durch Neutralsalze. Biochem Z. 1910; 24: 210.

[142] Cuatrecasas P, Wilchek M. Single-step purification of avidin from egg white by affinity chromatography on biocytin-Sepharose columns. Biochem Biophys Res Co. 1968; 33(2): 235–239.

[143] Grönwall C, Ståhl S. Engineered affinity proteins – Generation and applications. J Biotechnol. 2009; 140(34): 254–269.

[144] Kastner M. Immobilized metal ion affinity chromatography. In: Michael K, editor. Protein liquid chromatography. Volume 61 of Journal of chromatography library. Elsevier; 2000. p. 301–383.

[145] Cheung RCF, Wong JH, Ng TB. Immobilized metal ion affinity chromatography: A review on its applications. Applied Microbiol Biot. 2012; 96(6): 1411–1420.

[146] Block H, Maertens B, Spriestersbach A, Brinker N, Kubicek J, Fabis R, et al. Immobilized-metal affinity chromatography (IMAC): A review. Methods in enzymology. 2009; 463: 439–473.

[147] Feldman PA, Bradbury PI, Williams JD, Sims GE, McPhee JW, Pinnell MA, et al. Large-scale preparation and biochemical characterization of a new high purity factor IX concentrate prepared by metal chelate affinity chromatography. Blood Coagul Fibrinolysis. 1994; 5(6): 939–948.

[148] Barros de Genaro AC, Tamagawa RE, Azzoni AR, Alves Bueno SM, Miranda EA. Recovery and purification of aprotinin from industrial insulin-processing effluent by immobilized chymotrypsin and negative IMAC chromatographies. Process Biochem. 2002; 37(12): 1413–1420.

[149] Yang Y, Geng X. Mixed-mode chromatography and its applications to biopolymers. J Chromatogr A. 2011; 1218(49): 8813–8825.

[150] Zhang K, Liu X. Mixed-mode chromatography in pharmaceutical and biopharmaceutical applications. J Pharmaceut Biomed. 2016; 128: 73–88.

[151] Zhao G, Dong XY, Sun Y. Ligands for mixed-mode protein chromatography: Principles, characteristics and design. J Biotechnol. 2009; 144(1): 3–11.

[152] Khalaf R, Forrer N, Buffolino G, Butt A, Morbidelli M. Model-based description of peptide retention on doped reversed-phase media. J Chromatogr A. 2015; 1407: 169–175.

[153] Deppert WR, Lukačin R. Hydroxyapatite chromatography. In: Michael K, editor. Protein liquid chromatography. Volume 61 of Journal of chromatography library. Elsevier; 2000. p. 271–299.

[154] Cummings LJ, Snyder MA, Brisack K. Protein chromatography on hydroxyapatite columns. In: Richard RB, Murray PD, editors. Methods in enzymology. Volume 463. Academic Press; 2009. p. 387–404.

[155] Kaleas KA, Schmelzer CH, Pizarro SA. Industrial case study: Evaluation of a mixed-mode resin for selective capture of a human growth factor recombinantly expressed in *E. coli*. J Chromatogr A. 2010; 1217(2): 235–242.

[156] Pezzini J, Joucla G, Gantier R, Toueille M, Lomenech AM, Le Sénéchal C, et al. Antibody capture by mixed-mode chromatography: A comprehensive study from determination of optimal purification conditions to identification of contaminating host cell proteins. J Chromatogr A. 2011; 1218(45): 8197–8208.

[157] Hilbrig F, Freitag R. Isolation and purification of recombinant proteins, antibodies and plasmid DNA with hydroxyapatite chromatography. Biotechnol J. 2012; 7(1): 90–102.

[158] Nicoud RM. Chromatographic processes: Modeling, simulation and design. Cambridge University Press; 2015.

[159] Schmidt-Traub H, Schulte M, Seidel-Morgenstern A. Preparative chromatography. John Wiley & Sons; 2012.

[160] Guiochon G, Felinger A, Shirazi DG, Katti A. Fundamentals of preparative and nonlinear chromatography. Second edition. Elsevier; 2006.

[161] Ruthven DM. Principles of adsorption and adsorption processes. John Wiley & Sons; 1984.

[162] Gu T, Tsai GJ, Tsao G. Modeling of nonlinear multicomponent chromatography. Adv Biochem Eng Biot. 1993; 49: 45–71.

[163] DeGroot MH. A conversation with George Box. Stat Sci. 1987; 2(3): 239–258.

[164] Coquebert de Neuville B, Tarafder A, Morbidelli M. Distributed pore model for bio-molecule chromatography. J Chromatogr A. 2013; 1298: 26–34.

[165] Stickel JJ, Fotopoulos A. Pressure-flow relationships for packed beds of compressible chromatography media at laboratory and production Scale. Biotechnol Prog. 2001; 17(4): 744–751.

[166] Flodin P. Methodological aspects of gel filtration with special reference to desalting operations. J Chromatogr A. 1961; 5(2): 103–115.

[167] Hagel L. Gel filtration: Size exclusion chromatography. In: Janson JC, editor. Protein purification: Principles, high resolution methods, and applications. Third edition. John Wiley & Sons, Inc.; 2011. p. 51–91.

[168] DePhillips P, Lenhoff AM. Pore size distributions of cation-exchange adsorbents determined by inverse size-exclusion chromatography. J Chromatogr A. 2000; 883(1–2): 39–54.

[169] Kuga S. Pore-size distribution analyis of gel substances by size exclusion chromatography. J Chromatogr A. 1981; 206(3): 449–461.

[170] Giddings JC, Kucera E, Russell CP, Myers MN. Statistical theory for the equilibrium distribution of rigid molecules in inert porous networks. Exclusion chromatography. J Phys Chem. 1968; 72(13): 4397–4408.

[171] Knox JH, Scott HP. Theoretical-models for size-exclusion chromatography and calculation of pore-size distribution from size exclusion chromatography data. J Chromatogr A. 1984; 316(DEC): 311–332.

[172] Kvaalen E, Tondeur D. Constraints on phase equilibrium equations. Chem Eng Sci. 1988; 43(4): 803–810.

[173] Kopaciewicz W, Rounds MA, Fausnaugh J, Regnier FE. Second international symposium on high-performance liquid chromatography of proteins, peptides and polynucleotides retention model for high-performance ion-exchange chromatography. J Chromatogr A. 1983; 266: 3–21.

[174] Brooks CA, Cramer SM. Steric mass-action ion exchange – Displacement profiles and induced salt gradients. AIChe J. 1992; 38(12): 1969–1978.

[175] Hunter AK, Carta G. Protein adsorption on novel acrylamido-based polymeric ion-exchangers I. Morphology and equilibrium adsorption. J Chromatogr A. 2000; 897(1–2): 65–80.

[176] Mollerup JM. The thermodynamic principles of ligand binding in chromatography and biology. J Biotechnol. 2007; 132(2): 187–195.

[177] Mollerup JM. A review of the thermodynamics of protein association to ligands, protein adsorption, and adsorption isotherms. Chem Eng Technol. 2008; 31(6): 864–874.

[178] Poole CF, Schuette SA. Contemporary practice of chromatography. Elsevier; 1984.

[179] Melander W, Horvath C. Salt effects on hydrophobic interactions in precipitation and chromatography of proteins – Interpretation of lyotropic series. Arch Biochem Biophys. 1977; 183(1): 200–215.

[180] Danckwerts PV. Continuous flow systems. Chem Eng Sci. 1953; 2(1): 1–13.

[181] Martin AJP, Synge RLM. A new form of chromatogram employing two liquid phases: A theory of chromatography. 2. Application to the micro-determination of the higher monoamino-acids in proteins. Biochem J. 1941; 35(12): 1358–1368.

[182] Kozeny J. Uber kapillare Leitung der Wasser in Boden. Sitzungsber Akad Wiss Wien. 1927; 136: 271–306.

[183] Carman PC. Fluid flow through granular beds. Trans Ins Chem Eng. 1937; 15: 150–167.

[184] Taylor R, Krishna R. Multicomponent mass transfer. John Wiley & Sons, Inc; 1993.

[185] Krishna R, Wesselingh J. The Maxwell–Stefan approach to mass transfer. Chem Eng Sci. 1997; 52(6): 861–911.

[186] Wakao N, Funazkri T. Effect of fluid dispersion coefficients on particle-to-fluid mass transfer coefficients in packed beds. Chem Eng Sci. 1978; 33(10): 1375–1384.

[187] Carberry JJ. A boundary-layer model of fluid-particle mass transfer in fixed beds. AIChE J. 1960; 6(3): 460–463.

[188] Wilson EJ, Geankoplis CJ. Liquid mass transfer at very low Reynolds numbers in packed beds. Ind Eng Chem Fund. 1966; 5(1): 9–14.

[189] Einstein A. Über die von der molekularkinetischen Theorie der Wärme geforderte Bewegung von in ruhenden Flüssigkeiten suspendierten Teilchen. Ann d Physik. 1905; 322(8): 549–560.

[190] Polson A. The some aspects of diffusion in solution and a definition of a colloidal particle. J Phys Colloid Chem. 1950; 54(5): 649–652.

[191] Wang CK, Northfield SE, Swedberg JE, Harvey PJ, Mathiowetz AM, Price DA, et al. Translational diffusion of cyclic peptides measured using pulsed-field gradient NMR. J Phys Chem B. 2014; 118(38): 11129–11136.

[192] García De La Torre J, Huertas ML, Carrasco B. Calculation of hydrodynamic properties of globular proteins from their atomic-level structure. Biophys J. 2000; 78(2): 719–730.

[193] Flury M, Gimmi TF. Solute diffusion. In: Dane HD, Topp CG, editors. Methods of soil analysis: Part 4 – physical methods. vol. 5 of Soil science society of America book series. Soil Science Society of America; 2002. p. 1323–1351.

[194] Cunningham AB, Lennox JE, Ross RJ. Biofilm: The hypertextbook. National Science Foundation; 2011.

[195] van Deemter JJ, Zuiderweg FJ, Klinkenberg A. Longitudinal diffusion and resistance to mass transfer as causes of nonideality in chromatography. Chem Eng Sci. 1956; 5(6): 271–289.

[196] Carta G, Ubiera AR, Pabst TM. Protein mass transfer kinetics in ion exchange media: Measurements and interpretations. Chem Eng Technol. 2005; 28(11): 1252–1264.

[197] Carta G, Rodrigues AE. Diffusion and convection in chromatographic processes using permeable supports with a bidisperse pore structure. Chem Eng Sci. 1993; 48(23): 3927–3935.

[198] Rodrigues AE. Permeable packings and perfusion chromatography in protein separation. J Chromatogr B Biomed Sci Appl. 1997; 699(1–2): 47–61.

[199] Wu Y, Abraham D, Carta G. Particle size effects on protein and virus-like particle adsorption on perfusion chromatography media. J Chromatogr A. 2015; 1375: 92–100.

[200] Wu Y, Simons J, Hooson S, Abraham D, Carta G. Protein and virus-like particle adsorption on perfusion chromatography media. J Chromatogr A. 2013; 1297: 96–105.

[201] Coquebert de Neuville B, Lamprou A, Morbidelli M, Soos M. Perfusive ion-exchange chromatographic materials with high capacity. J Chromatogr A. 2014; 1374: 180–188.

[202] Hahn R, Bauerhansl P, Shimahara K, Wizniewski C, Tscheliessnig A, Jungbauer A. Comparison of protein A affinity sorbents: II. Mass transfer properties. J Chromatogr A. 2005; 1093(12): 98–110.

[203] Nfor BK, Noverraz M, Chilamkurthi S, Verhaert PDEM, van der Wielen LAM, Ottens M. High-throughput isotherm determination and thermodynamic modeling of protein adsorption on mixed mode adsorbents. J Chromatogr A. 2010; 1217(44): 6829–6850.

[204] Pakiman N, Isa NH, Abol Hassan MA, Walter JK, Abdullah N. Comparison of binding capacity and affinity of monoclonal antibody towards different affinity resins using high-throughput chromatography method. J Appl Sci. 2012; 12(11): 1136.

[205] Baur D, Angarita M, Müller-Späth T, Morbidelli M. Optimal model-based design of the twin-column CaptureSMB process improves capacity utilization and productivity in protein A affinity capture. Biotechnol J. 2016; 11(1): 135–145.

[206] Melter L, Ströhlein G, Butté A, Morbidelli M. Adsorption of monoclonal antibody variants on analytical cation-exchange resin. J Chromatogr A. 2007; 1154(12): 121–131.

[207] Yamamoto S, Nakanishi K, Matsuno R. Ion-exchange chromatography of proteins. vol. 43 of Chromatographic Science; 1988.

[208] Ishihara T, Kadoya T, Yamamoto S. Application of a chromatography model with linear gradient elution experimental data to the rapid scale-up in ion-exchange process chromatography of proteins. J Chromatogr A. 2007; 1162(1): 34–40.

[209] Pfister D, Steinebach F, Morbidelli M. Linear isotherm determination from linear gradient elution experiments. J Chromatogr A. 2015; 1375: 33–41.

[210] Vajda J, Conze W, Muller E. Kinetic plots in aqueous size exclusion chromatography of monoclonal antibodies and virus particles. J Chromatogr A. 2015; 1426: 118–125.

[211] Schmidt-Nielsen K. Countercurrent systems in animals. Sci Am. 1981; 244(5): 118–129.

[212] Gottschalk CW, Mylle M. Micropuncture study of the mammalian urinary concentrating mechanism: Evidence for the countercurrent hypothesis. Am J Physiol-Legacy Content. 1959; 196(4): 927–936.

[213] Midtgård U. The rete tibiotarsale and arteriovenous association in the hind limb of birds: A compartive morphological study on countercurrent heat exchange systems. Acta Zool. 1981; 62(2): 67–87.

[214] Taylor CR, Lyman CP. Heat storage in running antelopes: Independence of brain and body temperatures. Am J Physiol-Legacy Content. 1972; 222(1): 114–117.

[215] Biegler LT, Grossman IE, Westerberg AW. Systematic methods of chemical process design. Prentice Hall; 1997.

[216] Ruthven DM, Ching CB. Counter-current and simulated counter-current adsorption separation processes. Chem Eng Sci. 1989; 44(5): 1011–1038.

[217] Migliorini C, Mazzotti M, Morbidelli M. Continuous chromatographic separation through simulated moving beds under linear and nonlinear conditions. J Chromatogr A. 1998; 827(2): 161–173.

[218] Mazzotti M, Storti G, Morbidelli M. Optimal operation of simulated moving bed units for nonlinear chromatographic separations. J Chromatogr A. 1997; 769(1): 3–24.

[219] Berg C. The hypersorption process for gas fractionation. The Hague; 1951. Third World Petroleum Congress.

[220] Rodrigues AE. Simulated moving bed technology. Principles, design and process applications. Butterworth-Heinemann; 2015.

[221] Mazzotti M. Equilibrium theory based design of simulated moving bed processes for a generalized Langmuir isotherm. J Chromatogr A. 2006; 1126(1–2): 311–322.

[222] Mazzotti M. Design of simulated moving bed separations: Generalized Langmuir isotherm. Ind Eng Chem Res. 2006; 45(18): 6311–6324.

[223] Ma Z, Wang NHL. Standing wave analysis of SMB chromatography: Linear systems. AIChE J. 1997; 43(10): 2488–2508.

[224] Xie Y, Wu D, Ma Z, Wang NHL. Extended standing wave design method for simulated moving bed chromatography: Linear systems. Ind Eng Chem Res. 2000; 39(6): 1993–2005.

[225] Mallmann T, Burris BD, Ma Z, Wang NHL. Standing wave design of nonlinear SMB systems for fructose purification. AIChE J. 1998; 44(12): 2628–2646.

[226] Siitonen J, Sainio T. Unified design of chromatographic separation processes. Chem Eng Sci. 2015; 122: 436–451.

[227] Charton F, Nicoud RM. Complete design of a simulated moving bed. J Chromatogr A. 1995; 702(1): 97–112.

[228] Biressi G, Ludemann-Hombourger O, Mazzotti M, Nicoud RM, Morbidelli M. Design and optimisation of a simulated moving bed unit: Role of deviations from equilibrium theory. J Chromatogr A. 2000; 876(1): 3–15.

[229] David L, Yun J, Nicoud RM. Comparing multi-column chromatographic processes for purifying monosaccharides part I: A simplified approach. Adsorption. 2017; p. 577–591.

[230] Shukla AA, Hinckley P. Host cell protein clearance during protein A chromatography: Development of an improved column wash step. Biotechnol Prog. 2008; 24(5): 1115–1121.

[231] Nogal B, Chhiba K, Emery JC. Select host cell proteins coelute with monoclonal antibodies in protein A chromatography. Biotechnol Prog. 2012; 28(2): 454–458.

[232] Hahn R, Schlegel R, Jungbauer A. Comparison of protein A affinity sorbents. J Chromatogr B. 2003; 790(1): 35–51.

[233] Perez-Almodovar EX, Carta G. IgG adsorption on a new protein A adsorbent based on macroporous hydrophilic polymers. I. Adsorption equilibrium and kinetics. J Chromatogr A. 2009; 1216(47): 8339–8347.

[234] Urh M, Simpson D, Zhao K. Affinity chromatography: General methods. In: Richard RB, Murray PD, editors. Methods in enzymology. Volume 463. Academic Press; 2009. p. 417–438.

[235] Wrzosek K, Gramblička M, Tóthová D, Antošová M, Polakovič M. Impact of ionic strength on adsorption capacity of chromatographic particles employed in separation of monoclonal antibodies. Chem Pap. 2010; 64(4): 461–468.

[236] Gramblička M, Tóthová D, Antošová M, Polakovič M. Influence of pH on adsorption of human immunoglobulin gamma, human serum albumin and horse myoglobin by commercial chromatographic materials designed for downstream processing of monoclonal antibodies. Acta Chimica Slovaca. 2008; 1(1): 85–94.

[237] Shalliker RA, Catchpoole HJ, Dennis GR, Guiochon G. Visualising viscous fingering in chromatography columns: High viscosity solute plug. J Chromatogr A. 2007; 1142(1): 48–55.

[238] Stone MC, Carta G. Patterns of protein adsorption in chromatographic particles visualized by optical microscopy. J Chromatogr A. 2007; 1160(1–2): 206–214.

[239] Tao Y, Carta G. Rapid monoclonal antibody adsorption on dextran-grafted agarose media for ion-exchange chromatography. J Chromatogr A. 2008; 1211(1–2): 70–79.

[240] Zhu M, Carta G. Protein adsorption equilibrium and kinetics in multimodal cation exchange resins. Adsorption. 2016; 22(2): 165–179.

[241] Spahn H, Schlünder EU. The scale-up of activated carbon columns for water purification, based on results from batch tests – I: Theoretical and experimental determination of adsorption rates of single organic solutes in batch tests. Chem Eng Sci. 1975; 30(5): 529–537.

[242] Steinebach F, Angarita M, Karst DJ, Müller-Späth T, Morbidelli M. Model based adaptive control of a continuous capture process for monoclonal antibodies production. J Chromatogr A. 2016; 1444: 50–56.

[243] Angarita M, Müller-Späth T, Baur D, Lievrouw R, Lissens G, Morbidelli M. Twin-column CaptureSMB: A novel cyclic process for protein A affinity chromatography. J Chromatogr A. 2015; 1389: 85–95.

[244] Gottschlich N, Kasche V. Purification of monoclonal antibodies by simulated moving-bed chromatography. J Chromatogr A. 1997; 765(2): 201–206.

[245] Bryntesson M, Hall M, Lacki K, inventors; Chromatography method. WO 2008153472 A1; 2008, filing date.

[246] Mierendorf R, Paaren H, Grabski A, Wilke A, Baier J, Oroskar A, inventors; Continuous Isocratic Affinity Chromatography. US20080053901 A1; 2007, filing date.

[247] Müller-Späth T, Aumann L, Ströhlein G, Bavand M, inventors; Method for identification and purification of multi-specific polypeptides. EP 2500073 A1; 2011, filing date.

[248] Ransohoff TC, Bisschops MAT, inventors; Continuous processing methods for biological products. WO 2012078677 A3; 2011, filing date.

[249] Valery E, Morey C, inventors; Method and device for separating fractions of a mixture. WO 2007101944 A2; 2007, filing date.

[250] Gjoka X, Rogler K, Martino RA, Gantier R, Schofield M. A straightforward methodology for designing continuous monoclonal antibody capture multi-column chromatography processes. J Chromatogr A. 2015; 1416: 38–46.

[251] Pfister D, David L, Holzer M, Nicoud RM. Designing affinity chromatographic processes for the capture of antibodies. Part I: A simplified approach. J Chromatogr A. 2017; 1494: 27–39.

[252] Angelo J, Pagano J, Müller-Späth T, Mihlbacher K, Chollangi S, Xu X, et al. Scale-up of twin-column periodic counter-current chromatography for mAb purification. BioProcess International. 2018; 16(4).

[253] Baur D, Angarita M, Müller-Späth T, Steinebach F, Morbidelli M. Comparison of batch and continuous multicolumn protein A capture processes by optimal design. Biotechnol J. 2016; 11: 920–931.

[254] Ng CK, Osuna-Sanchez H, Valéry E, Sørensen E, Bracewell DG. Design of high productivity antibody capture by protein A chromatography using an integrated experimental and modeling approach. J Chromatogr B. 2012; 899: 116–126.

[255] Ng CK, Rousset F, Valéry E, Bracewell DG, Sørensen E. Design of high productivity sequential multi-column chromatography for antibody capture. Food Bioprod Process. 2014; 92(2): 233–241.

[256] Corriou JP. Process control: Theory and applications. Springer; 2004.

[257] Bequette BW. Process control: Modeling, design and simulation. Prentice Hall; 2003.

[258] Engell S, Toumi A. Optimisation and control of chromatography. Comp Chem Eng. 2005; 29(6): 1243–1252.

[259] Morari M, Zafiriou E. Robust process control. Prentice Hall; 1989.

[260] Borrelli F, Bemporad A, Morari M. Predictive control for linear and hybrid systems. Cambridge University Press; 2017.

[261] Karst DJ, Steinebach F, Soos M, Morbidelli M. Process performance and product quality in an integrated continuous antibody production process. Biotechnol Bioeng. 2017; 114(2): 298–307.

[262] Mathiasson L, Åkerblom A, Skoglar H, Blom H, Łącki K. The use of dynamic control in periodic counter-current chromatography. GE Healthcare Life Sciences; 2015.

[263] Chromacon. Contichrom Twin-column FPLC Chromatography. Process Control and Optimization by AutomAb; 2016.

[264] Müller-Späth T, Ulmer N, Aumann L, Ströhlein G, Bavand M. Automated process development and control of a twin-column counter-current process (CaptureSMB) for affinity capture. Prep Symposium. Boston; 2014.

[265] Gottschalk U. Process scale purification of antibodies. John Wiley & Sons; 2011.

[266] Kelley BD, Tobler SA, Brown P, Coffman JL, Godavarti R, Iskra T, et al. Weak partitioning chromatography for anion exchange purification of monoclonal antibodies. Biotechnol Bioeng. 2008; 101(3): 553–566.

[267] Fahrner RL, Lester PM, Blank GS, Reifsnyder DH. Non-flammable preparative reversed-phase liquid chromatography of recombinant human insulin-like growth factor-I. J Chromatogr A. 1999; 830(1): 127–134.

[268] Garke G, Hartmann R, Papamichael N, Deckwer WD, Anspach F. The influence of protein size on adsorption kinetics and equilibria in ion-exchange chromatography. Sep Sci Technol. 1999; 34(13): 2521–2538.

[269] Walton KS, Sholl DS. Predicting multicomponent adsorption: 50 years of the ideal adsorbed solution theory. AIChE J. 2015; 61(9): 2757–2762.

[270] Myers A, Prausnitz JM. Thermodynamics of mixed-gas adsorption. AIChE J. 1965; 11(1): 121–127.

[271] Radke CJ, Prausnitz JM. Thermodynamics of multi-solute adsorption from dilute liquid solutions. AIChE J. 1972; 18(4): 761–768.

[272] Guélat B, Ströhlein G, Lattuada M, Morbidelli M. Electrostatic model for protein adsorption in ion-exchange chromatography and application to monoclonal antibodies, lysozyme and chymotrypsinogen A. J Chromatogr A. 2010; 1217(35): 5610–5621.

[273] Guélat B, Khalaf R, Lattuada M, Costioli M, Morbidelli M. Protein adsorption on ion exchange resins and monoclonal antibody charge variant modulation. J Chromatogr A. 2016; 1447: 82–91.

[274] Guélat B, Ströhlein G, Lattuada M, Delegrange L, Valax P, Morbidelli M. Simulation model for overloaded monoclonal antibody variants separations in ion-exchange chromatography. J Chromatogr A. 2012; 1253: 32–43.

[275] Pfister D, Morbidelli M, Nicoud RM. A continuum theory for multicomponent chromatography modeling. J Chromatogr A. 2016; 1446: 50–58.

[276] Carta G, Lewus RK. Film model approximation for multicomponent adsorption. Adsorption. 2000; 6(1): 5–13.

[277] Tao Y, Chen N, Carta G, Ferreira G, Robbins D. Modeling multicomponent adsorption of monoclonal antibody charge variants in cation exchange columns. AIChE J. 2012; 58(8): 2503–2511.

[278] Forrer N, Butte A, Morbidelli M. Chromatographic behavior of a polyclonal antibody mixture on a strong cation exchanger column. Part I: Adsorption characterization. J Chromatogr A. 2008; 1214(1–2): 59–70.

[279] Forrer N, Butte A, Morbidelli M. Chromatographic behavior of a polyclonal antibody mixture on a strong cation exchanger column. Part II: Adsorption modelling. J Chromatogr A. 2008; 1214(1–2): 71–80.

[280] Nicoud RM, Clavier JY, Perrut M. Preparative SFC: Basics and applications. In: Caude M, Thiébaut D, editors. Practical supercritical fluid chromatography and extraction. vol. 2; 1999. p. 397.

[281] Mazzotti M, Storti G, Morbidelli M. Supercritical fluid simulated moving bed chromatography. J Chromatogr A. 1997; 786(2): 309–320.

[282] Jensen TB, Billiet HAH, Van DWLAM, inventors; Method of substantially continuously separating two compounds using a moving bed or a simulated moving bed. WO2000033934 A1; 1999, filing date.

[283] Abel S, Mazzotti M, Morbidelli M. Solvent gradient operation of simulated moving beds: II. Langmuir isotherms. J Chromatogr B. 2004; 1026(1): 47–55.

[284] Keßler LC, Gueorguieva L, Rinas U, Seidel-Morgenstern A. Step gradients in 3-zone simulated moving bed chromatography: Application to the purification of antibodies and bone morphogenetic protein-2. J Chromatogr A. 2007; 1176(12): 69–78.

[285] Martínez Cristancho CA, Seidel-Morgenstern A. Purification of single-chain antibody fragments exploiting pH-gradients in simulated moving bed chromatography. J Chromatogr A. 2016; 1434: 29–38.

[286] Song SM, Kim IH. A three-zone simulated moving-bed for separation of immunoglobulin Y. Korean J Chem Eng. 2013; 30(8): 1527–1532.

[287] Li P, Xiu G, Rodrigues AE. Proteins separation and purification by salt gradient ion-exchange SMB. AIChE J. 2007; 53(9): 2419–2431.

[288] Seidel-Morgenstern A, Keßler LC, Kaspereit M. New developments in simulated moving bed chromatography. Chem Eng Technol. 2008; 31(6): 826–837.

[289] Agrawal G, Kawajiri Y. Comparison of various ternary simulated moving bed separation schemes by multi-objective optimization. J Chromatogr A. 2012; 1238: 105–113.

[290] Nicolaos A, Muhr L, Gotteland P, Nicoud RM, Bailly M. Application of equilibrium theory to ternary moving bed configurations (four+four, five+four, eight and nine zones): I. Linear case. J Chromatogr A. 2001; 908(12): 71–86.

[291] Xie Y, Mun S, Chin CY, Wang NHL. Simulated moving bed technologies for producing high purity biochemicals and pharmaceuticals. In: Hwang NHC, Woo SLY, editors. Frontiers in Biomedical Engineering: Proceedings of the World Congress for Chinese Biomedical Engineers. Springer US; 2003. p. 507–527.

[292] Xie Y, Mun S, Kim J, Wang NHL. Standing wave design and experimental validation of a tandem simulated moving bed process for insulin purification. Biotechnol Prog. 2002; 18(6): 1332–1344.

[293] Voigt U, Kinkel J, Hempel R, Nicoud RM, inventors; Chromatographic process for obtaining highly purified cyclosporin A and related cyclosporins. US006306306B1; 1997, filing date.

[294] Hotier G, Toussaint JM, Terneuil G, Lonchamp D, inventors; Continuous process and apparatus for chromatographic separation of a mixture of at least three constituents into three purified effluents using two solvents. US005093004A; 1990, filing date.

[295] Abel S, Bäbler MU, Arpagaus C, Mazzotti M, Stadler J. Two-fraction and three-fraction continuous simulated moving bed separation of nucleosides. J Chromatogr A. 2004; 1043(2): 201–210.

[296] Paredes G, Abel S, Mazzotti M, Morbidelli M, Stadler J. Analysis of a simulated moving bed operation for three-fraction separations (3F-SMB). Ind Eng Chem Res. 2004; 43(19): 6157–6167.

[297] Wei F, Li M, Huang F, Chen M, Jiang H, Zhao Y. A novel pseudo simulated moving bed with solvent gradient for ternary separations. J Chromatogr A. 2011; 1218(20): 2906–2911.

[298] Wei F, Shen B, Chen M, Zhao Y. Study on a pseudo-simulated moving bed with solvent gradient for ternary separations. J Chromatogr A. 2012; 1225: 99–106.

[299] Aumann L, Morbidelli M. A continuous multi-column counter-current solvent gradient purification (MCSGP) process. Biotechnol Bioeng. 2007; 98(5): 1043–1055.

[300] Aumann L, Stroehlein G, Morbidelli M. Parametric study of a 6-column counter-current solvent gradient purification (MCSGP) unit. Biotechnol Bioeng. 2007; 98(5): 1029–1042.

[301] Müller-Späth T, Aumann L, Morbidelli M. Role of cleaning-in-place in the purification of mAb supernatants using continuous cation exchange chromatography. Sep Sci Technol. 2009; 44(1): 1–26.

[302] Krättli M, Müller-Späth T, Morbidelli M. Multifraction separation in counter-current chromatography (MCSGP). Biotechnol Bioeng. 2013; 110(9): 2436–2444.

[303] Silva RJS, Rodrigues RCR, Osuna-Sanchez H, Bailly M, Valéry E, Mota JPB. A new multi-column, open-loop process for center-cut separation by solvent-gradient chromatography. J Chromatogr A. 2010; 1217(52): 8257–8269.

[304] Müller-Späth T, Ströhlein G, Aumann L, Kornmann H, Valax P, Delegrange L, et al. Model simulation and experimental verification of a cation-exchange IgG capture step in batch and continuous chromatography. J Chromatogr A. 2011; 1218(31): 5195–5204.

[305] Ströhlein G, Aumann L, Mazzotti M, Morbidelli M. A continuous, counter-current multi-column chromatographic process incorporating modifier gradients for ternary separations. J Chromatogr A. 2006; 1126(1): 338–346.

[306] Steinebach F, Krättli M, Storti G, Morbidelli M. Equilibrium theory based design space for the multicolumn countercurrent solvent gradient purification process. Indus Eng Chem Res. 2017; 56(45): 13482–13489.

[307] Aumann L, Morbidelli M. A semicontinuous 3-column counter-current solvent gradient purification (MCSGP) process. Biotechnol Bioeng. 2008; 99(3): 728–733.
[308] Steinebach F, Ulmer N, Decker L, Aumann L, Morbidelli M. Experimental design of a twin-column counter-current gradient purification process. J Chromatogr A. 2017; 1492: 19–26.
[309] Müller-Späth T, Krättli M, Aumann L, Ströhlein G, Morbidelli M. Increasing the activity of monoclonal antibody therapeutics by continuous chromatography (MCSGP). Biotechnol Bioeng. 2010; 107(4): 652–662.
[310] Müller-Späth T, Aumann L, Melter L, Ströhlein G, Morbidelli M. Chromatographic separation of three monoclonal antibody variants using multi-column counter-current solvent gradient purification (MCSGP). Biotechnol Bioeng. 2008; 100(6): 1166–1177.
[311] Müller-Späth T, Ströhlein G, Lyngberg O, Maclean D. Enabling high purities and yields in therapeutic peptide purification using multicolumn countercurrent solvent gradient purification. Chim Oggi Chem Today. 2013; 31(5): 56–60.
[312] Krättli M, Steinebach F, Morbidelli M. Online control of the twin-column counter-current solvent gradient process for biochromatography. J Chromatogr A. 2013; 1293: 51–59.
[313] Hendriks LJA, de Kruif J, Throsby M, Bakker ABH, Müller-Späth T, Ulmer N, et al. Purifying common light-chain bispecific antibodies. Bioprocess Int. 2013; 11(5): 36–45.
[314] Müller-Späth T, Aumann L, Ströhlein G, Kornmann H, Valax P, Delegrange L, et al. Two step capture and purification of IgG2 using multi-column counter-current solvent gradient purification (MCSGP). Biotechnol Bioeng. 2010; 107(6): 974–984.
[315] Ulmer N, Steinebach F, Morbidelli M. Short course on continuous chromatography for biotherapeutics. Labcourse on the continuous purification of PEGylated lysozyme. ETH Zurich; 2015.
[316] Erdem G, Abel S, Morari M, Mazzotti M, Morbidelli M, Lee JH. Automatic control of simulated moving beds. Indus Eng Chem Res. 2004; 43(2): 405–421.
[317] Erdem G, Abel S, Morari M, Mazzotti M, Morbidelli M. Automatic control of simulated moving beds II: Nonlinear isotherm. Indus Eng Chem Res. 2004; 43(14): 3895–3907.
[318] Grossmann C, Ströhlein G, Morari M, Morbidelli M. Optimizing model predictive control of the chromatographic multi-column solvent gradient purification (MCSGP) process. J Process Contr. 2010; 20(5): 618–629.
[319] Papathanasiou MM, Avraamidou S, Oberdieck R, Mantalaris A, Steinebach F, Morbidelli M, et al. Advanced control strategies for the multi-column counter-current solvent gradient purification process. AIChE J. 2016; 62(7): 2341–2357.
[320] Birch JR, Onakunle Y. Biopharmaceutical proteins. vol. 308 of Therapeutic proteins. Humana Press; 2005.
[321] Dembowsky K, Stadler P. Novel therapeutic proteins: Selected case studies. Wiley-VCH; 2001.
[322] Farkaš P, Bystrický S. Chemical conjugation of biomacromolecules: A mini-review. Chemical Pap. 2010; 64(6): 683–695.
[323] Shu JY, Panganiban B, Xu T. Peptide-polymer conjugates: From fundamental science to application. Annu Rev Phys Chem. 2013; 64: 631–657.
[324] Wong SS. Chemistry of protein conjugation and cross-linking. CRC Press; 1991.
[325] Kalia J, Raines RT. Advances in bioconjugation. Current Org Chem. 2010; 14(2): 138–147.
[326] Dean KM, Palmer AE. Advances in fluorescence labeling strategies for dynamic cellular imaging. Nat Chem Biol. 2014; 10(7): 512–523.
[327] Hermanson GT. Bioconjugate techniques. Third edition. Academic Press; 2013.

[328] Fee CJ. Protein conjugates purification and characterization. In: PEGylated protein drugs: Basic science and clinical applications. Springer; 2009. p. 113–125.

[329] Pfister D, Ulmer N, Klaue A, Ingold O, Morbidelli M. Modeling the kinetics of protein conjugation reactions. Chem Ing Tech. 2016; 88(11): 1598–1608.

[330] Perez HL, Cardarelli PM, Deshpande S, Gangwar S, Schroeder GM, Vite GD, et al. Antibody-drug conjugates: Current status and future directions. Drug Discov Today. 2014; 19(7): 869–881.

[331] Dozier JK, Distefano MD. Site-specific PEGylation of therapeutic proteins. Int J Mol Sci. 2015; 16(10): 25831–25864.

[332] Schumacher D, Hackenberger CPR. More than add-on: Chemoselective reactions for the synthesis of functional peptides and proteins. Curr Opin Chem Biol. 2014; 22: 62–69.

[333] Veronese FM, Mero A, Pasut G. Protein PEGylation, basic science and biological applications. In: PEGylated Protein Drugs: Basic Science and Clinical Applications. Springer; 2009. p. 11–31.

[334] Brocchini S, Balan S, Godwin A, Choi JW, Zloh M, Shaunak S. PEGylation of native disulfide bonds in proteins. Nature Protoc. 2006; 1(5): 2241–2252.

[335] Brocchini S, Godwin A, Balan S, Choi JW, Zloh M, Shaunak S. Disulfide bridge based PEGylation of proteins. Adv Drug Deliver Rev. 2008; 60(1): 3–12.

[336] Fontana A, Spolaore B, Mero A, Veronese FM. Site-specific modification and PEGylation of pharmaceutical proteins mediated by transglutaminase. Adv Drug Deliver Rev. 2008; 60(1): 13–28.

[337] Pfister D, Bourgeaux E, Morbidelli M. Kinetic modeling of protein PEGylation. Chem Eng Sci. 2015; 137: 816–827.

[338] Fee CJ, Van Alstine JM. Prediction of the viscosity radius and the size exclusion chromatography behavior of PEGylated proteins. Bioconjugate Chem. 2004; 15(6): 1304–1313.

[339] Birdsall RE, Shion H, Kotch FW, Xu A, Porter TJ, Chen W. A rapid on-line method for mass spectrometric confirmation of a cysteine-conjugated antibody-drug-conjugate structure using multidimensional chromatography. MAbs. 2015; 7(6): 1036–1044.

[340] Le LN, Moore JM, Ouyang J, Chen X, Nguyen MD, Galush WJ. Profiling antibody drug conjugate positional isomers: A system-of-equations approach. Anal Chem. 2012; 84(17): 7479–7486.

[341] Monkarsh SP, Spence C, Porter JE, Palleroni A, Nalin C, Rosen P, et al. Isolation of positional isomers of mono-poly(ethylene glycol)ylated interferon α-2a and the determination of their biochemical and biological characteristics. In: Harris JL, Zalipsky S, editors. Poly(ethylene glycol). Chemistry and biological applications. vol. 680 of ACS Symposium Series. American Chemical Society; 1997. p. 207–216.

[342] Maiser B, Kroner F, Dismer F, Brenner-Weiss G, Hubbuch J. Isoform separation and binding site determination of mono-PEGylated lysozyme with pH gradient chromatography. J Chromatogr A. 2012; 1268: 102–108.

[343] ATTO-TEC. Recommended procedures for labeling; 2016. Accessible from: https://www.atto-tec.com/.

[344] Pfister D, Ingold O, Morbidelli M. Model-based development of an on-column PEGylation process. React Chem Eng. 2016; 1: 204–217.

[345] Pfister D, Morbidelli M. Integrated process for high conversion and high yield protein PEGylation. Biotechnol Bioeng. 2016; 113(8): 1711–1718.

[346] Wang X, Hu J, Pan D, Teng H, Xiu Z. PEGylation kinetics of recombinant hirudin and its application for the production of PEGylated HV2 species. Biochem Eng J. 2014; 85: 38–48.

[347] Turecek PL, Bossard MJ, Schoetens F, Ivens IA. PEGylation of biopharmaceuticals: A review of chemistry and nonclinical safety information of approved drugs. J Pharm Sci. 2016; 105(2): 460–475.

[348] Grace M, Youngster S, Gitlin G, Sydor W, Xie L, Westreich L, et al. Structural and biologic characterization of pegylated recombinant IFN α-2b. J Interf Cytok Res. 2001; 21(12): 1103–1115.

[349] Marcoux J, Champion T, Colas O, Wagner-Rousset E, Corva N, Van Dorsselaer A, et al. Native mass spectrometry and ion mobility characterization of trastuzumab emtansine, a lysine-linked antibody drug conjugate. Protein Sci. 2015; 24(8): 1210–1223.

[350] Shen BQ, Xu K, Liu L, Raab H, Bhakta S, Kenrick M, et al. Conjugation site modulates the *in vivo* stability and therapeutic activity of antibody-drug conjugates. Nature Biotechnol. 2012; 30(2): 184–189.

[351] Molek J, Ruanjaikaen K, Zydney AL. Effect of electrostatic interactions on transmission of PEGylated proteins through charged ultrafiltration membranes. J Membrane Sci. 2010; 353(1–2): 60–69.

[352] Molek JR, Zydney AL. Separation of PEGylated αLactalbumin from unreacted precursors and byproducts using ultrafiltration. Biotechnol Prog. 2007; 23(6): 1417–1424.

[353] Ruanjaikaen K, Zydney AL. Purification of singly PEGylated α-lactalbumin using charged ultrafiltration membranes. Biotechnol Bioeng. 2011; 108(4): 822–829.

[354] Rohrer T. Consideration for the safe and effective manufacturing of antibody drug conjugates. Chem Oggi/Chem Today. 2012; 30: 76–79.

[355] Zeman LJ, Zydney AL. Microfiltration and Ultrafiltration. Principles and Applications. Marcel Dekker; 1996.

[356] Molek KS. Ultrafiltration of PEGylated proteins. Pennsylvania State University; 2008.

[357] Rodriguez-Aller M, Guillarme D, Beck A, Fekete S. Practical method development for the separation of monoclonal antibodies and antibody-drug-conjugate species in hydrophobic interaction chromatography. Part 1: Optimization of the mobile phase. J Pharmaceut Biomed. 2016; 118: 393–403.

[358] Pfister D. Processes for the production of PEGylated proteins. ETH Zurich; 2015.

[359] Thayer AM. Building antibody-drug conjugates. Chem Eng News. 2014; 92(3): 13–20.

[360] Seely JE, Richey CW. Use of ion-exchange chromatography and hydrophobic interaction chromatography in the preparation and recovery of polyethylene glycol-linked proteins. J Chromatogr A. 2001; 908(1–2): 235–241.

[361] Moosmann A, Christel J, Boettinger H, Mueller E. Analytical and preparative separation of PEGylated lysozyme for the characterization of chromatography media. J Chromatogr A. 2010; 1217(2): 209–215.

[362] Maiser B, Dismer F, Hubbuch J. Optimization of random PEGylation reactions by means of high throughput screening. Biotechnol Bioeng. 2014; 111(1): 104–114.

[363] Moosmann A, Blath J, Lindner R, Muller E, Bottingert H. Aldehyde PEGylation kinetics: A Standard protein versus a pharmaceutically relevant single chain variable fragment. Bioconjugate Chem. 2011; 22(8): 1545–1558.

[364] Chavez DM, Orpiszewski J, inventors; Apparatuses and processes for increasing protein pegylation reaction yields. US20050089952 A1; 2004, filing date.

[365] Rathore A, Bhambure R, Ghare V. Process analytical technology (PAT) for biopharmaceutical products. Anal Bioanal Chem. 2010; 398(1): 137–154.

[366] Cromwell MEM, Hilario E, Jacobson F. Protein aggregation and bioprocessing. AAPS J. 2006; 8(3): E572–E579.

[367] Wang W, Nema S, Teagarden D. Protein aggregation – Pathways and influencing factors. Int J Pharm. 2010; 390(2): 89–99.

[368] Shire SJ, Shahrokh Z, Liu J. Challenges in the development of high protein concentration formulations. J Pharm Sci. 2004; 93(6): 1390–1402.

[369] Rathore A, Anderson S, Sharma A. Filter clogging issues in sterile filtration. BioPharm Int; 2008.

[370] Narhi LO, Schmit J, Bechtold-Peters K, Sharma D. Classification of protein aggregates. J Pharm Sci. 2012; 101(2): 493–498.

[371] Sharma VK, Kalonia DS. Experimental detection and characterization of protein aggregates. In: Wang W, Roberts CJ, editors. Aggregation of therapeutic proteins. John Wiley & Sons, Inc.; 2010. p. 205–256.

[372] Cole JL, Lary JW, Moody T, Laue TM. Analytical ultracentrifugation: Sedimentation velocity and sedimentation equilibrium. Method Cell Biol. 2008; 84: 143–179.

[373] Wahlund KG, Nilsson L. Flow FFF – Basics and key applications. In: Williams RSK, Caldwell DK, editors. Field-flow fractionation in biopolymer analysis. Springer Vienna; 2012. p. 1–21.

[374] Sahin E, Roberts CJ. Size-exclusion chromatography with multi-angle light scattering for elucidating protein aggregation mechanisms. In: Voynov V, Caravella AJ, editors. Therapeutic proteins: Methods and protocols. Humana Press; 2012. p. 403–423.

[375] Li Y, Weiss WFt, Roberts CJ. Characterization of high-molecular-weight non-native aggregates and aggregation kinetics by size exclusion chromatography with inline multi-angle laser light scattering. J Pharm Sci. 2009; 98(11): 3997–4016.

[376] Nicoud L, Arosio P, Sozo M, Yates A, Norrant E, Morbidelli M. Kinetic analysis of the multistep aggregation mechanism of monoclonal antibodies. J Phys Chem B. 2014; 118(36): 10595–606.

[377] Amin S, Barnett GV, Pathak JA, Roberts CJ, Sarangapani PS. Protein aggregation, particle formation, characterization & rheology. Curr Opin Colloid Interface Sci. 2014; 19(5): 438–449.

[378] Das TK. Protein particulate detection issues in biotherapeutics development – Current status. AAPS PharmSciTech. 2012; 13(2): 732–746.

[379] Paul R, Graff-Meyer A, Stahlberg H, Lauer ME, Rufer AC, Beck H, et al. Structure and function of purified monoclonal antibody dimers induced by different stress conditions. Pharm Res. 2012; 29(8): 2047–2059.

[380] Esue O, Kanai S, Liu J, Patapoff TW, Shire SJ. Carboxylate-dependent gelation of a monoclonal antibody. Pharm Res. 2009; 26(11): 2478–2485.

[381] Brych SR, Gokarn YR, Hultgen H, Stevenson RJ, Rajan R, Matsumura M. Characterization of antibody aggregation: Role of buried, unpaired cysteines in particle formation. J Pharm Sci. 2010; 99(2): 764–781.

[382] Telikepalli SN, Kumru OS, Kalonia C, Esfandiary R, Joshi SB, Middaugh CR, et al. Structural characterization of IgG1 mAb aggregates and particles generated under various stress conditions. J Pharm Sci. 2014; 103(3): 796–809.

[383] Nicoud L, Owczarz M, Arosio P, Morbidelli M. A multiscale view of therapeutic protein aggregation: A colloid science perspective. Biotechnol J. 2015; 10(3): 367–378.

[384] Lazzari S, Nicoud L, Jaquet B, Lattuada M, Morbidelli M. Fractal-like structures in colloid science. Adv Colloid Interface Sci. 2016; 235: 1–13.

[385] Nicoud L, Lattuada M, Yates A, Morbidelli M. Impact of aggregate formation on the viscosity of protein solutions. Soft Matter. 2015; 11(27): 5513–5522.

[386] Nicoud L, Jagielski J, Pfister D, Lazzari S, Massant J, Lattuada M, et al. Kinetics of monoclonal antibody aggregation from dilute toward concentrated conditions. J Phys Chem B. 2016; 120(13): 3267–3280.

[387] Chi EY, Krishnan S, Randolph TW, Carpenter JF. Physical stability of proteins in aqueous solution: Mechanism and driving forces in non-native protein aggregation. Pharm Res. 2003; 20(9): 1325–1336.

[388] Saito S, Hasegawa J, Kobayashi N, Tomitsuka T, Uchiyama S, Fukui K. Effects of ionic strength and sugars on the aggregation propensity of monoclonal antibodies: Influence of colloidal and conformational stabilities. Pharm Res. 2013; 30(5): 1263–1280.

[389] Lehermayr C, Mähler HC, Mäder K, Fischer S. Assessment of net charge and protein–protein interactions of different monoclonal antibodies. J Pharm Sci. 2011; 100(7): 2551–2562.

[390] Arzensek D, Kuzman D, Podgornik R. Colloidal interactions between monoclonal antibodies in aqueous solutions. J Colloid Interface Sci. 2012; 384(1): 207–216.

[391] Roberts D, Keeling R, Tracka M, van der Walle CF, Uddin S, Warwicker J, et al. The role of electrostatics in protein–protein interactions of a monoclonal antibody. Mol Pharm. 2014; 11(7): 2475–2489.

[392] Saluja A, Fesinmeyer RM, Hogan S, Brems DN, Gokarn YR. Diffusion and sedimentation interaction parameters for measuring the second virial coefficient and their utility as predictors of protein aggregation. Biophys J. 2010; 99(8): 2657–2665.

[393] Chennamsetty N, Voynov V, Kayser V, Helk B, Trout BL. Design of therapeutic proteins with enhanced stability. Proc Natl Acad Sci USA. 2009; 106(29): 11937–11942.

[394] Chennamsetty N, Voynov V, Kayser V, Helk B, Trout BL. Prediction of aggregation prone regions of therapeutic proteins. J Phys Chem B. 2010; 114(19): 6614–6624.

[395] Agrawal NJ, Kumar S, Wang XL, Helk B, Singh SK, Trout BL. Aggregation in protein-based biotherapeutics: Computational studies and tools to identify aggregation-prone regions. J Pharm Sci. 2011; 100(12): 5081–5095.

[396] Neal BL, Asthagiri D, Lenhoff AM. Molecular origins of osmotic second virial coefficients of proteins. Biophys J. 1998; 75(5): 2469–2477.

[397] Curtis RA, Lue L. A molecular approach to bioseparations: Protein–protein and protein–salt interactions. Chem Eng Sci. 2006; 61(3): 907–923.

[398] Fischer H, Polikarpov I, Craievich AF. Average protein density is a molecular-weight-dependent function. Protein Sci. 2004; 13(10): 2825–2828.

[399] Tessier PM, Lenhoff AM, Sandler SI. Rapid measurement of protein osmotic second virial coefficients by self-interaction chromatography. Biophys J. 2002; 82(3): 1620–1631.

[400] Chi EY, Krishnan S, Kendrick BS, Chang BS, Carpenter JF, Randolph TW. Roles of conformational stability and colloidal stability in the aggregation of recombinant human granulocyte colony-stimulating factor. Protein Sci. 2003; 12(5): 903–913.

[401] Kelly SM, Price NC. The use of circular dichroism in the investigation of protein structure and function. Curr Protein Pept Sci. 2000; 1(4): 349–384.

[402] Garidel P, Hegyi M, Bassarab S, Weichel M. A rapid, sensitive and economical assessment of monoclonal antibody conformational stability by intrinsic tryptophan fluorescence spectroscopy. Biotechnol J. 2008; 3(9-10): 1201–1211.

[403] Hawe A, Sutter M, Jiskoot W. Extrinsic fluorescent dyes as tools for protein characterization. Pharm Res. 2008; 25(7): 1487–1499.

[404] Barth A, Zscherp C. What vibrations tell us about proteins. Q Rev Biophys. 2002; 35(4): 369–430.

[405] Wen ZQ. Raman spectroscopy of protein pharmaceuticals. J Pharm Sci. 2007; 96(11): 2861–2878.

[406] Angarita M, Arosio P, Müller-Späth T, Baur D, Falkenstein R, Kuhne W, et al. Role of urea on recombinant Apo A-I stability and its utilization in anion exchange chromatography. J Chromatogr A. 2014; 1354: 18–25.

[407] Sahin E, Weiss WF, Kroetsch AM, King KR, Kessler RK, Das TK, et al. Aggregation and pH-temperature phase behavior for aggregates of an IgG2 antibody. J Pharm Sci. 2012; 101(5): 1678–1687.

[408] Johnson CM. Differential scanning calorimetry as a tool for protein folding and stability. Arch Biochem Biophys. 2013; 531(1–2): 100–9.

[409] Vermeer AWP, Norde W. The thermal stability of immunoglobulin: Unfolding and aggregation of a multi-domain protein. Biophys J. 2000; 78(1): 394–404.

[410] He F, Becker GW, Litowski JR, Narhi LO, Brems DN, Razinkov VI. High-throughput dynamic light scattering method for measuring viscosity of concentrated protein solutions. Anal Biochem. 2010; 399(1): 141–143.

[411] Connolly BD, Petry C, Yadav S, Demeule B, Ciaccio N, Moore JMR, et al. Weak interactions govern the viscosity of concentrated antibody solutions: High-throughput analysis using the diffusion interaction parameter. Biophys J. 2012; 103(1): 69–78.

[412] Chari R, Jerath K, Badkar AV, Kalonia DS. Long- and short-range electrostatic interactions affect the rheology of highly concentrated antibody solutions. Pharm Res. 2009; 26(12): 2607–2618.

[413] Yadav S, Shire SJ, Kalonia DS. Viscosity behavior of high-concentration monoclonal antibody solutions: Correlation with interaction parameter and electroviscous effects. J Pharm Sci. 2012; 101(3): 998–1011.

[414] He F, Woods CE, Litowski JR, Roschen LA, Gadgil HS, Razinkov VI, et al. Effect of sugar molecules on the viscosity of high concentration monoclonal antibody solutions. Pharm Res. 2011; 28(7): 1552–1560.

[415] Chiti F, Stefani M, Taddei N, Ramponi G, Dobson CM. Rationalization of the effects of mutations on peptide and protein aggregation rates. Nature. 2003; 424(6950): 805–808.

[416] Hiemenz PC. Principles of colloid and surface chemistry. Dekker; 1986.

[417] Russel WB, Saville DA, Schowalter WR. Colloidal dispersions. Cambridge University Press; 1989.

[418] Andrews JM, Roberts CJ. A Lumry-Eyring nucleated polymerization model of protein aggregation kinetics: 1. Aggregation with pre-equilibrated unfolding. J Phys Chem B. 2007; 111(27): 7897–7913.

[419] Li Y, Roberts CJ. A Lumry-Eyring nucleated-polymerization model of protein aggregation kinetics. 2. Competing growth via condensation and chain polymerization. J Phys Chem B. 2009; 113(19): 7020–7032.

[420] Sahin E, Grillo AO, Perkins MD, Roberts CJ. Comparative effects of pH and ionic strength on protein-protein interactions, unfolding, and aggregation for IgG1 antibodies. J Pharm Sci. 2010; 99(12): 4830–4848.

[421] Campioni S, Carret G, Jordens S, Nicoud L, Mezzenga R, Riek R. The presence of an air–water interface affects formation and elongation of alpha-synuclein fibrils. J Am Chem Soc. 2014; 136(7): 2866–2875.

[422] Arosio P, Beeg M, Nicoud L, Morbidelli M. Time evolution of amyloid fibril length distribution described by a population balance model. Chem Eng Sci. 2012; 78: 21–32.

[423] Nicoud L, Lazzari S, Balderas Barragn D, Morbidelli M. Fragmentation of amyloid fibrils occurs in preferential positions depending on the environmental conditions. J Phys Chem B. 2015; 119(13): 4644–4652.

[424] Roberts CJ. Non-native protein aggregation kinetics. Biotechnol Bioeng. 2007; 98(5): 927–938.

[425] Pinholt C, Hartvig RA, Medlicott NJ, Jorgensen L. The importance of interfaces in protein drug delivery – Why is protein adsorption of interest in pharmaceutical formulations? Expert Opin Drug Deliv. 2011; 8(7): 949–964.

[426] Nicoud L, Sozo M, Arosio P, Yates A, Norrant E, Morbidelli M. Role of cosolutes in the aggregation kinetics of monoclonal antibodies. J Phys Chem B. 2014; 118(41): 11921–11930.

[427] Arosio P, Jaquet B, Wu H, Morbidelli M. On the role of salt type and concentration on the stability behavior of a monoclonal antibody solution. Biophys Chem. 2012; 168–169: 19–27.

[428] Timasheff SN. The control of protein stability and association by weak interactions with water: How do solvents affect these processes? Annu Rev Biophys Biomol Struct. 1993; 22: 67–97.

[429] Vagenende V, Yap MGS, Trout BL. Molecular anatomy of preferential interaction coefficients by elucidating protein solvation in mixed solvents: Methodology and application for lysozyme in aqueous glycerol. J Phys Chem B. 2009; 113(34): 11743–11753.

[430] Nicoud L, Cohrs N, Arosio P, Norrant E, Morbidelli M. Effect of polyol sugars on the stabilization of monoclonal antibodies. Biophys Chem. 2015; 197(0): 40–46.

[431] Arosio P, Rima S, Lattuada M, Morbidelli M. Population balance modeling of antibodies aggregation kinetics. J Phys Chem B. 2012; 116(24): 7066–7075.

[432] Wang W, Roberts CJ. Non-Arrhenius protein aggregation. AAPS J. 2013; 15(3): 840–851.

[433] Kayser V, Chennamsetty N, Voynov V, Helk B, Forrer K, Trout BL. Evaluation of a non-Arrhenius model for therapeutic monoclonal antibody aggregation. J Pharm Sci. 2011; 100(7): 2526–2542.

[434] Wu H, Truncali K, Ritchie J, Kroe-Barrett R, Singh S, Robinson AS, et al. Weak protein interactions and pH- and temperature-dependent aggregation of human Fc1. MAbs. 2015; 7(6): 1072–1083.

[435] Weiss WFI, Young TM, Roberts CJ. Principles, approaches, and challenges for predicting protein aggregation rates and shelf life. J Pharm Sci. 2009; 98(4): 1246–1277.

[436] Privalov PL. Cold denaturation of proteins. Crit Rev Biochem Mol Biol. 1990; 25(5): 281–306.

[437] Ghosh R, Calero-Rubio C, Saluja A, Roberts CJ. Relating protein-protein interactions and aggregation rates from low to high concentrations. J Pharm Sci. 2016; 105(3): 1086–1096.

[438] Lilyestrom WG, Yadav S, Shire SJ, Scherer TM. Monoclonal antibody self-association, cluster formation, and rheology at high concentrations. J Phys Chem B. 2013; 117(21): 6373–6384.

[439] Nicoud L, Lattuada M, Lazzari S, Morbidelli M. Viscosity scaling in concentrated dispersions and its impact on colloidal aggregation. Phys Chem Chem Phys. 2015; 17(37): 24392–2402.

[440] Wiesbauer J, Prassl R, Nidetzky B. Renewal of the air–water interface as a critical system parameter of protein stability: Aggregation of the human growth hormone and its prevention by surface-active compounds. Langmuir. 2013; 29(49): 15240–15250.

[441] Mazzer AR, Perraud X, Halley J, OHara J, Bracewell DG. Protein A chromatography increases monoclonal antibody aggregation rate during subsequent low pH virus inactivation hold. J Chromatogr A. 2015; 1415: 83–90.

[442] Gagnon P, Nian R, Leong D, Hoi A. Transient conformational modification of immunoglobulin G during purification by protein A affinity chromatography. J Chromatogr A. 2015; 1395: 136–142.

[443] Shukla AA, Gupta P, Han X. Protein aggregation kinetics during protein A chromatography: Case study for an Fc fusion protein. J Chromatogr A. 2007; 1171(12): 22–28.

[444] Arosio P, Rima S, Morbidelli M. Aggregation mechanism of an IgG2 and two IgG1 monoclonal antibodies at low pH: From oligomers to larger aggregates. Pharm Res. 2013; 30(3): 641–654.

[445] Yu L, Zhang L, Sun Y. Protein behavior at surfaces: Orientation, conformational transitions and transport. J Chromatogr A. 2015; 1382: 118–134.

[446] Gillespie R, Nguyen T, Macneil S, Jones L, Crampton S, Vunnum S. Cation exchange surface-mediated denaturation of an aglycosylated immunoglobulin (IgG1). J Chromatogr A. 2012; 1251: 101–110.

[447] Guo J, Carta G. Unfolding and aggregation of monoclonal antibodies on cation exchange columns: Effects of resin type, load buffer, and protein stability. J Chromatogr A. 2015; 1388: 184–194.

[448] Gospodarek AM, Hiser DE, O'Connell JP, Fernandez EJ. Unfolding of a model protein on ion exchange and mixed mode chromatography surfaces. J Chromatogr A. 2014; 1355: 238–252.

[449] Guo J, Zhang S, Carta G. Unfolding and aggregation of a glycosylated monoclonal antibody on a cation exchange column. Part I. Chromatographic elution and batch adsorption behavior. J Chromatogr A. 2014; 1356: 117–128.

[450] Guo J, Carta G. Unfolding and aggregation of a glycosylated monoclonal antibody on a cation exchange column. Part II. Protein structure effects by hydrogen deuterium exchange mass spectrometry. J Chromatogr A. 2014; 1356: 129–137.

[451] Muca R, Marek W, Piatkowski W, Antos D. Influence of the sample-solvent on protein retention, mass transfer and unfolding kinetics in hydrophobic interaction chromatography. J Chromatogr A. 2010; 1217(17): 2812–2820.

[452] Jungbauer A, Machold C, Hahn R. Hydrophobic interaction chromatography of proteins: III. Unfolding of proteins upon adsorption. J Chromatogr A. 2005; 1079(12): 221–228.

[453] Fogle JL, Fernandez EJ. Amide hydrogen-deuterium exchange: A fast tool for screening protein stabilities in chromatography. LCGC North America. 2006; p. 96–101.

[454] Benedek K, Dong S, Karger BL. Kinetics of unfolding of proteins on hydrophobic surfaces in reversed-phase liquid chromatography. J Chromatogr. 1984; 317: 227–243.

[455] McNay JL, Fernandez EJ. Protein unfolding during reversed-phase chromatography: I. Effect of surface properties and duration of adsorption. Biotechnol Bioeng. 2001; 76(3): 224–232.

[456] McNay JL, O'Connell JP, Fernandez EJ. Protein unfolding during reversed-phase chromatography: II. Role of salt type and ionic strength. Biotechnol Bioeng. 2001; 76(3): 233–240.

[457] Xiao Y, Rathore A, O'Connell JP, Fernandez EJ. Generalizing a two-conformation model for describing salt and temperature effects on protein retention and stability in hydrophobic interaction chromatography. J Chromatogr A. 2007; 1157(12): 197–206.

[458] Arosio P, Barolo G, Müller-Späth T, Wu H, Morbidelli M. Aggregation stability of a monoclonal antibody during downstream processing. Pharm Res. 2011; 28(8): 1884–1894.

[459] Ghosh R. Protein bioseparation using ultrafiltration: Theory, applications and new developments. Imperial College Press; 2003.

[460] Rosenberg E, Hepbildikler S, Kuhne W, Winter G. Ultrafiltration concentration of monoclonal antibody solutions: Development of an optimized method minimizing aggregation. J Membrane Sci. 2009; 342(1–2): 50–59.

[461] Wang W. Protein aggregation and its inhibition in biopharmaceutics. Int J Pharm. 2005; 289(12): 1–30.

[462] Roberts CJ, Das TK, Sahin E. Predicting solution aggregation rates for therapeutic proteins: Approaches and challenges. Int J Pharm. 2011; 418(2): 318–333.

[463] Rouet R, Lowe D, Christ D. Stability engineering of the human antibody repertoire. FEBS Lett. 2014; 588(2): 269–277.

[464] Rajan RS, Li T, Aras M, Sloey C, Sutherland W, Arai H, et al. Modulation of protein aggregation by polyethylene glycol conjugation: GCSF as a case study. Protein Sci. 2006; 15(5): 1063–1075.

[465] Kamerzell TJ, Esfandiary R, Joshi SB, Middaugh CR, Volkin DB. Protein-excipient interactions: Mechanisms and biophysical characterization applied to protein formulation development. Adv Drug Deliver Rev. 2011; 63(13): 1118–1159.

[466] Shukla AA, Hinckley PJ, Gupta P, Yigzaw Y, Hubbard B. Strategies to address aggregation during protein A chromatography. Bioprocess Int. 2005; 3: 36–44.

[467] Arakawa T, Philo JS, Tsumoto K, Yumioka R, Ejima D. Elution of antibodies from a protein-A column by aqueous arginine solutions. Protein Expr Purif. 2004; 36(2): 244–248.

[468] Turecek P, Siekmann J. Factor VIII polymer conjugates. EP2318050B1; 2009.

[469] Bankston TE, Carta G. Apolipoprotein A-I-Milano anion exchange chromatography: Self association and adsorption equilibrium. Biotechnol J. 2010; 5(10): 1028–1039.

[470] Arakawa T, Timasheff SN. Stabilization of protein structure by sugars. Biochemistry. 1982; 21(25): 6536–6544.

[471] Gao D, Wang LL, Lin DQ, Yao SJ. Evaluating antibody monomer separation from associated aggregates using mixed-mode chromatography. J Chromatogr A. 2013; 1294: 70–75.

[472] Chmielowski RA, Meissner S, Roush D, Linden TO, Glowacki E, Konietzko J, et al. Resolution of heterogeneous charged antibody aggregates via multimodal chromatography: A comparison to conventional approaches. Biotechnol Prog. 2014; 30(3): 636–645.

[473] Aldington S, Bonnerjea J. Scale-up of monoclonal antibody purification processes. J Chromatogr B. 2007; 848(1): 64–78.

[474] Wellhoefer M, Sprinzl W, Hahn R, Jungbauer A. Continuous processing of recombinant proteins: Integration of refolding and purification using simulated moving bed size-exclusion chromatography with buffer recycling. J Chromatogr A. 2014; 1337: 48–56.

[475] Park BJ, Lee CH, Koo YM. Development of novel protein refolding using simulated moving bed chromatography. Korean J Chem Eng. 2005; 22(3): 425–432.

Index

Printed in the United Kingdom by TJ Clays Ltd.